S0-ASV-178

BEYOND ENVIRONMENTALISM

BEYOND ENVIRONMENTALISM

A Philosophy of Nature

Jeffrey E. Foss
University of Victoria
Department of Philosophy
Victoria, British Columbia
Canada

WILEY

A JOHN WILEY & SONS, INC., PUBLICATION

Copyright © 2009 by John Wiley & Sons, Inc. All rights reserved.

Published by John Wiley & Sons, Inc., Hoboken, New Jersey.
Published simultaneously in Canada.

No part of this publication may be reproduced, stored in a retrieval system, or transmitted in any form or
by any means, electronic, mechanical, photocopying, recording, scanning or otherwise, except as
permitted under Sections 107 or 108 of the 1976 United States Copyright Act, without either the prior
written permission of the Publisher, or authorization through payment of the appropriate per-copy fee to
the Copyright Clearance Center, Inc., 222 Rosewood Drive, Danvers, MA 01923, (978) 750-8400, fax
(978) 750-4470, or on the web at www.copyright.com. Requests to the Publisher for permission should
be addressed to the Permissions Department, John Wiley & Sons, Inc., 111 River Street, Hoboken, NJ
07030, (201) 748-6011, fax (201) 748-6008, or online at http://www.wiley.com/go/permission.

Limit of Liability/Disclaimer of Warranty: While the publisher and author have used their best efforts in
preparing this book, they make no representations or warranties with respect to the accuracy or
completeness of the contents of this book and specifically disclaim any implied warranties of
merchantability or fitness for a particular purpose. No warranty may be created or extended by sales
representatives or written sales materials. The advice and strategies contained herein may not be suitable
for your situation. You should consult with a professional where appropriate. Neither the publisher nor
author shall be liable for any loss of profit or any other commercial damages, including but not limited to
special, incidental, consequential, or other damages.

For general information on our other products and services or for technical support, please contact our
Customer Care Department within the United States at (800) 762-2974, outside the United States at (317)
572-3993 or fax (317) 572-4002.

Wiley also publishes its books in a variety of electronic formats. Some content that appears in print may
not be available in electronic formats. For more information about Wiley products, visit our web site at
www.wiley.com.

Library of Congress Cataloging-in-Publication Data:

Foss, Jeffrey E., 1948–
    Beyond environmentalism : a philosophy of nature / Jeffrey E. Foss
        p. cm.
    Includes bibliographical references and index.
    ISBN 978-0-470-17941-3 (pbk.)
    1. Environmentalism.   2. Naturalness (Environmental sciences)   3. Philosophy
of nature.   I. Title.
    GE195.F676   2008
    333.72–dc22

                                                            2008022815
Printed in the United States of America

10   9   8   7   6   5   4   3   2   1

To life, consciousness, and caring

CONTENTS

PREFACE

This book presents a philosophy of nature that goes beyond the limitations and mistakes of environmentalism to provide a better understanding of our place in the natural world and our responsibilities concerning it. Although critical of popular environmentalism, it nevertheless has deep sympathies with its sense that nature is sacred and that we should protect its health and beauty. But, unlike environmentalism, which begins with a dichotomy between human beings and their natural environment, this philosophy of nature is based on the fact that humankind is an integral part of the natural world. Although humankind should—indeed, must—seek its own destiny, it cannot flourish unless nature flourishes as a whole. However, we are not obligated to make ourselves as small as possible in order that nature take the path it would have followed had we never been here and had agriculture and technology never been developed. On the other hand, the natural world is not simply a resource that we should use and manage solely in order to pursue our own goals. There is instead a balance, a symbiosis, that we should seek to realize.

Because it is crucial for humankind to achieve a proper relationship with the rest of the natural world, this book has been designed to serve the purposes of both the general reader and college and university students. The main text is reader-friendly and easily accessible to the reading public at large. The footnotes contain further discussion of technicalities and references to scholarly works that support the claims of the main text. These two levels develop side by side, providing material of interest to both the popular and the academic audience. Case studies of such things as global warming, the "sixth extinction," the precautionary principle, pollution, recycling, and so on, provide food for thought for the casual reader as well as the basis for class discussion and further study by students. Feature boxes provide introductions to topics such as eco-sabotage, the Gaia hypothesis, the urban heat-island effect, the logic of carbon offsets, pastoralism, and the fact–value dichotomy. Graphics are used to illustrate facts and arguments as required. By employing these different elements, the book provides various levels of interest for each reader.

Balance, fair-mindedness, and objectivity are in short supply in much of the global debate concerning environmental issues, and this book aims to restore them so that the debate may reach a successful conclusion. Environmental issues come to the fore in countless classrooms in virtually every subject, just as they do in countless discussions in the popular media and in people's living rooms. This book is designed to inform these discussions scientifically, to enlighten them philosophically, and thus to play a role in bringing about a sound and happy relationship between *Homo sapiens* and the rest of the natural order.

ACKNOWLEDGMENTS

I have relied upon the help and encouragement of too many people to list here, and beg the forgiveness of anyone who does not find the help they provided me acknowledged below.

I would like to thank my wife, Christina Foss, for correcting countless grammatical and stylistic errors, and for her steadfast belief in this book during its darkest hours. For keeping me in touch with the wilderness, I am grateful to my brother, John Foss, who charted our annual hikes in the Canadian Rockies. Thank you, too, to my daughter, Sylvie Foss, who created the artistic rendering of our front yard that appears on the cover, and made my computer behave. Fellow philosopher Steven Davis encouraged me to persist in my analysis of recycling, while fellow academic, the economist Paul Geddes, supported an economic approach to this analysis. I owe a debt of gratitude to my University of Victoria (UVic) colleague David Johnston, who kindly took it upon himself to keep me informed of breaking news stories on the environment and environmentalism. My friend, fellow musician, and UVic colleague Matt James introduced me to the works of Ulrich Beck. I am specially thankful to Dr. Richard Lindzen, the renowned MIT and Harvard climate scientist, for providing me invaluable discussions by email and phone of global warming theory. Paul MacRae gave me my first taste of published support in 2000 as editorial writer for the *Victoria Times-Colonist*, and is now a colleague, in more senses than one, at UVic. Ross McKitrick encouraged me to keep working on this book even when publication seemed doomed by the pro-environmental tenor of the times, and educated me via his work and personal communications about many aspects of the global warming debate raging just under the surface of the united front presented by the popular media. Thanks to Reg Mitchell, chemist and UVic colleague, who gave me timely support in print in my debate with IPCC scientist Andrew Weaver. Fellow philosopher Jan Narveson taught me by example to stand up to the sometimes impolite barrage of criticism leveled at those who dare question environmentalism. My UVic colleague John Newcomb has kindly kept me informed of news events concerning environmental activists. James Gary Shelton, famed for his knowledge of bears and bear attacks on humans, reaffirmed for me the value of firsthand, as opposed to theoretical, knowledge of nature. MIT climate scientist Willie Soon gave me valuable insights into variations in Arctic climate and the relevance of solar activity. Finally, my friend and fellow philosopher James Young led me to investigate the link between the ban on DDT and increased deaths due to malaria. Thank you to all.

JEFFREY E. FOSS

Introduction: Why I Wrote This Book

As a university professor, I have attended many lectures, and presented many. Usually, such affairs are models of scholarly decorum, with kind introductions at the start and polite applause at the finish. Not so for one talk I gave, entitled, "Why I Am Not an Environmentalist."* A few days before the talk I was contacted by an officer of the campus security force who warned me that threats of physical violence against me were circulating on Internet sites devoted to environmentalist causes. I was advised to be cautious in my movements and was reassured that security would be quietly present at the talk. Violence? What sort of violence? Well, he couldn't really say. I was surprised--in fact, a bit shocked. It had never once occurred to me that the police would ever be required to protect my academic freedom! I had a hard time believing what was being said. I am a *philosophy* professor, I told him, nobody cares that much about what I say. Well, he said, apparently *some* people cared enough to try to stop me from saying whatever it was I wanted to say.

At least somebody *cared* about the topic, I told myself, something not always guaranteed, and gave the matter no more thought. I was to learn that not just some people, but a lot of people, cared, really cared, about environmentalism. Usually, philosophical lectures draw small audiences, but my talk had to be moved to a larger room, twice. Still, on the day of the talk the room was filled to overflowing. Microphones from a radio station were in front of the lectern, along with some unidentified tape recorders. A newspaper reporter claimed ownership of one of them and informed me that he would be taking notes. Apparently, the mere existence of an academic who was not an environmentalist, and willing to say so, was actually *news*. I did realize that academics tend to be uniformly pro-environmentalist. In fact, budging the complacent academic orthodoxy on this issue was why I had written my lecture in the first place, but I didn't realize that the world at large had noticed the pro-environmentalist orthodoxy in the universities and colleges—or cared.

Knowing that environmental issues are heavily laden with emotion, I had tried to craft my talk to be as modest as possible without abandoning my central thesis: Nature is not in a state of crisis; environmentalism is based on a mistake, and is becoming more a matter of faith than of reason. It began well enough, but met ever-louder groans of displeasure from members of the audience who disagreed with me. They didn't merely disagree, they were *offended*. When the time came for discussion,

---

* University of Victoria, Canada, December 3, 1999.

Beyond Environmentalism: A Philosophy of Nature, By Jeffrey E. Foss
Copyright © 2009 John Wiley & Sons, Inc.

many of the questions were, more accurately speaking, accusations, and my replies were met with disdain rather than reason. Did I not know that every bowl of cereal I ate contributed to the most rapid decline in biodiversity in the history of the planet? Did I really think that riding a motorcycle to school would clear me of guilt when the globe overheated? Did I not know that nature was being *devastated?*

"Devastation," I repeated the word out loud, while turning it over in my mind. "Devastation. Is that really, *literally*, the right word?"

"Totally!" my interrogator replied, as his fellow environmentalists voiced a chorus of agreement. Where was this devastation occurring, I asked. The entire city! they said. The surrounding fields! The very air we breathed! These all were scenes of devastation. When I pointed out that scientific measures show air, water, and soil pollution to have been steadily decreasing for nearly half a century, I met simple disbelief. I must have gotten my facts wrong. The military–industrial complex must have paid their hired scientists to fix the figures.

Maybe, I thought, direct observation would work where reason and data had failed. So I went to the windows along one side of the room, and looked out. The audience looked out too, and calmed down a bit. It was a beautiful, serene autumn evening, the last rays of sunlight fading on the brightly colored autumn leaves of the campus.

"If this is devastation, why is it beautiful?" I asked.

This was met by groans of disbelief.

But, really, it *was* beautiful! That is what I thought then, and I still do.

Faced with so many people who could not see the beauty that lay right before their eyes, I decided then and there to persist in my philosophical investigation of the environment and our relationship to it. As I investigated further, I became more and more convinced that contemporary environmentalism has blinded many of us to the facts. I became convinced that it is a powerful ideology that has—usually out of the best of motives—created much confusion and misunderstanding. So I was gradually persuaded that it was time for a *new* environmental philosophy.

For a few years, however, I avoided publication of my ideas. This was due mainly to the fact that there was an enormous amount I wanted to learn about environmentalism—and the environment—before saying anything more about it, but it was also due partly to my fear that engaging in a public debate with environmentalists might destroy the peace and happiness of my family life. In the days just after I had presented my lecture I was interviewed by local papers and radio stations, and soon found myself under a forceful media attack. On the one side I was reproached by some scientists* for my refusal to embrace environmentalism, while on the other side I was chastised by environmentalists for embracing science as though it were a modern-day religion. Up the middle charged local representatives of environmentalist groups such as the Sierra Club and the Western Canada Wilderness Committee, who saw me as an enemy of the very planet itself. Meanwhile I began to receive lots of mails, e-mails, and even personal visits. Most of the e-mails tore strips off

---

* Notable among these is Andrew Weaver, a climatologist and one of the lead authors of the vastly influential Assessment Reports of the Intergovernmental Panel on Climate Change.

me, although I recall some of a very different sort. Apparently, I had become not just a convenient target for environmentalists, but a spark of hope for those who took a different view and, in some cases, dare I say it, felt morally victimized by environmentalism.

One batch of e-mails came from some people working inside the national park bureaucracies of the United States and Canada, who were distressed by on-going programs in which entire lakes, streams, and even long stretches of rivers were being poisoned to kill all of their fish. If this sounds like a strange thing for national park personnel to do, you are not up to speed on the latest environmental thinking. Strong poisons were being used to kill *all* of the fish so that these bodies of water could be restocked by "native" species, species that were there before the arrival of Europeans on the continent. The "whistle-blowers" e-mailing me about these pro-grams were too frightened to say anything publicly, but they thought perhaps I might stand up and do so. Another group of mails and e-mails was from people whose land had been appropriated so that it could return to a state of nature as part of a park or nature preserve. They were also hoping that I could publicize their plight, since the media just seemed to be ignoring them.

Of special interest was a letter from a geophysicist along with a sample of his pub-lications. He explained that his research as a federal government scientist addressing the issue of climate change showed that removal of forest by forestry and agriculture significantly raised surface air temperatures.* Because this result could explain much of the observed rise in global temperatures, it challenged the view that global warm-ing is caused by the carbon dioxide emissions produced by humans. Although he was a contributor to the 1995 report of the Intergovernmental Panel on Climate Change, his work was now ignored, as were his proposals for combating climate change. This was no surprise to him, since he had been aware since the late 1970s that "environ-mental activists" had taken over the agenda of his governmental department. "Here I, a physical scientist, was completely out of place," he wrote. "These people did not want the facts, . . . they just wanted any environmental concerns, unsupported by data, to be put foremost. I was astonished!" There was a movement within climate science to create and enforce a consensus around the carbon dioxide view, so his research had gradually been stopped, its funding denied. He asked me to keep this subject in the public eye.

---

* His research (Lewis 1992; Lewis and Wang 1992, 1998; Lewis et al. 1993) involved measuring the temperature at various levels below the surface of the ground using boreholes. As is well known, the surface temperature of the ground is transmitted slowly downward through the underlying soil and rocks to yield a temperature record. The average surface temperatures going back for decades can be recovered from such measurements. His research showed that a sudden and permanent warming of a few degrees occurs when the ground cover, typically forest, is removed. The amount of warming correlates well with the heat energy that would have been used in transpiration by the trees that have been removed. This warming cannot be explained by global greenhouse warming, since it occurs all at once, at the time of deforestation for the different sites. Since most temperature records come from places where widespread changes in ground cover have occurred, these findings could explain temperature rises in these records on a basis other than anthropogenic global warming.

Then I was visited by a very distinguished professor of forestry who had recently retired, who conveyed a similar message. He had fled Hungary to escape its takeover by Russian communists, only to find himself many years later fleeing the scientific discipline of forestry itself as it was gradually taken over by ever more radical environmentalists opposed to any human use of the forest whatsoever. The amount of forest lost to logging, he told me, was only a fraction of what could sustainably be cut. The science on this was perfectly clear, but he was mischaracterized as an agent of the logging industry by the new environmentalist breed of forest scientists.[*]

I heard a quite different story with the same theme from a fellow who was not a licensed scientist, but an expert nevertheless. He had devoted his life to bears,[†] studying them both in the library and in the wild. He ran a well-known and well-attended school teaching those who worked in the wilderness how to handle bears. He wrote me to say that the environmentalists' often repeated claim that logging reduced bear habitat was the very opposite of the truth. There is nothing for bears to eat in forests, he told me. Bears live on the edges of forests, in clearings made by fire or logging, where they can find food. The false claims are just environmentalist propaganda, he said, invented by a coterie of biologists who never leave the comfort of their university offices and laboratories to see what is actually going on in the woods. His message had a special resonance for me. As someone who has spent some time in the wilderness, I too had noticed bears' extraordinary fondness for areas recently cleared by logging. You can hike through the woods and meadows for days without ever seeing a bear or any signs of bears, only to find both in great numbers—along with deer, hares, grass, and berries—in and around patches that have been forested. Not that such anecdotal evidence proves anything by itself, but it does provide food for thought.

Among the many other communications I received, one other should be mentioned. This was not from a scientist but from a woman who had retired with her husband on Galiano Island, off the west coast of British Columbia, where they had created their own version of paradise on a 23-acre lot. They had built a house and settled down to enjoy their senior years together, shepherding a small flock of sheep as a pastime. Because of the sheep, their lot was designated as agricultural land. This caught the unfortunate attention of the local nature conservancy group, which was enamored with "restoration ecology," and with the backing of the local government, proceeded to have their land confiscated to return it to its "natural state." Apparently, getting rid of the flock of sheep would do them no good. They were desperate for help from any quarter, and thought that perhaps I might assist them—somehow. The last I heard from them, they were on the point of being evicted. Their case needs to be mentioned because it brings home the point that more is at stake than just the truth—important as the truth is. Environmentalism has many victims (I believe that this is the correct word here), people who through no fault of their own have had their lives uprooted, ruined, even lost as they were caught up in the machinations of the environmentalist movement. This couple was living the very same pastoral dream, the dream of a life

---

[*] In subsequent chapters, especially Chapter 5, we will have opportunity to explore further the rise of "environmental science" and its implications for both environmentalism and science.

[†] James Gary Shelton; see his book *Bear Attacks* (1998).

lived close to the land in a preindustrial agricultural lifestyle, that inspires so many environmentalists. Nevertheless, they became victims of environmentalism.

While environmentalists tried to silence me with words or threats of physical violence, many other people encouraged me to continue with my investigations. I listened politely to both sides and waited for the dust to settle, concerned that the peace and quiet of my home life not be interrupted by environmental activists or media intrusion. But I quietly continued my program of research into the phenomenon of environmentalism. This book is the result.

I have always loved nature, yet I am not an environmentalist. From the time I was a small child, I loved animals, loved the outdoors, loved the spectacle of sunshine and storm. Among my fondest memories is that of my first trip, at the age of 6, to the Rocky Mountains with my parents. I remember shrinking in my sleeping bag as the bears sniffed around the edges of the tent at night. I remember drinking crystal water straight from mountain streams. I remember sitting on a mountainside looking down on so many pine and spruce trees that they looked smooth as a carpet down below. Growing up a city boy with sidewalks underfoot, I dreamed of living in a cabin in the woods. That dream was just a dream, and my daily reality is still the city, still the sidewalks. But my love of nature, of living things, remains. I still go to the mountains, and into the woods, whenever I get the chance.

So why am I not an environmentalist? Do we not hear every day of the devastation of nature, of the pollution of our air and water, of the extinction of species, of global warming? Do we not see these things for ourselves? Isn't nature in a state of crisis? Having studied this question in some depth and breadth, I can confidently say: No. It has suffered some shocks, all right, but it's hardly in a state of crisis. There are many things that remain to be done, and should be done, to improve the health of the environment. Nevertheless, long after you and I have passed from this world, long after the human population has shrunk to its preindustrial levels, long after the highways and cabins and ski lifts have returned to earth, nature will still be here, thriving and beautiful.

I am someone who hates waste and excess. When I take a shower I turn down the water as low as I can; I write on both sides of a sheet of paper; my cars have been chosen for fuel efficiency and whenever feasible are parked in favor of motorcycle, bicycle, or walking. That's just the way I am—I guess I have absorbed the ethics of my parents, who lived through the Great Depression of the 1930s: Waste not, want not. Whatever the reason, I am struck and saddened by the profligacy and waste I see all around me, the lights kept blazing all night long, the buildings heated though no one is in them, the hordes of huge vehicles rushing everywhere at once on the roads. It would be really persuasive to say to everyone, "Cut back, or there will be an environmental disaster!" But the facts simply do not support this contention. Nice as it would be to pull out this trump card in an argument, I will instead stick to the truth. When all is said and done, it is more important to save the truth than it is to save material goods.

Environmental science—which is not to be confused with environmentalism—reveals nature to be very robust. It is not weak, no matter how many times you may have heard it described as fragile, sensitive, delicate, as though it were a structure of sticks and straws held together by spider webs. Certainly, we humans have had an effect on nature, but that effect is largely on our own sensibilities, and concerns our feelings about nature rather than the health of nature as such. When we see a clear-cut forest, our feelings are aroused; we see the death of the trees as a wound we have inflicted on the tender body of nature herself—but what the retired professor of forestry who visited me said is indeed true. In fact, nature destroys more trees every year by lightning-sparked fires than loggers mow down with their chain saws. Indeed, nature would destroy much more forest than it does now if it were not for our efforts to spot and put out forest fires.

If you have ever wandered through a burned forest, you will witness a far more thorough desolation than anything you will see in a clear-cut. After a forest fire, the trees, the shrubs, the grasses, the very soil itself is burned away, right up to the edges of streams and lakes. Nothing stirs. This is what nature does as part of her everyday business. By comparison, a logged forest is teeming with life, with rabbits, deer, and elk that come to munch on the newly verdant ground cover that had previously been suppressed under the dark forest canopy. But despite the desolation of the burned forest, despite the mournfulness we feel among the charred trunks that stand like lonely sentinels, forest fires play a role in the natural cycle of death and renewal.

It is difficult for us to accept death. Millions of years of evolution have ingrained in us a love of life and a horror of death. But we must remind ourselves that death is just the other side of the very same natural process that gives us life. Ephemeral creatures such as human beings tend to overlook nature's big picture: Whether it falls to the logger or to lightning, the forest grows back, at least until the next ice age crushes it under miles of ice. And even then, after a few hundreds of thousands of years, the ice will retreat and the forests march toward the poles once again. There is no wonder that we fail to notice this big picture, given that our lives occur in a flash against the almost inconceivably huge background it presents.

Every summer I return to one of the most beautiful—and busiest—national parks on the planet. The marks of human beings are many. Sprawling towns have sprung up to house, feed, and entertain the legions of vacationers. The highways are clamorous rivers of cars, trucks, trailers, and recreational vehicles. The people in these vehicles wish the other vehicles were not there. They wish that they had the highway all to themselves, that they had nature all to themselves. They complain about the traffic, about the noise, about the crowds of people—and about the destruction of nature. Yes, the people in these vehicles have environmentalist sympathies. They don't like the traffic that they themselves create. They would like to see nature untamed rather than under their own wheels. They would like to see nature reign rather than subdued by human beings—all from the comfort of their front seats.

Yet, despite the highway, the motels, the towns, the many marks of humankind, nature does, in fact, still reign. An hour's walk into the woods beside the highway takes you to a place few people witness, where lodgepole pine perfumes the air, where purple fireweed blooms unseen, where grizzlies and black bears are the local authorities. By the time you have reached the ridge of the mountain a few miles back from the highway, the highway is invisible, and nothing can be heard but the

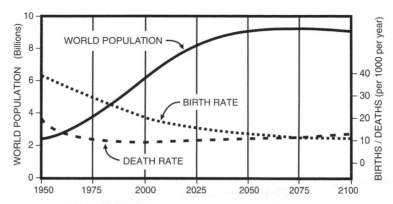

Figure I.1 World population peak. The United Nations estimates that world population will peak at just over 9 billion people in 2075, the year when the birth rate falls below the death rate. Because we are living longer due to steady increases in life expectancy, the population peaks very gradually, with the old heavily outnumbering the young. Birth rates continue to fall slowly and death rates continue to rise slowly up to 2100, causing accelerating population shrinkage. (Based on UNDESA 2004, pp. 5–17.)

wind, the birds, the insects, the distant whisper of rushing water—the same sounds that have filled these mountain passes for thousands of years. Over the ridge you see range after range of mountains invisible from the highway. They are topped with glaciers glinting in the sun, the remnants of the last ice age, and down below you see valley after cool green valley. These realms are open to those few people who take the time, the trouble, the energy, to walk—not to mention the risk of crossing paths with an angry or hungry bear. This place, and millions like it, was here through all the innumerable wars and armistices, all the loves and losses, all the comings and goings of human history, and was untouched by them.

The bottom line of this book is good news: This place on the mountain ridge, this lovely wilderness, will remain. I cannot give you a guarantee, for no one knows the future with certainty. Still, we can, and do, rely on reasonable expectations for the future—as we must. To live today we must have some belief about the morrow. And the vast range of evidence indicates that most human development will happen not in the wilderness, but along the highways, on the farms, in the towns, and in the cities. Not every one of the millions of places like this one on the mountain ridge will remain untouched, but most will. And by the end of this century, when human population is in the beginning of its long decline (Figure I.1), this place on the mountain ridge will be visited even less by human beings.

Eventually, inevitably, wilderness will begin to reclaim territory borrowed from it for awhile by humankind, and once again nothing will be heard but the wind, the birds, the insects, and the whisper of distant waterfalls. That is the good news.

If you cannot accept it, ask yourself why not. If you truly love nature, is it not cynical to reject this good news absolutely? If, on the other hand, you are willing to admit at least the possibility of good news, then please read on.

1 The Need for a Philosophy of Nature

I would have ... the scholar finely sift all things with discretion, and have him harbor in his head nothing by mere authority....

—Michel de Montaigne (1575)

There are many reasons that I am not an environmentalist, even though I believe it obviously necessary that we, the human species, develop the right relationship with the natural world. Environmentalism is a movement that has sprung up spontaneously from the soil of human concern and conviction, so it suffers from the weaknesses that afflict popular ideologies. It is not a system of thought, but a loose collection of putative facts, questionable creeds, and hastily conceived calls for action—fortified throughout with plain truths, worthy ideals, and sound plans. Because environmentalism contains so much that is right, it must be analyzed and evaluated if we are to have any hope of salvaging what is right and sound, even as what is wrong and unsound heads for its inevitable collision with reality.

I am not an environmentalist, but neither am I an anti-environmentalist. I am a non-environmentalist who would like to help found a philosophy of nature that recognizes environmental facts and values. Among these is the fact that we are entering a crucial century in the history of humankind, one where our own population will peak at about 9 billion (see Figure I.1), while ever larger proportions of us come to enjoy the benefits of modern life. This will be a crucial, epoch-making process, one filled with hazards for the natural world and for us, just as the environmentalists warn—but also one filled with promise. What will the facts be when this century has passed? That is largely up to us. As for values, I think everyone, environmentalist or not, would like to see nature, including humankind, emerge from the century whole, healthy, and flourishing. This is no time to rely on the accidental accumulation of ideas and programs that make up environmentalism. The time has come to resurvey our place in nature, as well as the place of nature in our lives. The time has come to identify the essential facts, to formulate the central values, and to develop a basis for discussing policy.

The time has come, that is, for a philosophy of nature. A philosophy of nature differs from environmental philosophy in that it is not based on the concept of the

Beyond Environmentalism: A Philosophy of Nature, By Jeffrey E. Foss
Copyright © 2009 John Wiley & Sons, Inc.

environment. Every organism is distinct and separate from its environment, and human beings are no exception. So to speak of the environment is to set up a conflict between us and the natural world. It is less divisive and more accurate to begin with nature as a whole, which includes us as an integral part. Nature has given birth to *Homo sapiens* just as it has given birth to every other organism on Earth, and we cannot exist except within nature.* Our body chemistry is engaged with the chemistry of the planet. The very air that flows in and out of your body to keep you alive as you read these words is a part of nature that untold billions of plants helped create by breathing out oxygen over millions of years. The fit between hand and glove is crude compared to the fit between us and nature: We are leaves on the tree of life. To speak of the environment is to speak about that tree and everything else that is not us. But we cannot solve "environmental" problems until we recognize human nature within nature, and nature within our human nature. So let us begin with nature, the natural world as a whole, and seek wisdom about it, our role within it, and its role within us. In other words, given that philosophy is the pursuit of wisdom, let us begin the philosophy of nature.

The failings of environmentalism are so numerous and diverse that a complete catalog would be impractical. Instead I have identified, in the following chapters, the main types of mistake. These I will describe starkly, in order to make them clear. There is a danger that this approach will alienate the very people who most need to be aware of what I will portray: namely, environmentalists themselves. If you would describe yourself as an environmentalist, I urge you to keep reading. Remember, I want what you want: that we find the right relationship with the rest of the living world, as well as with the nonliving world that supports it.

I grant—in fact I happily attest—that were it not for environmentalists, we would probably be in the midst of a severe (or more severe) environmental problem at this very moment. We cannot advance toward our common goal—humankind in the right relationship with the natural world in which and through which it finds its existence—unless our debt to spontaneous environmentalism is acknowledged. However, we also cannot advance toward our common goal until we also recognize where environmentalist thinking has gone wrong. I assume that everyone who is still reading this book at this page has some sense of concern about the environment. Your concern for the environment is invaluable, but you do not want to waste it on poorly conceived projects, or worse, have it discredited by investing it in ideas that cannot withstand the test of time. Bear with me, and I think you will see the obstacles more clearly—and the way forward.

---

* Some humans have left the Earth on relatively brief trips into outer space, which has been accomplished by storing oxygen, water, and food for use during the trip. Conceivably, we could live for longer periods away from Earth, so this is a qualification to the claim that we cannot exist except within nature. Interestingly enough, however, all engineering visions of human colonies or multigenerational travel far from Earth involve bringing sufficiently many plants and animals along to provide the needed food and oxygen by biological processes—so it is a qualified qualification, so to speak. In any case, at the present time the truth is that virtually every one of us cannot exist anywhere but within the natural world here on Earth.

1.1 THE GOAL OF THE BOOK

The ultimate goal of this book is not critical, but constructive: laying foundations for the philosophy of nature. Foundations must begin with clearing the ground. This is a big job in itself, given the large number of environmentalist constructions already in place. In the proper sense of the word, environmentalism is not a philosophy at all, because it is unexamined, unsystematic, and inconsistent—more a matter of practice than of theory. It is not an edifice of ideas, but a rough camp of converging interests, ranging from conservative lovers of wilderness on the right to radical haters of big business on the left. Philosophy is essentially an exploration of ideas, so our investigation of environmentalism will focus on the ideas that sustain it. As a popular movement, and as a collection of overlapping interests, environmentalism involves passion, faith, and history. Philosophy addresses the ideas that incite this passion, express this faith, and motivate this history, critically evaluating them for their logic, their truth, and their coherence. This is not to exclude passion, faith, and history from our investigation, but rather, to consider them from the point of view of the concepts that animate them.

1.2 THE STRUCTURE OF THE BOOK

The discussion will proceed along five parallel tracks:

1. The *primary text* that you are reading now, which is intended for the nonspecialist.
2. The *footnotes*, which will involve scholarly details and scientific technicalities. All scientific references cited as authoritative support in this discussion will be from proper, well-respected, scientific publications, although we will consider a range of important environmental views regardless of their source.
3. *Theses*, conclusions that are sufficiently important that they are marked as reference points in the philosophy of nature. These are not presented as proven truths nor as articles of faith, but rather, as focuses for ongoing study and discussion within the philosophy of nature.
4. *Case studies* of current issues of special interest within environmentalism or the philosophy of nature. Since the devil is always lurking in the details, in particular the scientific details, we must delve into them. When it comes right down to the real nitty-gritty, we need to know the numbers. But even when the debate seems to come down strongly on one side or the other of the issue in question, we should remember that case studies depend in part upon states of affairs that may change, upon ongoing scientific investigation, or both.
5. *Feature boxes* will designate detachable features of the ongoing discussion that are not essential but which fill out the discussion in relevant and valuable ways, in roughly the way that a detachable hood is a feature on a jacket or a detachable basket is a feature on a bicycle.

The advantage of having parallel tracks is that when you find yourself reading a section that has become, perhaps, too technical for your liking, or which is perhaps unappealing to you for other reasons, you can simply skip over it and go on to the next section, case study, box, and so on. Since the tracks are parallel, they all lead to the same destination in the end.

1.3   THE NEED FOR OBJECTIVITY

Every day in the news, on television, in the newspapers, and on the World Wide Web, environmentalists criticize people and businesses for their greed and carelessness in harming the environment. Every now and then we also hear from someone who is critical of the environmentalists. When an environmental "watch dog group" puts forward a speaker who condemns the use of genetically modified (GM) crops, predicting the collapse of ecosystems and human agricultural production as a result of their use, we sometimes also hear from a farmer who points out that the GM crop requires less pesticide, less fertilizer, or less water, and so also has an environmental benefit. The very next day, we will again hear from environmentalists who will claim that the farmer was a friend of, or in the pocket of, the seed companies that sell the seed for the GM crop. This is supposed to be damning criticism of what the farmer has said. Similarly, when an environmentalist critic comes forward to chastise people who want to harvest some trees for lumber, we may, although it is not very likely, hear from a logger who points out that the amount of forest in the United States has remained constant since 1920 (USDA 2000, p. 2) and that the amount of forest over the entire Earth has remained constant since 1948 (United Nations' Food and Agriculture data, 1949–1995), even as the population and agricultural output for both North America and the world have increased dramatically. The next day we will hear from an environmentalist who will criticize what the logger said on the grounds that loggers have vested interests in logging. This is supposed to discredit completely what the logger has said.

Critics of environmentalism are castigated and shunned because they are, supposedly, supported by business, by corporations, by unions, by right-wing organizations, or by other groups that have an interest in resisting some environmental claims and programs. Such people, we are told, are to be ignored because they have anti-environmental interests, and so are *biased*. Their views, unlike those of the environmentalists, are therefore said to lack credibility. Yet the people who make these claims are representatives of environmental organizations such as Greenpeace, Friends of the Earth, or the Sierra Club, groups that have an obvious and clearly stated interest in the issue—or even executives of these groups who draw their salaries from them. Why is it that the views of environmentalists are not thought to be biased given their obvious anti-GM, antilogging interests and the money they receive from the environmentalist organizations that support them? Apparently, it is assumed immediately that the interests of businesspeople are less trustworthy than those of environmentalists. For the environmentalist that is axiomatic, since business, particularly in the form of the dreaded multinational corporation, is seen as the main enemy of the environment.

Of course, it is just as quickly assumed from the other side, the side of business, that every environmentalist is against business, if not against the material progress of humankind itself.

But a moment's reflection reveals that both sides of this contest are interested parties who have deeply entrenched views prior to their appearance in the news. Both sides are funded by organizations with deep pockets. Reflection also reveals that attacking the people who make the arguments, instead of criticizing the arguments they make, is illogical. To attack the arguer rather than the argument is to commit the logical fallacy of *ad hominem*. A dentist can make a perfectly sound argument in favor of dental care—or against it for that matter. A thief can make a perfectly good argument against theft, and an alcoholic can soundly inveigh against the dangers of drink. We are told by pundits to "follow the money," to find a person's sources of income in order to see what makes him or her say and think the things that he or she thinks. This is somehow supposed to tell us whether or not to believe or take seriously the things that they say. But truth is not a football in a team sport. Truth is not won by picking a side and then judging whatever anyone says simply by figuring out whether or not they are on your side. Objectivity is the only guide to the truth. Objectivity requires listening to *what* is said and judging it on the basis of its evidence and logic, not on the basis of whether you are for or against the speaker.

I do not mean to suggest that we should ignore people's sense of right and wrong, their ethics, personal commitments, intellectual affiliations, political orientations, or to sum it up in a single word, their *values*. Indeed, I believe that one of the quickest, most efficient routes to discovering a person's thoughts—one's own or someone else's—is by searching out that person's values. What I *am* denying is that the soundness of a person's arguments and therefore the truth of their conclusions can be determined by the identification of their values. Truth and falsehood are determined by the facts, not by the values of the person speaking. To put it another way, I am committed to logic as the means of evaluating the soundness of argument. The fundamental principles of logic will be my guide in what follows, and their canons will sound whenever they must. Logic, the creation of the originators of the philosophical tradition I follow,* measures the soundness of an argument solely in terms of whether or not it can establish the truth of its conclusion. Nothing else is relevant. There are only two relevant issues logically speaking: (1) Would the premises of an argument entail the conclusion if they were true? and (2) Are the premises true?†

My intention is to evaluate all claims solely on the basis of the arguments that support them and to evaluate all arguments solely on the basis of logic. Of course, arguments that people put forward are a function of their values. Surely we must take an interest—an objective, philosophical interest—in the values that motivate arguments, including our own. Sometimes it is very difficult, if not actually impossible, to

---

* My training and expertise is in (and limited to) the Western philosophical tradition that is commonly thought to begin with the so-called pre-Socratic philosophers (Thales, Anaximander, Anaximenes, Heraclitus, etc.,) and has developed into today's Continental and analytic traditions. The study of logic itself is present in this tradition from the beginning, very obviously in the works of Heraclitus and Parmenides, and as a specifically identified philosophical subdiscipline by the time of Aristotle.

† Those with training in logic will recognize that the first issue concerns the validity of the argument, and the second, the soundness of the argument, assuming that it is valid.

really understand an argument without understanding what motivates it. The motivations of an argument are completely irrelevant to the soundness of the argument from a logical point of view, but they are not irrelevant to understanding a person's views, values, system of thought: in a word, a person's philosophy. But where the objective spirit reigns, no one should be embarrassed to admit their views, their commitments, their affiliations. We should take it as a mundane matter of fact that people do have views, do have strong likes and dislikes, do have passionately held values. There is nothing wrong with that. There is nothing wrong with professional environmentalists drawing their living from the billions of dollars raised by environmentalist groups, just as there is nothing wrong with their critics finding their source of income elsewhere. In this spirit of objectivity, we should have nothing to hide.

I have been cheered by what appears to be the gradual opening of minds on all sides. Perhaps this is more illusory than real—it is difficult to say. The mere fact that my research has found a publisher—to whom I am extremely grateful—is a bit of evidence of this gradual broadening of the discussion surrounding our relationship with the rest of nature. I welcome any move toward openness and objectivity when it comes to any issue, including those identified and popularized by environmentalism. The questions should not be closed, and those who want to have a look at them should not be branded as deniers or tree-huggers or enemies of the planet or superannuated hippies. Both sides need to be heard in any serious decision, no matter how obvious things may look at the outset. That's the only fair and impartial procedure. If truth and falsehood really are obvious when it comes to nature, this will still be the case after we have had a closer look. We take a second look before we cross the street. It cannot hurt to have a second look before we transform our values, policies, and economics to meet environmentalist objectives.

1.4   THE NEED FOR PHILOSOPHY

Objectivity, impartiality, fairness, freedom from prejudice, an open mind, value-neutrality—by whatever name you call it, this is the first necessity of philosophy. Unfortunately, there are so many unsound ideas circulating within popular environmentalism that at this time, objectivity can only be achieved by revealing their flaws. So I may give the impression of bias as I criticize one popular environmentalist conviction after another. If I do, however, it is merely an impression created by our place in the history of humankind's environmental thinking, where further progress demands that we set aside old ideas to make room for the new.* These new ideas will themselves be subjected to the same relentless criticism as their predecessors. Some of them will be identified explicitly and numbered for easy reference. Their

---

* Professional philosophers, at least, will not have to be reminded how typical it is for works in philosophy to begin with criticism of extant doctrine. Perhaps the most famous example in the modern era is Descartes' *Meditations on First Philosophy* (1641), which begins, in the first meditation, with "I was convinced of the necessity of undertaking once in my life to rid myself of all the opinions I had adopted, and of commencing anew the work of building from the foundation." While I intend nothing as radical here as wholesale rejection of all environmentalist opinion, I do share Descartes' conviction that we must at least clear sufficient space to begin from new foundations.

sole claim on our assent is their ability to withstand this criticism. No doubt they, too, will eventually be outgrown and swept aside by ideas of greater accuracy and deeper understanding. This is to be welcomed. Philosophers do not expect to find truth descending from the mountain, written in stone, but instead hope to take the first tentative steps toward wisdom.

Philosophy takes both the close view and the long view. Professional philosophers are, on the one hand, trained in the logical details, the hair-splitting minutiae of argument, and on the other, in the grand historical sweep of ideas. When people think of religion, for example, they think of the religions of those they see around themselves. But when philosophers think of religion, they think of the detailed arguments for and against God's existence, the justification of faith, and so on, on the one hand, and the great religious traditions down through history on the other. When people think of science, they think of the science they were taught in school and hear about in the news, whereas philosophers think about the gradual emergence of the current logic of scientific evidence and theory, alongside the thousands of years of scientific tradition that has gradually led to today's sophisticated science along what can only be called an indirect route. Philosophers have learned that the way to understand what people agree or disagree about is to reveal its logic and to trace its historical origins. Environmentalism is no exception: It will be understood, if it is to be understood at all, by attending to its logic and its historical development.

The philosopher's dual focus on argument and the history of ideas can bring some much needed clarity to the current struggles between environmentalists and their opponents, even if this does not automatically provide solutions. Because of its long historical perspective and focus on logic, the philosophical point of view is a bit above the fray. The idea is to shine a well-focused light on the subject, one that might even be called harsh, in order to get the clearest, most detailed view of it. Philosophy has its own project, one that is now well into its third millennium, a project to bring reason and light into people's lives. Presuppositions and assumptions that are seldom if ever stated explicitly act as unquestioned and invisible rulers of our thoughts, and thereby of our lives themselves. Philosophy's first task is not programming but deprogramming. Its goal is to free our thoughts and lives of these unacknowledged rulers. The rulers of our thoughts, like the rulers of our countries, should be chosen freely, with open eyes.

A philosophical work asks its readers to ask themselves why they do what they do, why they say what they say, and why they think what they think. Meanwhile, the philosophical writer must be asking these questions of himself or herself. The task is a communal one, taking the form of a discussion. So, I will be using the word "we" to refer to this philosophical grouping consisting of you, dear reader, and me. I expect that you will be asking questions about what I am saying as you read along. That is a good thing. I do not by any means expect that you will agree with me at every step—indeed, I expect that you will disagree at many points. I do not even expect that everything I am about to assert is indeed true, since I, like any human being, make mistakes. Happily, the real gold in philosophy is not the assertions, but the questions. I will have done my job as a philosopher if I have provided you with good questions—ones that must be addressed to the unacknowledged rulers of your thought and life. Our aim is to open a discussion, not close it, a discussion about nature and our lives and thoughts insofar as they have to do with nature.

1.5   THE RISE OF ENVIRONMENTALISM

We are animals of the species *Homo sapiens*, a type of large ape that arose by the process of natural selection. That process is, in fact, nothing other than the struggle for existence, a struggle that every organism wages with its natural competitors. Nature giveth, and nature also taketh away. Although we revere nature for giving us life, we must also fear it as the inevitable destroyer of life. In our struggle for existence we came to have a few features not available to our evolutionary ancestors: notably, a bigger brain, craftier hands, and a highly communicative tongue. These features permitted us to become involved in an enhanced form of cultural evolution not available to other organisms, and our cultural evolution in turn begat agriculture, civilization, science, and technology. This gave us a unique tactical advantage in our struggle within the natural domain. To an extent unprecedented in the biological world, we have begun, occasionally, to get the upper hand in this struggle. In fact, we have begun to enjoy the first beginnings of liberation from the struggle itself. Liberation is a good thing—if we realize that we are liberated only to the extent that we reject the struggle itself.

Unfortunately, our partial liberation from the struggle is all too easily confused with absolute victory and the unconditional surrender of nature to us. Our historical struggle with nature is not one that we can actually win, for if we were to defeat nature itself, we would defeat the very source of our own existence. We are animals, and we therefore depend on nature for our existence. Unfortunately, in our battles we have in too many cases overshot the mark: like the Romans, who were not content with merely defeating their historical enemy, the Carthaginians, but went on to erase Carthage itself from the face of the Earth. But even the Romans would never have dreamed of defeating the Earth itself—that would only be self-defeating in the end.

Our natural instinct for survival has often carried our struggle for existence too far. Although we romanticize our hunter-gatherer ancestors, they were the first to go too far in this way. As the renowned environmental scientist Niles Eldredge notes (1998, p. 35): "We modern humans were clearly like bulls in a china shop, disrupting ecosystems wherever we went." Eldredge attributes the extinction of huge numbers of species of larger mammals, birds, and reptiles to our hunter-gatherer forefathers, including our cousins, the Neanderthals.* We also romanticize the pastoral life of small-scale agriculture long before tractors, chemical fertilizers, pesticides, and

---

* This is the start of the famous "sixth extinction" that he, Paul Ehrlich, E. O. Wilson, and other environmental scientists describe. In Eldredge's words, "Modern humans ... reached Australia about 40,000 years ago, triggering a die-off of the larger native species of Australian mammals and lizards.... [J]ust a little over 12,000 years ago, humans first crossed the Bering Land Bridge.... Immediately the big hairies—the woolly mammoth, the mastodon, the giant bison, the woolly rhinoceros—became extinct." Species vanished in droves in South America, the Caribbean, Madagascar, in short, wherever we went (Eldredge 1998, p. 35).

genetically modified crops, but primitive agriculture only intensified our assault on the environment. All through Asia Minor, Greece, the Balkans, we repeatedly cleared the valleys and hillsides for pasture and crops, and then abandoned them when the soil washed away. It was only with the rise of our scientific–technological civilization that we began to notice the harm we were doing.

Fortunately, a crisis of extinction and habitat loss was only narrowly avoided by a sudden growth spurt in environmental consciousness spurred on by the activism of environmentalists—for which we owe them a debt of gratitude. It was as if the human race were in a gigantic airliner that was screaming down toward an inevitable crash with the natural world below. My experience of those days was much like that of people who lived in other American or European cities at the time. Although I grew up near the edge of the civilized world (there were endless miles of wilderness to the north and the west of my small city), under my feet there was concrete and in my nostrils there was the reek of automobile exhaust and the sulfur released by the cheap and plentiful natural gas that heated our houses. I played on the banks of a river into which raw sewage poured from numerous outlets of the city sewage system, where it blended with the effluent from numerous refineries and chemical plants that was launched into the faster-moving waters farther from shore by huge overhanging pipes. A few miles downstream the tea-dark waters flowed dank and foul-smelling over the gray mud and odious slime of the river bottom.

We children were told in school that rivers naturally cleansed themselves in about 10 miles. This magical self-cleansing property didn't seem plausible to me, or borne out by the evidence I could see with my own eyes. To top it all off, clouds of radioactive fallout released by aboveground nuclear weapons tests (particularly from the huge USSR devices) would sometimes swirl about us for weeks, with radiation levels so high they caused babies to be born prematurely. Even as children we suspected that the airliner of human civilization was screaming toward the Earth. Our generation grew up with the imminent possibility of the end of the world itself by nuclear holocaust. The idea that humankind was not quite sane, or possible suicidal, was ingrained in us every time we ducked under our desks when the nuclear air raid sirens sounded. The founding and now senior members of the environmentalist community come from this nightmarish background.

Fortunately, the airliner of human civilization pulled up at the last minute, engines howling, and collision with the Earth was narrowly avoided. By the time I was a young man, the false reassurances of my childhood that the environment would clean itself up were replaced by stern scientific warnings that it would not, and the process of cleaning up pollution began in earnest. Not only the rivers, but all the waters, the land, and the air as well became focal points of a general spring cleaning. The word *pollution*, which was scarcely heard in the days of Elvis Presley, became the universally recognized name of the curse of the modern world by the time of the Beatles. "Pollution" became the environmental imperative, the sharp administrative prod that spurred us to clean up the mess we had been making. The airliner gradually gained altitude and began to cruise at a safer distance from the ground. Today, the waters of the river of my childhood flow through the city nearly unblemished. The fish have returned, and people catch and eat them without hazard. The animals of the countryside have begun to reclaim their former territories in what is now urban territory. Deer graze in the parks and gardens, pursued by coyotes and, once again, the odd mountain lion.

Now the newly industrializing countries of the world are repeating the same mistakes that led to the environmental problems I witnessed when young. Once again they are determined to gain the advantages of industrialization, mindless of the pollution it creates. Once again the air, the waterways, and the soil are being used as dumping grounds. Once again the level of toxins in the air, drinking water, and foods is rising. They seem to be at the first stage of a pattern that is repeating itself.

1. In the first stage, nature is seen as both friend and as adversary.* The first imperative of all organisms is the struggle to survive, a battle that is waged with other agents and elements within the natural world itself. We struggle with disease, pestilence, famine, predators, wind, rain, and storm. The beauty of nature is recognized, especially in terms of the abundance of game or the fertility of the ground, but so is the destructiveness and fearsomeness of nature. Nature is both loved and hated. It is in the first stage that agriculture, civilization, and industrialization are eventually attained. In this first stage, environmentalism does not exist.

2. In the second stage the by-products of agriculture, civilization, and industrialization, such as deforestation, depleted soil, and pollution, begin to harm people, and nature is seen as vengeful. This seems to be the stage at which many industrializing peoples find themselves today. When we in the industrialized world found ourselves in this second stage, we took the steps necessary to clean up the mess we had made. We hope—and even with some confidence expect—that industrializing peoples will see their way of life, their economy, screaming toward Earth, and that they also will pull up before they crash. In the second stage, environmentalists reverse the received system of values in which humans were placed above the environment (i.e., nonhuman nature) and, instead, place the environment above humans.†

3. If we look carefully, we can see a third stage emerging in which nature is no longer seen as the adversary, but as our partner. Nature has taken a bit of a beating in the last century of our struggle with it, and we have begun to realize that it is no longer quite the fearsome adversary that has struck us down with

---

* In the Old Testament, which is shared by at least three great religions (Judaism, Christianity, and Islam), humankind is instructed by God to "subdue the earth." This expression of the first stage of environmental attitudes is by no means unique to this text and these religions. Indeed, it is a virtually universal aspect of cultures that live close to the Earth, as hunter-gatherers or low-technology agriculturists.

† In the terminology of environmentalist ethics, the second stage of environmental awareness adopted by people in general corresponds to "shallow ecology," in which the environment is valued but only in terms of its relevance to human beings. In this stage, the environmentalist instead adopts "deep ecology," in which the environment is viewed as intrinsically valuable (valuable in itself, regardless of human values). We return to these topics in later chapters.

disease, famine, and pestilence so many times in the past. We and nature are now seen as more evenly matched. So we now do not need to view nature as the adversary in our historical struggle for survival. Even while the struggle goes on, we can recognize that nature is our ally—indeed, the source of our existence—in this struggle. Our struggle is not merely with the other agents and elements within nature, but for nature as a whole as well.

For the last few decades we have been cast by environmentalists as the aggressor, and the environment has been cast as the victim. This is the second stage, in which our relationship to nature is still conceived as a struggle in which we and the environment are opposed. Environmentalism reverses the attitudes of the first stage by siding with the environment (i.e., nonhuman nature) instead of with humankind. In the environmentalist campaign for the triumph of nature over humankind, "environments" are viewed as sensitive, species are thought of as endangered, and nature is portrayed as weak and wounded. In the industrialized world, at least, we no longer fear that nature will strike us down with disease or famine. Instead, we fear that we may strike down nature itself. Environmentalists cry out that we must save the planet—*from ourselves*. Instead of viewing the environment with fear and awe, we view it with guilt and self-constraint.

As I show more clearly in the rest of the book, this is an overreaction to the pollution crisis of the mid-twentieth century. In fact, nature still holds the upper hand. Sure, over the last few centuries we have had a series of victories in our struggle to survive, but we are still just a tiny event against the background of cosmic changes that have brought the world to the present moment. So far we have been lucky. The plagues have been relatively mild, the famines relatively short, and the meteorites relatively small. But nature, the Earth, *has not been subdued*. There is another ice age coming, and there is, so far as we know, nothing we can do about it. Eventually, the sun will expand into a red giant and swallow up the Earth in its nuclear fire. It is only arrogance on our part to view ourselves as giants who will bring nature itself—the universe by another name—to its knees if we do not control ourselves. Worse, to take this view maintains the outdated emotional dynamic of our struggle with nature—it is to remain at the second stage. We have an opportunity now to move on to the third stage, where we see ourselves as partners with nature—junior partners, to be sure, but partners nevertheless. Diagramming the basis of this new relationship, this new environmental philosophy, is the final goal of this book.

We have just recently avoided a collision between humankind and the Earth itself and should not be surprised that a chorus of warning voices has been raised. Since the near-collision was unexpected and terrifying, our emotional response to it has been protracted and exaggerated. Our sense of alarm verges on panic even though the crisis has passed in the industrialized world and looks like it will be avoided in the industrializing world as well. It is not surprising that an unforeseen brush with disaster has created a compensating overcorrection in the direction of caution. But it is time to calm down, catch our collective breath, and take stock of the situation. If we do, perhaps we can develop a sound philosophy of nature that will enable it, including us as an essential part, to flourish. We should not fool ourselves: Nature will always have the upper hand and the last word. This is not a partnership of equals. Eventually, our species will be gone and nature will still be here. In the meantime, we can achieve a bit of freedom from the age old struggle for existence by working with nature instead of against it.

1.6   PHILOSOPHY OF NATURE AS A PATH

In our search for wisdom about nature we must be guided—and cheered—by a sense of shared purpose. I share with environmentalists the deep conviction that human beings must radically reconceive their relationship with nature if they are ever to make that relationship truly healthy and beneficial for all involved. This relationship can only be achieved if we are just as ready to reject unsound ideas as to welcome the sound. Environmentalism has done the world a great favor by bringing the natural world into our thinking and our day-to-day business. Unfortunately, given the urgency of environmental problems that the world faced in the midst of the twentieth century, when environmentalism was born as a powerful social force, our consciousness was raised not by calm thought and discussion, but by relentless environmentalist *advocacy*. Perhaps that was the strategy that the times dictated—it seemed crucial to get results right away—but nevertheless, it brought in its wake the one-sidedness that advocacy entails.

And so it was that environmentalism simplified complex issues. It exaggerated one side of them at the expense of the other. It focused on the dark clouds and ignored the silver linings. It created ringing campaign slogans: Recycle! Protect the wilderness! Save the planet! It grabbed media attention. Who hasn't seen the images of baby seals about to be clubbed to death for their fur? Or belching smokestacks? Or tiny rubber boats challenging huge tankers? Or people chaining themselves to trees? In this way, popular environmentalism was able to emerge from backstage to take a starring role in our daily lives. The danger is that the starring actor has begun to believe that her advertising copy is the truth, the whole truth, and nothing but the truth. Now that nature has gained the attention it so clearly deserves, we have to move beyond slogans to a deeper, truer understanding of our relationship to nature.

THESIS 1:   We need to go beyond environmentalism and establish a new discipline, the philosophy of nature, dedicated to a deeper understanding of the relationship between humankind and the rest of nature.

A philosophy is not the sort of thing that can be built in a single season, under the direction of a single architect. The goal of this book is to start a discussion—not end one—about "environmental" facts, values, and action, a discussion based on the traditional philosophical principles of logic, evidence, and fair-mindedness.

Achieving a fair beginning is ambition enough. When it comes to many-splendored nature, thorough study and complete arguments are out of the reach of a single book. The huge scope and massive complexity of the topic, along with its multifaceted importance to us, defy any attempt at completeness in so short a span. All we can hope to do is present a clear vision of the fundamental facts and arguments, and then trust that people's good sense will guide them to see their way to the truth.

2 Environmentalism's Apocalyptic Assumption

Defects of empirical knowledge have less to do with the ways we go wrong in philosophy than with defects of character: such things as the simple inability to shut up; determination to be thought deep; hunger for power; fear, especially the fear of an indifferent universe.

—David Stove (1990)

2.1 THE ENVIRONMENTALIST WORLDVIEW

Recently, a couple of cheerful, but nevertheless gravely serious, Greenpeace fundraisers came to my door asking if they could count on my monetary help in Greenpeace's struggle to "save the planet." When I asked what the planet needed to be saved from, they almost rolled their eyes at my woeful ignorance, and began to recite solemnly the litany of humankind's environmental sins, which, to be frank, I had already heard many times before: extinctions of species; destruction of habitat; increasingly dangerous levels of toxins in the air, the soil, the water, and mothers' milk; glowing dumps of nuclear wastes—in short, the defilement of Earth. To hear them, you would think that we should all be in bed, sick from the poisons we had been eating and breathing, and that nothing but ashes, smoke, and drizzling toxic rain should be visible outside.*

---

* Interestingly enough, a few years earlier some other Greenpeace fundraisers had sought money explicitly for the Greenpeace campaign to prevent nuclear testing by the French in the Muroroa Atoll in the South Pacific. I had then asked whether they thought preventing the French test would stabilize or destabilize the global balance of nuclear weapons and whether, therefore, it would increase or decrease the probability of nuclear war. They replied that Greenpeace had no political goals, and that its only concern was the good of the environment. When I suggested that Greenpeace's hindering of the French was inevitably political and must have political consequences, they were unmoved. Nuclear testing was bad for the environment, and that was good enough for them: Nuclear testing had to be stopped. They were sure that anything that was good for the environment would, as if guided by an invisible hand, work for the greater good politically, economically, socially. As unbelievable as this is once it is expressed plainly, the notion that environmental values magically transcend all other values is nevertheless a common article of faith for environmentalists. We return to this issue in Chapter 3.

Beyond Environmentalism: A Philosophy of Nature, By Jeffrey E. Foss
Copyright © 2009 John Wiley & Sons, Inc.

In fact, the suburb where I live is green, the trees are plentiful and filled with squirrels and birds (including owls, hawks, goshawks, and Steller's jays), and after dark the deer and raccoons wander and raid people's gardens. My suburb is not very special in this regard. The return of wild species to the cityscape is not unique to my city but a common feature of many North American and European cities these days. I have been in suburbs in three continents that have more or less as many trees, squirrels, fields, rabbits, walking trails, birds, and bird-watchers as mine does. The streams, brooks, and waterfronts in most urban areas have been restored from their sorry state a few decades ago, when they were apt to be used as sewers or dumps, so the return of wildlife includes fish as well as mammals and birds.

"It looks nice," the female said, "but it's not natural. What you see around you is not a natural environment. Anyway, it's not sustainable. We owe it to future generations to leave them a world that they can actually survive in, not a toxic wasteland. We are literally killing the planet."

"Suburbia is an illusion," he added. "It all runs on fossil fuels, which we can't keep burning forever. It's all got to come crashing down sometime. We'll be *extremely* lucky to make it to the middle of the century." I agreed that it did seem unlikely to me that we would still be burning gasoline in our cars in 50 years' time (mind you, I was picturing technological progress, not environmental collapse).

"Well, perhaps then, you would like to help us combat the *environmental crisis*," they said, hopefully. The belief that the environment is in a state of crisis is the most common article of faith among environmentalists. The title of a recent book by a distinguished biologist, Niles Eldredge (1998), shows what I mean: *Life in the Balance: Humanity and the Biodiversity Crisis*. In *Time to Change*, a biologist and television-environmentalist advocate, David Suzuki (1994), repeatedly uses the phrase "environmental crisis." This list could be extended indefinitely,* but there is no need. If you are an environmentalist, you will take it as given that the world is in an environmental crisis.

They said the crisis was urgent, and I invited them inside to tell me more. Of course, I had heard about the environmental crisis before. It was a concept that became popular after the publication by the best-selling environmental book ever, *The Limits to Growth* (Meadows 1972). This book of the Club of Rome, which sold some 30 million copies worldwide in some 30 languages, created a sensation when it was published in 1972. I remembered reading it with some alarm years before either of my fundraisers had been born. It had employed MIT-designed computer models (which in those days automatically implied scientific authority) to forecast the future of the planet. The future it predicted was dire: massive environmental degradation,

---

* A search of the online bookstore Amazon.com lists 6048 entries for "environmental crisis," while the Internet search engine Google.com yields a whopping 24,100,000 entries for the same phrase (August 1, 2007). Perhaps one of the earliest uses of the concept in a book title is *World Eco-Crisis* (Kay and Skolnikoff 1972), which, very interestingly, contains an introduction by Maurice F. Strong, then Secretary-General of the United Nations Conference on the Human Environment. He went on to become the architect of the Kyoto protocols as Secretary-General of the 1992 Rio Earth Summit. In his introduction he says that international agreements will be necessary to deal with the crisis, and that these must issue from the United Nations, where some 35 years later he still works to produce the necessary legal apparatus to deal with this supposed crisis.

undrinkable water, unbreathable air, unproductive farmland, illness, famine, despair. All of this was supposed to have come to pass well before the end of the century.*

I could not help but think that the urgency of the environmental crisis had a time scale unlike any we have seen before. It is continuous, ongoing urgency, ongoing for decades, in fact.[†] Although environmental apocalypse is upon us, or has even over-taken us, the crunch always remains in the near, but unspecified, future. This is a form of urgency that cheats the test of time. We are reminded here of old movies in which the heroine is tied to the railway tracks, screaming, while the train rushes up to her—over and over again. For nearly half a century, at least since Rachel Carson's *Silent Spring* (1962), we have been told to repent, that the end is nigh—told, in fact, that we are al-ready in the very midst of the final destruction—and yet things have not gotten worse.

How could I begin to talk with these Greenpeace fundraisers when they believed in a crisis that they had never seen? Suspecting that their concern for the environment might perhaps be the expression of concern for themselves, I suggested that instead of saving the planet from us, maybe they should really worry about saving us from the planet.[‡]

"When the planet gets tired of us," I said, "it will just brush us off like this," and I made the motion of brushing a crumb off my sleeve. I was thinking of the many species that had been removed from the face of the Earth before us. All it took was a new plague, a new global ice age,[§] or a comet racing toward us from the Oort cloud.

"No," they said, the planet would never hurt us, although we hurt it continually. They explained that the environment was naturally balanced and that only our species upset this balance of nature. Nature itself, they reassured me, was wisely benign toward every organism in the environment. The environment was intrinsically good, and would do good, and only good, if it were not for us environmentally wicked humans. Yet some 20 meters away from these petitioners were outcroppings of the scarred rocks which all over the midlatitudes of our planet bear mute testimony to the last ice age. Compared to it, all the works of man, all of our environmental

---

* The follow-up book, *Beyond the Limits: Confronting Global Collapse* (Meadows et al. 1993), does not back away from the apparently false predictions of its predecessor, which, if they had come true, would have been obvious. Instead, it takes the view that the limits to growth can be passed without any obvious results, at least at first, just as a wild population of deer, say, can "overshoot" the limits of its grazing area without any obvious sign of danger, only to be decimated by a disastrous famine later on. Plausible or not, this transformation of global collapse from an observable to a theoretical state does constitute a stark change from the first report. To reaffirm the urgency of the environmental crisis, a third book has also been published in this series (Meadows et al. 2004).

† The following title from the prestigious journal *Science* is poignant in this regard: "Conservation: Tactics for a Constant Crisis" (Soulé 1991).

‡ Expression of fear and concern for humankind, particularly "future generations," is also a staple concept of environmentalism. It is a very unstable concept for environmentalists: As we shall see in more detail in a later chapter, there is an antihuman element in environmentalism as well. Since we are the sole source of environmental harm, it would be better for the environment if we did not exist. Since environmentalists' guiding ideal is the good of the environment, they must see their own species in a very unflattering light. The following book title expresses another concept that is common among environmentalists: *Nature's Revenge* (Johnson et al. 2006). The idea is that nature will give us grief, and that we deserve whatever we get: Nature will be getting vengeance, we will be getting our just deserts.

§ The evidence is mounting that once or twice the entire Earth has been covered by glaciers in extreme ice-age conditions.

effects since we were spawned by evolution, pale to insignificance. During the ice ages, forest and field, lake and marsh, were frozen solid and bulldozed south, the soil itself scoured off the surface of the ground, and the very rocks below deeply scarred. Environmentalists talk about environmental destruction and yet seem not to have heard about the ice ages. The miles-thick crust of ice actually weighed large portions of Earth's land surface itself below sea level (as it still does in parts of Antarctica and Greenland). It is quite likely that the glaciers will return and that the ponds and meadows and forests around us will be utterly destroyed, along with all of the animals that now make their homes in those ponds, meadows, and forests. The planet will brush us off—again.

"You must have heard about global greenhouse warming! That's something we're doing to the planet. It's not doing it to us." They launched into an explanation of the basic principles of the scientific model: We use fossil fuels, this releases carbon dioxide (CO_2) into the air, and CO_2 acts like the glass walls of a greenhouse, letting in the pure white light of the Sun while trapping the red light that Earth would otherwise excrete into outer space, warming the planet disastrously. I knew from experience that it would have been pointless to explain that greenhouses do not work by tapping infrared radiation but by interrupting circulation. That was a long story, and these people were too busy to spend that much time on someone like me, who was obviously an environmental dinosaur. They said gravely that unless we stopped using oil and coal, disaster upon disaster would descend upon not only us, but that millions of others species would be doomed to extinction. "Extinction is forever," she reminded me. Then she said the glaciers would melt and the seas would rise up to flood New York, London, Hong Kong, and Rio de Janeiro, along with great swathes of Bangladesh, wiping out millions of the poorest, most desperate people on the planet.

It would have seemed callous to interrupt them in their solemn delivery of this grave warning, but it seemed that they were laboring under the false impression created by Al Gore's film *An Inconvenient Truth*. In that film Gore shows coastal cities flooding under 30 feet of water, implying (although never saying) that this would be the immediate result of global warming—but the Intergovernmental Panel on Climate Change (IPCC),* the environmentalists' own recognized authority on the theory of greenhouse warming, had predicted only an 8-inch rise in sea level (20 cm plus or minus 10 cm) by 2100 (IPCC 2007a, p. 812, Fig. 10.31). An 8-inch rise in sea level does not spell disaster—it barely spells nuisance. In fact, this is less than an inch per century above the background rate of sea rise over the last few centuries (Hancock and Hayne 1996), so global warming has a negligible effect on sea-level rise.

Nor did I mention that for all we know, once again according to the IPCC itself, adding CO_2 to the air might actually result in half as much warming, or no warming, of the planet, as the IPCC itself admits. The IPCC states very clearly in its famed Third Assessment Report[†] that Earth's climate is "inherently *chaotic*" (IPCC 2001a,

---

* The IPCC is the scientific body created by the United Nations that has been warning the world about Global Greenhouse Warming and recommending immediate implementation of the Kyoto protocols to reduce the use of fossil fuels.

[†] This report is famous because in it the IPCC said that due to advances in climate modeling during the five years since its previous report, it had finally concluded that we human beings were the cause of the recent warming trend, mainly via our use of fossil fuels.

p. 773).* In a chaotic system, the IPCC said, "there are feedbacks that could potentially switch sign" (ibid.). What this means is that something which normally has one sort of effect may suddenly have the opposite effect. Since the warming itself was a feedback of the increasing CO_2, it followed that it, too, could even switch sign and so cause cooling rather than warming. In other words, global temperature is unpredictably wiggly—the signature of chaos in action. Indeed, nature as a whole, as revealed in all of its various measures, is unpredictably wiggly. Predicting the weather next week was difficult enough—what made anyone think that they could predict climate a century in advance? This was conceivable, but it hardly seemed likely. Could it possibly be well advised to stop using fossil fuels, and bring the economy to a virtual halt, on the basis of such a risky prediction?

NATURE IS UNPREDICTABLY WIGGLY

The concept of dynamical *chaos* is sophisticated, and its impact on science since its rediscovery in the 1980s (Henri Poincaré, 1854–1912, seems to have noticed it first, in the 1890s) has been profound. Fortunately, a simpler working idea of chaos is available: the *unpredictable wigglyness* of most natural measures. For example, if we look at a graph of global temperatures for the last 10,000 years (see Figure 2.1D). we see that the graph is wiggly, repeatedly reversing direction for no apparent reason.

From Figure 2.1A, 10,000 years ago (8000 B.C.E.) we see the Earth in the very last millennium of the last ice age. We can see that the temperatures climb 4 degrees or so in a herky-jerky sort of way over another thousand years to today's level, marked by the zero line. We also see that temperatures still continue to jump up and down abruptly by a degree or more. For example, there was a spike in temperatures 5000 years ago (3000 B.C.E.) of about 2 or 3 degrees that lasted about a century. Generally speaking, we do not know why the temperature jumps up and down this way, and these changes are referred to as *natural variability*. Chaos theory assumes that a reason does exist (it is a

---

* This is the statement made by the IPCC in its Third Assessment Report: "The climate system is particularly challenging since it is known that components in the system are inherently chaotic; there are feedbacks that could potentially switch sign, and there are central processes that affect the system in a complicated, nonlinear manner" (IPCC 2001, Section 14.2.2, Predictability in a Chaotic System, p. 773). It is vitally important to fully realize that the chaotic nature of climate is an inherent, hence inescapable, characteristic of climate, and one that renders prediction difficult *in principle*: "These complex, chaotic, nonlinear dynamics are an inherent aspect of the climate system. As the IPCC WGI Second Assessment Report (IPCC 1996) has previously noted: "Future unexpected, large and rapid climate system changes (as have occurred in the past) are, by their nature, difficult to predict" (ibid.). How difficult? It is impossible to say, precisely. In fact, it is quite possible that any prediction, no matter how certain it may seem, could be completely wrong: "This implies that future climate changes may also involve 'surprises'.... these arise from the nonlinear, chaotic nature of the climate system" (ibid.). In short (and as we shall see in more detail in what follows), we know on scientific grounds that we cannot know quite what the climate will do or why.

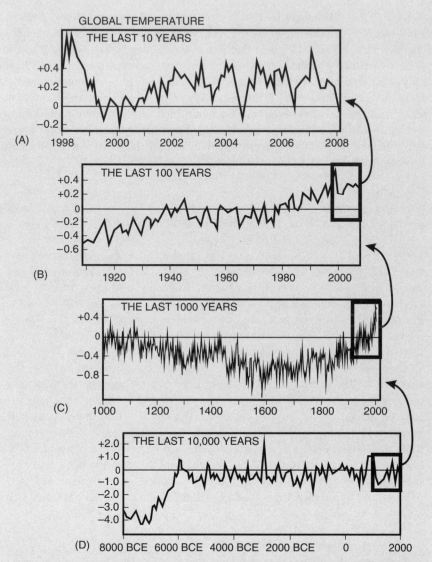

Figure 2.1 Global temperature is chaotic. One sign of deterministic chaos is irregular variation—or wigglyness—at all scales. Here we see global temperature estimates at 10, 100, 1000, and 10,000 years, and each shows wigglyness. Features that dominate at some scales, such as the famous "global warming" over the last 100 years, may disappear at other scales. In the last 10 years temperature has been falling. A thousand years ago temperatures were warmer than they are now. In 3000 B.C.E. temperatures were much higher than they are now, while in 8000 B.C.E. it was colder by several degrees, as Earth was still in the grip of the last ice age. (Based on Moberg et al. 2005, Brohan et al. 2006, Staniford 2006, and Carter 2008.)

deterministic theory), but proves that these jumps (because of the interminable complexity of the calculations) usually cannot be predicted.*

Part of the reason that these wiggles cannot be predicted is that they depend on even tinier changes that occur on a scale too small to even appear in this graph. One popular expression of chaos, the "butterfly effect," due to Edward Lorenz, the climatologist who rediscovered chaos in the 1960s (see Lorenz 1991), captures this feature: A butterfly flapping its wings in Brazil can cause a tornado several days later in Texas. Had the butterfly not flapped its wings, the tornado would not have happened, or would have happened at a different time or place. So a chaotic system is one that is *infinitely sensitive* to a range of variables: Tiny events can be amplified into enormous effects.

This means that to understand the large-scale changes we must turn to the small-scale changes. However, if we look at a graph at a smaller scale, as in the 1000-year graph just above the 10,000, we see that it is just as wiggly, just as full of unexplained natural variation. If we turn to yet smaller scales in the 100-year and 10-year graphs in the figure, we see the same wigglyness at these scales as well. Because the climate is a chaotic system, every smooth line turns out, on closer inspection, to be unpredictably wiggly. Ultimately, if we were to measure the temperature at any point in the atmosphere over a tiny fraction of a second, it, too, would still be jumping up and down unpredictably as molecules collided randomly with our measuring instrument.[†] This wigglyness at every scale is yet another characteristic of chaos.

Chaos means that our notions of temperature, global temperature, climate change, and so on, are merely human constructions, merely crude measures we place on a much finer-grained, infinitely complex reality. Weather and climate are prime examples of chaos that we witness every day. It is no accident that the rediscovery of chaos in the late twentieth century by Lorenz emerged in the study of computerized climate models. Chaos does not mean that prediction is impossible, only that it will always be difficult and imprecise. Chaotic systems may have various sorts of metastability, as it is sometimes called, which permit prediction using coarse measures over relatively short periods of time. For example, we know that it will generally be warmer in summer than in winter—but cannot reliably predict the temperature for any particular location for any particular day.

Since virtually every other natural measurement (e.g., populations of different plants and animals) is linked to climate, we will find that they too are unpredictably wiggly. Thus, nature itself is unpredictably wiggly—that is, nature is *chaotic*.

* See Foss (1992) for an accessible introduction to this topic.
[†] Thermometers themselves are instruments designed to dampen out variations in the kinetic energy of the atmosphere in order to measure its average kinetic energy (this being the definition of temperature). Without this dampening, we would see the temperature at a given point in the atmosphere jumping up and down unpredictably as this point suffers collisions with the individual molecules of the atmosphere.

Anyway, according to the IPCC's own predictions, even if we were to stop the use of fossil fuels within a few decades, this would have only a negligible effect on global warming in the next century: It would reduce by half a degree the warming of 3°F (2°C) predicted by 2100. Besides, it seemed quite unlikely to me that 100 years from now we would still be using mainly fossil fuels. In 1900, people still relied on horses for transportation and oxen for agriculture, while a century later they were jetting around the skies. Surely we should expect just as much change in the next 100 years as in the last. But I didn't get a chance to mention any of this, for my Greenpeace guests had moved on to predict terrible hurricanes that would "devastate" the world.

"Have you heard about Chris Landsea?" I asked. They had not.

"Haven't you heard of the Sixth Extinction?" my environmentalist fundraisers demanded. Indeed, I had, but according to the rules of polite conversation I had learned as a child, it was my turn to speak.

"Chris Landsea resigned from the IPCC because he believes that it has become politicized. He was their expert on hurricanes, and he concluded, as hurricane experts generally have, that global warming has only a tiny effect on hurricanes. The fellow in charge of writing the IPCC reports ignored his own experts, and instead gave interviews saying that global warming will cause hurricanes to become more intense, and that the very large number of hurricanes in recent years was due to anthropogenic greenhouse gases. Landsea repeatedly told him and the IPCC that this was not true—but they ignored him. So he was forced to resign."*

They sat silently, looking at me resentfully, as though a bad smell had been released into the room—as though the very idea that hurricanes would not devastate the world was unwelcome.

"It's all on the web," I said. "You can read it for yourself. Landsea made a formal statement."

"There is an overwhelming scientific consensus when it comes to climate change," the male Greenpeace member said.

"He must be funded by a multinational corporation with a vested interest in dumping megatons of carbon into the atmosphere," she chimed in.

My heart sank. I had often come to this point before with bright young environmentalists like these. They were good people, people who were ready, willing,

---

* Landsea's stated reason for resigning from the IPCC was "the misrepresentation of climate science while invoking the authority of the IPCC" by Dr. Trenberth, "Lead Author responsible for preparing the text on hurricanes" for the IPCC Fourth Assessment Report, along with the fact that "the IPCC leadership dismissed my concerns when I brought up the misrepresentation" (Landsea 2005). He points out that (1) "any impact in the future from global warming upon hurricanes will likely be quite small" (by 2100, assuming that global warming occurs as predicted, "hurricanes may have winds and rainfall about 5% more intense than today"); (2) "even this tiny change may be an exaggeration;" and (3) amazingly, the IPCC had admitted these conclusions in its previous reports. He concludes, on the basis of a lengthy interchange with the IPCC leadership (which he also made public), that it has, at least in part, "become politicized." As I will argue in Chapter 5, environmental science was the product of an environmentalist agenda from the start, so always had an element of political advocacy, inasmuch as it calls for action in the political domain.

and able to work for the greater good—which they took to be raising money for Greenpeace. I knew from the case of my own children that people of their age are taught environmentalist values and theories all the way through school. Their science texts and social studies texts lamented the environmental crisis, and professed environmentalist values as necessary to avoid environmental apocalypse. I recalled school projects in which my kids were instructed that the only sustainable form of human life would, in effect, repeal the industrial revolution. Farms would be small; no human-made fertilizers or pesticides would be used; only horses and oxen would be used for plowing and harvesting; people would live in small houses; cities would also be small, since population would be controlled and more people would have to live in the country to produce the needed food; long-distance trade would disappear; people would eat only local foods and use only local goods—and every person, animal, and plant would be happy, and every day would be sunny and bright. There were also villains in these lessons, and they were always impersonal: business, greed, and the dreaded "multinationals."

My Greenpeace alms-gatherers were automatically certain that anyone who doubted any part of the environmentalist doctrines they had absorbed as small children was a "denier" and probably in the service of dark powers such as the multinationals. They swiftly convicted deniers of falsehood because they were (supposedly) always funded by some (unspecified) multinational group with an interest in environmental issues. They did not realize that they had just falsely accused me of an intellectual crime, and they did not see the irony that they themselves were collecting money for a well-known multinational group with an interest in environmental issues and an annual budget of hundreds of millions of dollars, which it devotes solely to the promotion of its interests (not to mention the salaries of its executives and employees). Their minds were made up, their convictions enclosed completely in rhetorical armor. I asked them if they would look up Chris Landsea on the Internet. I might as well have asked them to start smoking. They looked at each other, and with a tacit nod, switched gears into a mode that I suspect had been taught them by their Greenpeace leaders as the proper move at this point.

"Well, it's obvious that you have a differing point of view," he said.

"Yes," she said, "we have to agree to disagree, and leave it at that." And so they headed for the door.

Environmentalism is a complex phenomenon, and not all environmentalists have the same views on every environmental issue. However, there are strong family resemblances among them. Mary Jones may have the Jones blue eyes and cheerfulness, while John Jones has the Jones chin and height, so that John and Mary are clearly Joneses although they do not share resemblances with each other. Similarly, each environmentalist shares enough characteristics with the others that his or her membership in the environmentalist family is clear. The principal set of environmental family resemblances seem to be these: that our form of life is unsustainable; that we are endangering life in general with our pollution; that our use of fossil fuels will cause runaway global warming; that we are causing a sixth great extinction; that we have upset the balance of nature—and that all of these things amount to an urgent environmental crisis.

2.2   IS THERE AN ENVIRONMENTAL CRISIS?

This is a very big question, but it seems that the onus is clearly on the environmentalist here. Obviously, we affect nature. Like any other species, we live by constant interaction with nature. Since our population is about to peak within a couple of generations (see Figure I.1 in the Introduction), our effect is unsurprisingly large at this point. But effect and harm are distinct. They are not the same. Peace and prosperity are at record highs right now, as is environmental awareness. Environmentalists are asking us to change our way of life radically: that is the intended effect of accepting their claim that there is an environmental crisis. Since acceptance of the claim would have such a disruptive effect, it is not to be taken lightly. Indeed, the claim of environmental crisis should be accepted only if the evidence for it is undeniable and its consequences grave, for the simple reason that changing our way of life radically would itself undeniably have grave consequences.

There is no way to settle this issue in a few pages, but we can begin to shine a light into some of the corners. The environmental crisis that environmentalists describe is the sum of a number of problems of various sorts: environmental, social, ethical, evaluative, and philosophical. We can hope to at most touch on the most important of these in the following pages. One of these, global warming, is so important, so timely, so complex that merely outlining the issues will require a thorough discussion, which we take up in Case Study 7.

Let's begin with pollution. How bad is it? In one of the worst cases of London smog, in December 1952, 4000 people met their untimely demise in just seven days (Lomborg 2001, p. 164). Will those days return?

CASE STUDY 1: IS POLLUTION GETTING WORSE?

Yes: In industrialized countries, allergies, asthma, and cancer have increased in recent decades, a sign of increasing pollution despite society's half-hearted efforts to control it. In newly industrializing countries such as China and the reindustrializing countries of the former USSR, where even the half-measures of the industrialized countries have not been taken, pollution is rampant. Pollution does not respect national boundaries, and in the absence of international laws cannot be controlled.

No: There is plenty of evidence that the trend in pollution is definitely downward. This is definitely true of air pollution, as can be seen in Figure 2.2. This graph is based on U.S. data, but they are typical of the developed nations. This graph considers the *criteria pollutants* as they are called, since they are the most toxic for life in general. Emissions of all of the criteria pollutants are trending downward and are generally at less than half their levels a few decades ago. The worst of them, lead, has been reduced the most, some 95% or so. In addition, countless measurements and constant

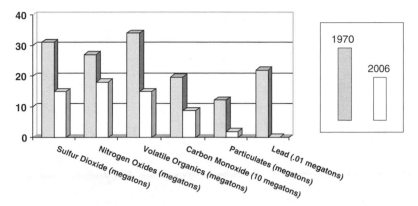

Figure 2.2   Air pollution trends. Air pollution in the United States has trended downward since 1970, during a period of steady growth in population, industrialization, and the economy. The sharpest decline has been in the most toxic pollutant, lead.

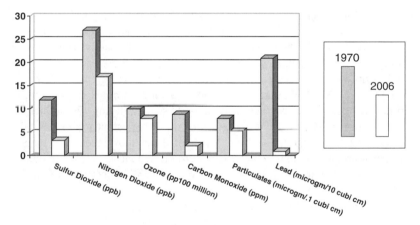

Figure 2.3   Air quality trends. Air quality in the United States has improved steadily since 1970, showing that reduction in emissions has paid off.

monitoring show that the reduction in emissions has reduced their levels in the atmosphere as well, as shown in Figure 2.3.*

Perhaps most significant of all is the evidence that shows the levels of pollutants are decreasing in the bodies of animals as well, as exemplified in Figure 2.4. This graph tracks the trend of a very troubling sort of pollutant, the persistent toxin, such as polychlorinated biphenyls (PCBs) and DDE (the metabolic product of DDT in the body) in the bodies of animals living in the wild (Granite Island is near the north

* Figures 2.2 and 2.3 are based upon U.S. Environmental Protection Agency data (EPA 2006, and EPA 2007a through EPA 2007g).

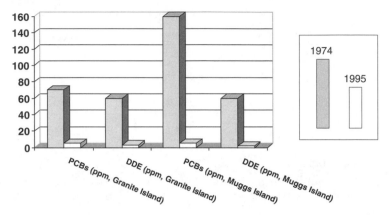

Figure 2.4 Persistent toxins trends. The levels of persistent toxins have fallen steeply in the place where it matters most: the tissues of animals. (Based on Ryckman et al. 2005.)

shore of Lake Superior) and in a highly industrialized area (Muggs Island is in Lake Ontario next to Toronto in the midst of the industrialized northeast corner of the United States). Toxins in the bodies of living things is ultimately what our concern about pollution is all about, so it is very reassuring that these levels have fallen so steeply. It is noteworthy that these measurements are of herring gulls, which are high in the food chain. Since they are at the top of a long complex food chain, toxins concentrate in their food and thus in their bodies, magnifying the ambient levels thousands of times. The fact that the levels of these persistent toxins has fallen so far and so fast tells us that the ambient levels of these substances must have fallen even farther than these graphs show. It is nice to know that our efforts to control the release of these substances has been so effective. It is also nice to be reminded just how robust and resilient nature can be.

So we can rest assured that when it comes to pollution, things are getting better, not worse, for ourselves and other living things, especially in the most developed nations. As for the increase in cancer rates, that is merely the downside of the fact that people are living longer. The cause of the increase in asthma and allergies is not known, but we do know it is not higher pollution levels, since the main increase is in the industrialized world, where pollution levels are down, not up.

Common Ground: More needs to be done, even in industrialized nations, since the levels of many pollutants are still harmful for humans and other living things, especially in cities during weather that provides poor air circulation. We do want to see the trend lines continue downward. Also, these data do not apply to industrializing nations, where pollution is greater than in the United States and even increasing. Something should be done about this. We should work to ensure that growing prosperity will lead to a higher level of environmental awareness (the second stage; see Section 1.5) as it did in the industrialized world.

One important aspect of the downward trend of pollution in the industrialized world is that it was achieved during an unprecedented rise in population, prosperity, and health. This counts against the environmentalist opinion that modern industrialized life is unsustainable. There is pretty persuasive evidence that human beings have caused massive species loss and habitat loss both in their hunter-gatherer form of life and in their preindustrial agricultural form. If that is true, the *only* thing that ever reversed our species' historical tendency to harm nature has been environmental standards within the context of increasingly sophisticated industrial methods. In any case, it is clear that we can improve environmental standards even as we improve our own health and happiness.

Do environmentalists dare accept this good news? Clearly, environmentalists take a very dim view of modern industrialized society. If this is because they think it is unsustainable, they should take a brighter view if it really is sustainable after all. If they do not, why not? Obviously, the belief in environmental crisis is crucial in pragmatic terms for maintaining the momentum of the environmentalist agenda and of the environmentalist establishment, including institutions such as Greenpeace. Certainly, we do not want to slide into complacency, but a steady state of alarm will eventually pall and the credibility of environmental concerns—as well as the credibility of environmentalism and environmentalist institutions—will be undermined. So to maintain morale and sow the seeds of agreement among all parties, we might consider the following points:

1. We are often told that industry is implacably opposed to any environmental programs since they cut into profits, that science (which gets over 90% of its funding from industrial sponsors) must follow suit, and that government cannot stand up against this coalition of interests, all of which stalls the fight against pollution. However, the production of ozone-depleting chlorofluoro-carbons (CFCs) has been phased out via the Montreal Protocol (1987). This was achieved by extensive consultation and cooperation among governments, scientists, and industries. Unless we assume that the success of the protocol is illusory, we are forced to conclude that these key groups can have a shared interest in working for the good of nature.

2. Over the last five decades the massive loss of forests due to acid rain has been predicted continually. However, the amount of forested land in the United States has remained steady since 1920, even as its population, agricultural production, and industrial production all increased dramatically (USDA 2000, p. 2). Only small losses of forest to acid rain have ever been documented. Unless we assume that acid rain is not a serious threat to forests, we are forced to conclude that the programs undertaken to reduce acid rain have been effective, and the progress reported in the statistical measures of the sort we have seen is indeed real.

3. We have often been warned that ongoing pollution threatens us with the collapse of agriculture. However, a few generations ago, chronic malnutrition afflicted more than half of the world's population, whereas only a tiny proportion of

humanity now faces this problem, even though population has more than doubled during the same period. The reason: Agricultural production has grown even faster than population. Unless we assume that pollution is not a serious threat to agriculture, we must conclude (1) that pollution has indeed been decreasing in the industrialized world, and (2) that there is still time to reverse the trend of increasing pollution in other parts of the world.

4. We have often heard dire predictions that toxic substances building up in the environment, in the air we breath, the water we drink, and the food we eat will result in an epidemic of disease and death. Fortunately, the average lifespan continues to grow, especially in the most industrialized countries. Unless we assume that pollution is not a serious threat to our health, we may conclude that pollution has indeed been decreasing in the industrialized world.

Let us turn now to the environmentalist belief that human beings have upset the balance of nature. This presumes that the balance of nature exists in the first place. However, although it is a central idea in popular environmentalists and still has some resonance even among environmental scientists, it has been officially rejected by biological science for some time. Nature is now generally recognized by scientists to be chaotic. As we have seen, the IPCC itself recognizes that the climate is chaotic, and it is difficult to see how climate could be chaotic without introducing chaos into the entire natural system.

Indeed, it now seems that evolution itself stagnates without nature's unpredictable changes. During long periods of natural stability, organisms become highly, even overly, adapted to local conditions, and thus more susceptible to harm given any sudden change. A plant that has genetically adapted to a precise range of temperatures will be less able to handle a sudden change of temperature than one that has adapted to a less stable regime. The great extinction events of the past seem to have been required for evolutionary progress. The dinosaurs ruled Earth for some 180 million years, an almost unimaginable span of time for ephemeral creatures such as us, whose species did not even exist much more than 100,000 years ago. The dinosaurs might have ruled forever were it not for natural chaos. As always, the disruption of natural stability caused a new blossoming of evolution, including the rise of mammals, including us.

Despite the apparent evolutionary necessity for unpredictable changes, environmentalists do not like them. Indeed, they ignore the beneficial effects of previous extinction events both large and small, and universally loathe any current species extinctions. It seems likely, therefore, that this environmentalist attitude is motivated not by biological science as such, but by rather personal affection for the current life-forms and ecosystems—a heartfelt *biological conservatism*, as it were. As we shall see in more detail in Chapter 5, environmental science has adopted and promotes a number of environmental values, of which the loathing of extinction is but one. However, regardless of the scientific status of the value, whether positive or negative, of species extinction, most of us would certainly be disturbed if we really were in the midst of a sixth great extinction event.

CASE STUDY 2: IS THE SIXTH GREAT EXTINCTION REAL?

There have been five great extinction events in the past,[*] each involving loss of 20 to 50% of all species. The sixth great extinction (6^{th}X) is supposed to be a human-caused loss of species that is just as massive. Niles Eldredge says: "As long ago as 1993, Harvard biologist E. O. Wilson estimated that Earth is currently losing something on the order of 30,000 species per year—which breaks down to the even more daunting statistic of some three species per hour. Some biologists have begun to feel that this biodiversity crisis—this 'Sixth Extinction'—is even more severe, and more imminent, than Wilson had supposed" (Eldredge 2001). He goes on to say: "We can divide the Sixth Extinction into two discrete phases. Phase One began when the first modern humans began to disperse to different parts of the world about 100,000 years ago. Phase Two began about 10,000 years ago when humans turned to agriculture" (ibid.). Thus, 6^{th}X is supposed to be a (1) rapid and massive extinction caused by humans that (2) began *very* long ago, and (3) is now accelerating. Is 6th X real?

No: The most relevant core of the doctrine for the environmentalist, point 3), is the claim that we are currently causing a massive acceleration in species extinction. This claim has a history of exaggerations that quickly ran aground on actual events. In 1979, the prestigious Norman Myers[†] estimated that 40,000 species became extinct every year (Myers 1979), and his number quickly became accepted wisdom among environmentalists. If Myers had been right, a million species would have disappeared by now, about 10 to 20% of all the species on Earth (scientific estimates of the total number of species are themselves pliable, with the most widely accepted numbers falling around the 5 million range). Thomas Lovejoy agreed: "Of the 3 to 10 million species now present on the earth, at least 500,000 to 600,000 will be extinguished during the next two decades" (Ray et al. 1993).[‡] Clearly nothing like this has happened; we could not help but notice if it had. Paul Ehrlich (1988) predicted a loss of up to 250,000 species per year over the same period, which would have resulted in the disappearance of 50 to 100% (!) of all species by now. E. O. Wilson still estimates 27,000 to 100,000 extinctions per annum. Meyers, ignoring his previous, and apparently false, prediction that some 800,000 extinctions should already have

---

[*] The five great (actual) extinctions occurred at the ends of the following eras: the Ordovician [440 million years ago (mya)], the Devonian (370 mya), the Permian (270 mya), the Triassic (245 mya), and the Cretaceous (65 mya).

[†] Myers' website at Oxford University begins with a list of his political appointments: "... scientific consultant and policy adviser to the White House, U.S. Departments of State and Defense, NASA, the World Bank, seven United Nations agencies, and the European Commission." It then goes on to list his administrative appointments in bodies that direct science and the arts: "He is a member of the U.S. National Academy of Sciences, the World Academy of Art and Science, the American Association for the Advancement of Science, and the Royal Society of Arts." Finally, there is reference to his scientific achievements: "Dr. Myers is the originator of the biodiversity 'hotspots' strategy, which has generated over $300 million for conservation activities," assuming that his "strategy" is a scientific achievement. Internet: http://www.green.ox.ac.uk/fellows.htm#Myers (15 March, 2004).

[‡] See http://www.nationalcenter.org/dos7127.htm (March 15, 2004).

occurred, still boldly predicts a loss of up to "one-third to two-thirds of all species now extant" within the "foreseeable future" (a rather vague temporal parameter from a scientific point of view; Myers and Knoll 2001). David Suzuki can claim to be the voice of moderation here, "Every year at least 20,000 species disappear forever, and the rate of extinction is speeding up" (1994, pp. 18–19), despite the absence of evidence for his conjecture.

Yes: These forecasts are based on known rates of habitat destruction, known densities of species in the various habitats that are being destroyed, and known sensitivities of these species to habitat loss. Most of the species are little known and do not appear on nature programs on television, so their loss is not being observed and documented. It does not follow that it is not occurring.

No: The International Union for the Conservation of Nature and Natural Resources (IUCN) maintains a database called the IUCN Red List (which is available online*). It lists a grand total of 684 extinctions since the year 1500.[†] Examination of this list reveals the sorry history of humans hunting various species out of existence. Most of the extinctions have occurred on islands, where local variants of pigs, ducks, squirrels, and so on, became recognized biological species. Because they were restricted to a small terrain, humans were able to hunt and kill every one of them. The rate of this sort of extinction has gone down steeply. Indeed, the Red List has catalogued only the tail end of a round of extinctions caused by human beings, one that climaxed with firearms and growing human populations, and has since subsided due to the environmental awakening of *Homo sapiens*. In fact, between 2004 and 2006, the Red List shows no extinctions at all—not one. So the evidence is clearly against point 3: Species loss is decelerating, not accelerating. In addition, no one ever suggested a massive extinction event 100,000 years ago. There is simply no evidence of any such thing. So the data are against point 2 as well.

Yes: There is a huge scientific literature in which various models of habitat-loss extinction are developed. Close examinations of relatively small areas of tropical rain forest will often reveal a very large number of species of plants and animals, some of which may never have been scientifically cataloged. Models can then take measures of the amount of tropical forest cleared annually to calculate loss of species.[‡] The models employ the concept of the *keystone species*, which if lost, as experiments have shown, cause further losses to other species. Resulting models show that species are connected to each other like the strands of a delicate spider web, so that after a few strands are cut, the other strands also fall.

No: Of course, if you picture the life on this planet as a delicate web, your scientific theory will predict that it will behave like a delicate web and will collapse under very

---

* http://www.iucnredlist.org (August 16, 2007).

[†] Of the extinct species, 260 are snails, 69 are insects, and 74 are mammals.

[‡] For an example, see E. O. Wilson's calculation leading to an estimated extinction rate of 27,000 species per year just within tropical forests (1992, p. 280). One cannot help but be awed by Wilson's confidence as he recreates the complex factors involved by means of complex formulas.

slight stress. Models can only show what your assumptions entail; they cannot show that your assumptions are correct. Only observation can tell us that. The observations tell a very different story, however. About 1.4 million species have been catalogued, and it is usually estimated that something like 10 times that many actually exist, say 10 to 20 million species. Since 684 extinctions have been observed in 507 years, even if we ignore the recent fall in the numbers of extinctions observed, we have observed fewer than two extinctions per annum. Even if we suppose that the actual number of extinctions might be 100 times as large as the number observed, it will still take tens of thousands of years for the proportion of species lost to approach that of the least of the great extinctions.

The fifth great extinction event, the one that killed off the dinosaurs, is thought to have been caused by a huge comet or meteorite striking Earth (Alvarez 1997), causing an explosion equivalent to billions of atomic bombs. That event was horrific to a degree we can barely comprehend. In the twinkling of an eye the atmosphere turned to fire, the ground heaved and opened in a global earthquake, and the sun was blotted out. Millions of species were lost within days. What, then, are we to make of such 6^{th}X claims as the following? "Sixty-five million years ago that extraterrestrial impact—through its sheer explosive power, followed immediately by its injections of so much debris into the upper reaches of the atmosphere that global temperatures plummeted and, most critically, photosynthesis was severely inhibited—wreaked havoc on the living systems of Earth. That is precisely what human beings are doing to the planet right now" (Eldredge 2001). Precisely? Surely Eldredge is not using the word in its scientific sense. Hyperbole is not precision.

In summary, 6^{th}X seems to be an exaggeration. Since the scientific *data* do not support 6^{th}X, it is hardly surprising that the 6^{th}X scientific literature bases it on models rather than or observation. Scientific theories are supposed to be based on evidence, but in the case of 6^{th}X, they are based on yet more theory, as illustrated in Eldredge's presentation of 6^{th}X above.*

2.3 OUR EMOTIONAL ENVIRONMENT

Should we be in a state of alarm about nature? It is an old joke that when the world fails to end on the prophesied date, prophets of doom simply regroup to name a later date. Unfortunately, when scientists engage in this maneuver, they give us promissory note science instead of real science. The prestigious *New York Times*, informed by environmentalist scientists, reported in 1991 (October 10) that rising seas caused by global warming would drive millions of people from their homes by 2000. This,

---

* For the scientist, a model is not an observable entity in the world but one through which the world is observed. Whereas one looks at a model airplane in the same way that one looks at the airplane of which it is a model, the scientist looks through his or her model at the things it models. For this reason, scientists will often confidently accept their own models even when others do not, for they see the evidence itself through the model. Of course, scientists will fight like tigers for their own models (or theories, as they are also known), a vital tendency which ensures that the model gets a chance to grow, develop, overcome hurdles, and show its potential. But it is the facts, not the fight, that determine whether a model is correct.

of course, never happened. And when it did not, the same scientists still warned that oceans will rise, and soon. But the oceans did not rise; instead, the prediction descended, at a very impressive rate, until it now rests at some 8 inches over the next century. This initial rapid exaggeration of danger, followed by a much slower surrender of our sense of alarm, is very human, very biological—and scientists are, after all, both human and biological. We have been equipped by evolution with the ability to anticipate bits of the future, because this has improved our potential to survive and reproduce. It is perfectly natural, then, that we instinctively become alarmed at perceived dangers, and only reluctantly let go of our sense of alarm. Better safe than sorry is the rule.

But if environmental apocalypse hasn't arrived yet, how long should we sustain our state of environmental alarm? How long can an imminent disaster be expected before we may conclude that it isn't really imminent? Our emotions have evolved to handle situations on a much shorter time scale than the changes that we have produced within nature in the past or can produce in the future. Since the environmental crisis is defined as being of our own making, it will not be sudden, like a meteor impact, but gradual, like the growth of civilization itself. Its time scale will also be orders of magnitude greater than our emotions are designed to accommodate. Does it really make sense for us to be alarmed—much less panicked?

To begin with the obvious, nature is in no danger whatever of being destroyed—at least not by us. It seems a bit ludicrous that this even has to be pointed out, but so many people accept the idea of the destruction of the environment, its devastation, complete collapse, and so on, that ludicrous or not, a few words must be said to lay this bogeyman to rest. Ever since Alvarez (1997) and others have shown beyond the shadow of a reasonable doubt that the era of the dinosaurs was brought to an end by a comet or meteorite striking the Earth in a cataclysmic impact, it has been clear that Earth's life is a very tough customer indeed. Cataclysmic changes are thought to be behind many, if not most, of the major species extinction events in the past, and there were many, including the five great extinctions. If repeated strikes by bolides (as these extraterrestrial bombshells are called) did not destroy Earth's life, we certainly will not. The history of Earth is full of catastrophic events: not just bolide impacts, but such things as 100-fold increases in cosmic rays as our solar system cycled through the spiral arms of the Milky Way galaxy, and the darkening of the sun as it encountered clouds of interstellar dust. These dire events have a cheerful lesson to teach us: The complex network of living creatures is not easily damaged, and when it is damaged, the many strands that remain regenerate a new net that is richer and more sophisticated than before.

The word *fragile* is almost always used to describe the environment. We are not to think of nature as a resilient network, but as a tenuous web. We are not to walk confidently on the face of Earth, but instead tentatively, apologetically, and fearfully. Environmentalists want us to see Earth's life as a house of cards, ready to topple given the least disturbance. But nothing could be farther from the truth. In many cases, their arguments employ equivocation between sensitivity and fragility. Living nature is sensitive—but sensitivity is a completely different thing from fragility. For example, a tiger is sensitive—it can sense things that would escape our notice entirely—but it is not fragile. Sensitivity is the capacity to respond to very subtle inputs. Yes, living nature is sensitive. It responds to the most subtle of inputs—often in very subtle and sophisticated ways. But sophistication of response is proof of resilience—the very opposite of fragility.

It is now undeniably clear that our dear Earth, our planet, our home, has survived dozens of crises, such as extraterrestrial impacts or massive climate change, with its precious cargo, the spark of life, intact. Life is tenacious, tough, adaptable, like the grass that inevitably springs up in cracks in the pavement. We now know that living things populate not only the land and the sea, but that extremophiles (Madigan and Morris 1997, D. Roberts 1998, Lubick 2002), as they are called, live in volcanic craters, under glaciers, even thousands of feet underneath the rocky crust of our beautiful Earth. Not one of the cataclysms experienced to date has been sufficient to remove all of the larger plants and animals, or even to threaten the extremophiles. Of course, the Alvarez event was horrible, and although not all flesh and flora were destroyed by it, we would never tolerate any action by our fellow human beings that would bring about anything remotely like it. Yet this positive fact must nevertheless be admitted: There is nothing that humankind could do, even if all of our bombs and poisons and slash-and-burn farmers were unleashed at once—and surely, more important, nothing that we will do—that would approximate what nature itself did in the Alvarez event.

Environmentalists should also meditate on the last ice age, to truly become conscious of the perpetual change of the environment, the incredible scale of these changes, and the fact that they are still going on—not with the sudden impact of a meteorite but at a pace difficult for creatures as ephemerally brief as human beings to notice. The last ice age (the last in a series of dozens of ice ages) is said to have come to an end a mere 7000 years ago, a twinkling of an eye in the 3.5-billion year history of life. We should also keep in mind that we have no reason to believe that the series of ice ages came to an end with the last period of glaciation. The glaciers are still retreating from the last ice age, and we are in the midst of an interglacial period, a sort of summer in a cycle of seasons on a geological time scale. In fact, summer has lingered unseasonably long.

Much as we may unconsciously drift toward the pleasant view that the ice ages are over, aided and abetted by our scientific doubts concerning the mechanism that causes them,* simple scientific induction tells us that another ice age is coming. Indeed, over the past few millions of years, ice ages have dominated the climatic picture. Interglacials have been brief, in the range of 10,000 to 20,000 years, while

---

* The most popular theory among scientists was first proposed persuasively by Milankovitch (1941): Changes in Earth's axial tilt in relation to its orbital eccentricity and the seasons cause the sequence of ice ages via alterations in solar heating (distance and angle), a theory that is typically called *orbital forcing*. Still, other theories, such as solar variability and changes in ocean currents, have their loyal supporters. The main problem with orbital forcing is that it should have been at work for billions of years, but the cycle of ice ages has appeared only during the last few millions of years, far less than 1% of the total. So some other mechanism is needed to support the orbital forcing hypothesis, and about this there is no general agreement. In fact, the Sun is supposed to have increased in brightness by some 30% since Earth was formed. The rapidity of the temperature changes involved in ice ages is another problem, since orbital changes are extremely gradual. Still, these and other problems are generally discounted given the generally good temporal agreement between orbital forcing and the ice ages.

the ice ages last for hundreds of thousands of years. Our current interglacial is reckoned to be 11,500 years old, about the length of the previous interglacial some 200,000 years ago. If this oft-repeated pattern is to be repeated, we are already on the brink of the next ice age. But soon or late, the next ice age is highly probable, and so is all the damage it will cause to living nature. Ice ages are not cataclysmic events such as meteor impacts but simply one stage in a very much slower cycle. Nevertheless, they cause enormous damage to living nature. Figure 4.1 (p. 60) is a snapshot of ice-age devastation.

Environmentalists may protest that since the ice age is natural, its effects cannot be called damage—but that is logically inconsistent. If human beings caused the same destruction of life and habitat, it would indeed be considered not just damage, but a travesty.

There was a time when it seemed reasonable to think that, just maybe, the tropical ecosystem remained intact during periods of glaciation, but we now know that this was not the case. Although paleometeorological and paleobiological studies of the Brazilian jungle and jungles in similar latitudes are yet fairly primitive, it does seem that the equatorial rain forests, just like the temperate ones, come and go with the cycle of ice ages. Although more study is needed to provide a really clear picture, the lands where the equatorial jungles now sit were much cooler and dryer during the long periods of ice. During those periods the plants and animals that now populate the more northerly temperate zones and highlands in and around the jungles moved to dominate the jungle areas themselves (see, e.g., N. Roberts 1998, pp. 73, 105).

Where did the countless jungle species go? Where did the corals and tropical warm-water fishes go? The fossil evidence indicates that many, perhaps most, were simply wiped out by cold and drought. Since there are far more species in the tropics than in more polar zones, we can only conclude that enormous numbers of species became extinct. The evidence also indicates that pockets of these tropical species survived, perhaps in rather different forms, and in smaller eco-niches, adapted to the then-prevailing conditions, in a manner similar to that in which temperate species become adapted to live at the treelines in high altitudes or high latitudes. It would seem that natural selection has favored plants and animals that have the ability to evolve rapidly, that is, to adapt over a few generations to changing conditions, such as periodic ice ages. But this much is certain: If past ice ages did not wipe them out, then, certainly, today's slash-and-burn farmers will not.

Have a little faith in nature—and in humankind!

THESIS 2:   Don't panic! Take a deep breath and look around. You will see that natural disaster is not just around the corner. You cannot think clearly if you panic.

3 Environmentalism's Transcendent Objective

Again I saw that under the sun the race is not to the swift, nor the battle to the strong, neither yet bread to the wise, nor yet riches to men of understanding, nor yet favor to men of skill; but time and chance happeneth to them all.

—Ecclesiastes, 9:11*

In its wisdom, the board in charge of a nearby nature sanctuary decided seven years ago that the Scotch broom and Himalayan blackberries that grew there had to be destroyed because they were nonnative species. If they were not native species, they must be *invasive species*, which as every environmentalist knows, are bad. Thus, they were doubly condemned, first for being part of the post-Columbian mass migration into North America, and second, because invasive species are thought to upset the balance of local ecosystems. This did not make perfect sense to me. Why should the broom and blackberries be unceremoniously exterminated, while we, the immigrant Europeans and Asians, stayed put? These plants were being used as scapegoats, it seemed. And were not all species invasive? All species expand into new territory when they get a chance; that is the only way that any plant or animal gets any living space anywhere. Life gets established on Earth by invasion. There would be no life, and no environments, without this invasion. I could understand getting upset when such an invasion causes plague, famine, or pestilence, but the broom and blackberries had been here for over a century, and were pretty meek.

When the teams of environmentalist volunteers arrived in the autumn with their tools sharpened for this new battlefront in the war to save the planet, I just swallowed hard and smiled. There was no point arguing. I would never change their minds, and the park board had made its decision. At least these environmentalists were not poisoning entire lakes or rivers to kill nonnative species of fish, as is now common practice, which would have been an uglier event. Not that beauty and ugliness are

---

* I do not intend this quotation from a book accepted by a number of modern religions to indicate an affiliation with any of them, nor that the quotation at the beginning of any chapter indicates an acceptance of anything beyond the poignancy of the idea that it expresses. In this passage from Ecclesiastes—which relevantly has another title, The Philosopher—we are reminded that virtue is not necessarily rewarded on Earth. Philosophers take note: This includes the virtue of wisdom itself, although the book goes on to say that wisdom serves us well if we so choose.

Beyond Environmentalism: A Philosophy of Nature, By Jeffrey E. Foss
Copyright © 2009 John Wiley & Sons, Inc.

the most important values environmentally speaking, I realize, but killing is ugly, say what you like. So the smaller plants were ripped up, the larger ones were cut down, and the dead plants were heaped untidily here and there.

Then next spring the quail returned, dozens and dozens of them, only to find their environment in ruins. They perched glumly in the spring rains around the graying stumps of their former homes. Gamely, they nested alongside the stacks of rotting broom and among the bits of cover that remained, without shelter from wind, rain, or sun. Soon they were spotted by a pair of hawks, for whom their misfortune was, alas, a windfall. The hawks began picking them off, and when the little ones of those that survived began foraging from the nest, the hawks picked them off as well. The hawks did so well by this unintended side effect of the environmentalists' work that they nested in some tall fir trees in the school ground next to the park. By the second spring, the hawks returned, and one of their children built a second nest, and together they raised two broods. But by the third spring, both nests were abandoned. The quail had now shared the fate of the broom and blackberry thickets, and their predators had moved on to better hunting grounds.

Paradoxically, these environmentalists had caused an environmental disaster from the point of view of the quail. Happily, from the quail's point of view, the authority and domain of these environmentalists is limited. Just yesterday in a blackberry thicket just outside the sanctuary boundary I saw a family of quail, a dozen or more chicks bouncing along behind their mother like loosely strung beads. As far as their species is concerned, relative to the challenges that nature has thrown at them over the millennia, these environmentalists and their nature sanctuary were only a minor inconvenience.

3.1   THE ENVIRONMENTALIST VALUE SYSTEM

The defining element of the environmentalist value system is its ultimate value: the good of the environment. In environmentalism as such, or *pure environmentalism*, the good of the environment is of value in and of itself, or intrinsically valuable, and all other values are valuable as means to it, or extrinsically valuable. Of course, environmentalists are people, hence complex, and generally do have values other than environmentalist values. Actual environmentalists tend to practice some form of *mixed environmentalism*, in which environmentalist values as such are just one set of values among others, a part of their value system that does not dominate the system as a whole.* What other values are blended with environmental values varies from person to person across all of the dimensions of human values, from religious values to those of atheistic humanism, from fascism to anarchism, and from misanthropy to the milk of human kindness. What all environmentalists have in common, whether pure or mixed, is a concern for the good of the environment.

---

* Pure and mixed environmentalism correspond pretty closely with radical and liberal environmentalism, although this is due largely to historical accident concerning which values tend to get mixed with environmentalism at this time. To avoid the misleading connotations of the terms *radical* and *liberal*, let us use the neutral terms *pure* and *mixed*. Pure and mixed environmentalism would match pure and applied environmentalism, except for the fact that pure environmentalism can be applied, and indeed is, in which case it becomes a transcendent objective, which is defined in Section 3.2.

The good of the environment is variously conceived. Religious people tend to value the environment as the work of God, while humanists are inclined to value it for its beauty and usefulness, while devout environmentalists are concerned for it for its own sake. Among biologists there is a spectrum from those who see nature mechanistically to those who see it as an emergent being in its own right, which is sometimes (perhaps fancifully) called *Gaia* (Abrams 1991; Lovelock 1991, 1995). The former take the good of the environment to be its optimum functioning (i.e., sufficiently large biomass, high biodiversity, etc.), while the latter tend to view it as the satisfaction of the environment's own goals or even its (perhaps metaphorical) happiness. Because these differences matter less for our purposes than what is held in common, I propose that we refer to the good of the environment (however it is individually conceived) as the *health* of the environment. This is, I think, a term that should be acceptable to all environmentalists.

If we want to understand environmentalism, we need to study it in its pure form. The pure environmentalist's values form a system inasmuch as all environmental values flow from the ultimate value, environmental health. More precisely, they depend upon environmental health in the same way that means depend upon ends. All means are defined in terms of some end. If we ask an environmental activist why she values the collapse of the global trading system, her answer will be that the system is bad for the environment, hence its collapse will be good. Thus, for this person the collapse of the global trading system is instrumentally good, while the health of the environment is good in itself, the ultimate good. Our activist might also value hard work and saving money in order that she can afford to travel to world capitals, where she can protest meetings of global traders, which would make work and savings valuable as a means to enable protest, which in turn is valuable as a means to environmental health. In the end, all environmental values come down to, or depend on, environmental health.

Environmental health, sad to say, is dangerously ill-defined. Part of the problem is that every goal, no matter what, is a function of values*—and there is no fact of the matter when it comes to values. That is, no fact or set of facts entails any value whatever, and vice versa. Facts and values are logically disparate. Values are freely chosen and subjective, whereas facts are determined by the world and objective. Thus, environmentalism is readily coopted by (i.e., mixed with) other value schemes or programs to change the world for the better. There is no way to prove, or demonstrate, a value. This in itself has some interesting consequences. One is that no one, and in particular no environmentalist, can speak authoritatively about environmental values There is no science of values, no evidence that can be adduced to support a value. As philosophers put it, "is" does not imply "ought" (see the box "The Fact–Value Dichotomy" below). Although we do recognize some

---

* Having a goal implies valuing that goal, viewing the goal as *good*, having a *positive attitude* about the goal, thinking that it is *right* (other things being equal) to pursue the goal, that the goal *ought* to be attained, that it is *moral* or *ethical* to seek the goal, that attaining the goal would be a *beautiful* thing, and so on. For the sake of brevity, let's lump all of these expressions or embodiments of evaluation under the single term, *value*. (Note to ethicists: I do not intend to imply that all of these concepts can be reduced to value but merely want to refer to them all efficiently via the single term *value*, in harmony with common usage.)

people's moral wisdom in some broad sense, we do not recognize anyone as qualified to dictate values to the rest of us. Values are a matter of choice, of judgment, of personal character, of freedom, and of each person's own vision of the good. So another consequence, and it is perhaps rather surprising, is that since environmental scientists, like the rest of us, are not moral experts, it follows that they, too, cannot dictate what is good or healthy for the environment.*

THE FACT–VALUE DICHOTOMY

Facts are bits of the way the world *is* (e.g., the fact that there are trees), whereas values are bits of the way that someone thinks it *should be* (e.g., that there should be more trees). There were facts before life evolved, before there was anyone around to know them, and before there were any values. But like beauty, values exist only in the eye of the beholder. Values exist only insofar as some being approves or disapproves of something in the world. The facts are the same for everybody. If it is a fact that the lynx is hunting the rabbit, it is a fact for both the lynx and the rabbit. But one and the same thing can have both a positive and a negative value, depending on one's point of view. Hunting the rabbit has a positive value for the lynx and simultaneously a negative value for the rabbit. Values may "contradict" each other in this way, but facts cannot, and this is one mark of the logical difference between them.

Values are neither true nor false. It is neither true nor false, for example, that you should not smoke cigarettes. On logical analysis, a value claim such as "You should not smoke cigarettes" has the same basic meaning as the imperative (command, or prescription) "Do not smoke cigarettes!" Use of the declarative sentence form for value claims (such as "smoking is wrong," "smoking is bad," "you ought not to smoke") disguises their true logical status, falsely making value claims look like descriptions. But value claims are imperatives, and imperatives are not descriptions, and hence are neither true nor false. Suppose that an officer tells a soldier, "Shut the door!" It would be meaningless for the soldier to answer, "That's false, sir." (or "That's true.").

Of course, one may cite facts to convince someone to quit smoking, but there is no logical connection between facts and the value claim itself (although there must be a psychological connection for the argument to work). The fact claim (or description) "Smoking is bad for your health" does not entail the value claim "You should not smoke." An additional premise is needed to support the entailment: namely, "You should not do things that are bad for your health," which is another value claim. Fact claims by themselves neither are, nor entail, value claims; in other words, "is" does not imply "ought." To make this mistaken form of inference is to commit the "naturalistic fallacy." Similarly, values do not entail facts. Thus, values and fact are logically independent, or are separated by a logical dichotomy.

---

* We will discuss this topic in more detail in what follows.

Although there is no way to prove, determine, or dictate what constitutes environmental health, there is nevertheless a group of environmental values that cluster together to vaguely define a widespread, popular vision of environmental health. One of these is the notion of environmental purity, or the *pristine* environment, into which no man or woman has entered: the virgin wilderness. The parallel here between environmental purity and sexual purity is obvious. To value pristine nature (and hence to revile sullied nature) is a form of environmental Puritanism, and simultaneously a form of antihumanism.

Presumably most people who value purity in nature are not radically puritanical and so would tolerate some human occupation in the wilderness as long as it was very small-scale, that is as long as it had a *small ecological footprint*. We are to picture a small dwelling on a small piece of land that takes few resources from its environment and puts little waste into it: that is, that has a *small environmental impact*. The purity of the wilderness is still lost, but at least the environmental insult of human occupation is kept to a minimum.* Human beings are thus environmentally damaging by their nature. Even in their primordial hunter-gatherer form, they have caused colossal species extinction. They must be upbraided, environmentally trained, and their instincts for speed, power, and consumption reined in, in order to be in minimal harmony with environmental health. Here there are obvious parallels with the religious notion of the fallen, or originally sinful, nature of human beings.

In any case, the evaluative focus of popular environmentalism is clear: Pristine wilderness untouched by humans is the ideal, and if we are not ready to remove ourselves from Earth altogether, we at least can try to make ourselves small, try to get as close to the ideal as we can, with a low-population, low-technology, low-consumption lifestyle that has the smallest possible effect on nature.† Popular environmentalism thus takes environmental health to be more important than human beings' actual values. Some human beings value big families, and most human beings value bread, books, television, electrical appliances, computers, and other products of our high-technology, high-consumption lifestyle. Environmentalists profess that environmental health must take precedence over these other values. In other words, popular environmentalism tends to treat environmental health as a *transcendent objective*.

---

* Note that this is merely to report the fact that environmentalists have this evaluative focus; it is not to be confused with the evaluative claim that anyone (even environmentalists) ought to have this value.

† Because knowing my environmental values may help you understand what I am saying, although they are not relevant to the soundness of my arguments (as noted in Chapter 1), I offer the following crude sketch, which is intended for the background, the notes, rather than the text of this book, until we take these issues up again in Chapters 7 through 11: I value living things over the nonliving, and of these the conscious over the nonconscious, and among the conscious I place the intelligent over the nonintelligent, and the kind over the cruel. Our ancestors, human and pre-human, struggled against an uncaring and often hostile universe to survive. The struggle for existence remains the main reality when it comes to environmental values, but our sympathies have broadened to include not only our kin, but all living things. Our existence was contained in the first instant of the universe as a potentiality, and although the purpose of the universe itself is either nonexistent or unknown, we have a right to be here. We now have the potential to become the nervous system of the planet, in the same way that neurons became the nervous systems of animals. This we should do for the good of the whole, which has struggled valiantly, but blindly, until now. We can give it sight.

3.2   ENVIRONMENTAL HEALTH AS A TRANSCENDENT OBJECTIVE

Because nature is fundamentally important, because each and every one of us, along with all other living things, depends on nature for our very existence, the value we place upon its health is naturally very high. Unfortunately, it is only a seemingly small step, then, to elevate the health of the environment to the status of a *transcendent objective*. By a transcendent objective I mean a value which is so important that it takes precedence over other all others. We all have objectives in life other than environmental health. For instance, we might want such things as a new car, to visit Beijing, or to have a family. However, none of these goals would be worth anything if the health of living nature collapsed. If the apocalyptic vision of environmental collapse that environmentalists predict were to happen, life would be nasty, brutish, and short, to re-deploy the memorable words of Thomas Hobbes (1651). Civilized society, as we know it, would collapse. There would be no gas for the car, Beijing would be an inaccessible wasteland, and one's family could only contemplate suffering and death. Therefore, it could be argued, and often is argued by environmentalists, that ensuring the health of the environment outweighs other objectives, and that they must give way before it. Just how far environmentalists are willing to go in this direction varies, but many do not or will not express or imply any limitation to their concern for nature, as we shall see. In other words, many environmentalists treat environmental health as a transcendent objective.*

Sensible as this may sound given the superficial argument above, it is really an extremely dangerous idea. Transcendent objectives have a hideous history. The most common examples are found in religious wars. Religions set up the transcendent goal of doing what God wants. What could possibly be more important than doing what God wants? Once again, one could argue, just as we did above for environmental health, that no other objective would be worth anything if we failed to do what God wants. Once again, the argument has enormous initial plausibility. Unfortunately, what God wants is a difficult issue, one obviously open to interpretation, and susceptible, moreover, to being twisted to serve other goals. The religious wars in Europe are a good example. After the Thirty Years War, one of the religious wars of the seventeenth century, travelers through some German territories saw only mile after mile of devastated cities and ruined villages, destroyed farms and burned fields, without a single living thing, only corpses, both human and animal. This devastation was caused not by an impersonal nuclear flash, but by the low-technological blades, nooses, bonfires, even the bare hands, of human beings like ourselves—inspired by a transcendent objective.

Moving ahead to the twentieth century, we find death and destruction still being perpetrated, but now in pursuit of transcendent objectives other than the will

---

* Pure environmentalists will, by definition, treat environmental health as a transcendent objective, but not conversely. Some environmentalists may let environmental values trump all other values, while nevertheless admitting that something other than the health of the environment is intrinsically valuable. For example, an environmentalist may take justice to be valuable in itself, but, nevertheless, prefer the good of the environment over justice should the two ever conflict.

of God, such as the international rule of communism. Stalin starved tens of millions to death in the Ukraine, and Pol Pot tortured and murdered millions in the killing fields of Cambodia, to name just two of the worst examples of the excesses of those seeking the transcendental objective of communism. Apparently, it is no easier to determine what the working class needs than it is to determine what God wants.

Just as it is difficult to say what God wants, it is difficult to say what the environment needs. As we have seen, environmentalists implicitly favor a particular paradigm of environmental health: the wilderness, or failing that, the near-wilderness into which humans have made only a very modest intrusion. Perhaps the early stages of the current interglacial period just prior to the rise of human agriculture, civilization, and population might roughly approximate this ideal, were it not for the fact that our ancestors had already, by then, exterminated several species and dislocated many others. In fact, it is not clear that our species ever lived in a manner fully consistent with the environmentalist ideal. This ideal of environmental health as *virgin* nature, nature untouched, or at least untrammeled, by human beings, has not been the product of environmental argument and debate. It has arisen as the counterimage to us as destroyers of nature, but no serious argument has been given for it, and none seems likely to be forthcoming. Moreover, we must notice that it has an implicitly antihuman bias, since the ideal of environmental health envisions human influence—or human "impact" as it is often called—reduced to an absolute minimum. The absolute minimum would imply that we disappear altogether.*

For now, let us just note that environmental health is a transcendent goal for many environmentalists. The transcendent goal, in its various guises, such as God's will, or the welfare of the working class, has served, and still serves, as the motivational weapon behind the sword and the bomb in humanity's perennial power struggles. Now we discover among us a transcendent goal in a new form, environmental health, vaguely conceived as the state of nature prior to the flourishing of the human race. Do we have any reason to think that in its new, environmental guise the transcendent objective will now assume a kinder and gentler role?

It is common to attribute war to hatred, but this superficially plausible explanation is far wide of the truth. Warfare requires the organized violence of thousands of individuals. This organization begins not with hatred but with the pursuit of an ultimate good. Warriors, with few exceptions, feel the same moral scruples as any of us when it comes to murdering and maiming others: the same horror, the same queasiness in their stomachs. But their scruples are overpowered, their horror overcome, and their stomachs hardened by the argument that their actions, disgusting as they may be, are necessary evils in the pursuit of the transcendent objective. A transcendent objective is like a trump card that overpowers all other cards, a moral principle that subjugates all other moral principles. Unfortunately, the health of the environment is a transcendent objective for environmentalists, a trump card when it comes to what ought to be done, whether by them or by others. It is therefore prey to the same vices as the other cases that we have considered.

___

* We return to the topic of environmentalism's antihuman bias in Chapter 4.

> ## CASE STUDY 3: ARE ECO-WARRIORS MOTIVATED BY IDEALISM OR ATTRACTED TO VIOLENCE?

Idealism: Eighteen-year-old Jake Sherman was converted to environmental activism out of idealism, and it was out of this idealism that he destroyed construction and logging equipment with homemade firebombs. Described as "a gifted, idealistic student" who "struck many of his fellow students and teachers as a particularly idealistic and spiritual person," Sherman was moved by the arguments of Tre Arrow, a militant environmentalist, to take violent action (C. Smith 2003). There appears to be nothing special about Sherman, or about Arrow, other than their high idealism. There is no evidence that they are unusually violent people. To the contrary, they were and are motivated by their conception of what is right and good. Arrow bravely perched on the high ledges of buildings of targeted businesses in order to bring media attention to what he saw as their environmental crimes. He once stayed on a ledge continuously for 11 days, obviously placing himself in harm's way, all in dedication to his environmental objective. One of his comrades was set ablaze in one of their arsonist attacks, an example of the risks of environmental activism and the necessity for the eco-warrior to be prepared for self-sacrifice. Arrow also perched in trees to prevent them from being cut down. In one case he became exhausted under the pressure of continual harassment from the police, and fell from a great height. Thus, he risked life and limb for trees, proof of environmental idealism. His disciple, Jake Sherman, was eventually sentenced to 41 months in federal prison following the firebombing of a construction firm, while Arrow at first escaped by going into hiding but was eventually arrested in Victoria, British Columbia. Heavy personal costs, yes, but like environmental warriors everywhere, they conceived of their actions as in service of a higher good.

In the mind of an eco-warrior the good of the environment transcends all other values. Risking one's own safety, freedom, and life itself can only be viewed as mere costs relative to this goal. When he was finally apprehended many months later, Arrow said, "I don't care about me. We're talking about ancient forest that doesn't grow back in a couple of years. We're talking about a planet that cannot be replaced," he said. "That to me is far more important than one person's individual case." Tre arrow sincerely described himself as a noble warrior on behalf of the environment: "As an activist, I stand tall. I hold my head high." He conceives of himself as one who was forced many years ago to make the stand of the warrior. In his own words, "I . . . had to put my own body between the chainsaws and the ancient forest."*

Violence: Note, however, the *logic* of warriors. Warriors say: I am willing to sacrifice myself. Assuming they do not think that others are worth more than they are, it follows that they are willing to sacrifice others as well. Arrow said that the forest is "far more important than one person's individual case"—he did not say that the forest was more important than his own case. Although this warrior hesitates to admit that he is willing to do harm to others, it nevertheless is implied by what he does admit and

---

* Associated Press, April 5, 2004; http://msnbc.msn.com/id/4668804.

what he actually does. The fundamental decision of the warrior is that violence will be his method and instrument. This is not a form of masochism but the sacrifice of basic human rights to a principle that is seen as higher. The readiness to resort to violence, to risk punishment, jail, bodily harm, and death, characterizes this denial, not of oneself, but of the value of human life itself.

The case of Patrick Moore, one of the founding members of Greenpeace, is interesting inasmuch as he has since come to recognize the flaws in the environmentalist movement, its logic, and the motivation of its activists (Bond 2000). Moore, who has a Ph.D. in ecology, was always troubled by the antiscientific attitude of environmentalists, who gave him the nickname "Dr. Truth" in response to his unwelcome attempts to bring facts to bear on environmental issues. Moore began to recognize too many similarities between eco-warriors and (in his own words) "Hitler youth"—in particular the substitution of indoctrination for understanding. He portrays eco-warriors as emotionally wed to activism itself and the warrior lifestyle. Moore eventually abandoned Greenpeace and founded a new group, Greenspirit, dedicated to a better informed, less violent approach to environmental health.*

Common Ground: People are both complex and different from one another. Different eco-warriors will have different motives and be moved by different arguments. Even a single eco-warrior may feel the pull of a number of different and conflicting motives and reasons. The question poses black-and-white alternatives, whereas the truth may consist of different and changing shades of gray.

On the one hand, we must recognize the nobility of idealistic sentiments. The willingness to make the ultimate sacrifice for one's convictions is the stuff of patriotism, legend, and song. On the other hand, we must recognize how dangerous these sentiments are to those things that we naturally hold dear, such as the right to go about our daily lives in safety—for there is scarcely any form of modern life that is not a crime in the mind of the environmental activist. Perhaps you work in a drugstore or department store—but then most of the products you sell are produced by large firms that are the stated enemies of the environmentalist. Perhaps you are a student—but then your university depends on the taxes provided by the modern environment-destroying economy. Perhaps you are a vegetarian who eats only organically produced foods—but only a fraction of those of us now living on the planet could survive on the yields of organic farming. Even if you are willing to tolerate the idea that most of your fellow human beings really should not be here if environmental health is to be achieved, what makes you think that you will be among those who should remain? Your organic vegetables were trucked to market in a fossil-fuel-burning vehicle, were they not?

No matter what we do, practically speaking, we get in the way of the transcendent ideal of environmental health. Really, this environmentalist goal *implies*— although

* See http://www.greenspirit.com, where Moore presents his history in his own words.

it almost never dares say—that it would be better for the environment if we were not here at all. The environment loses its purity, is no longer pristine, the second we enter it, touch it, or use it.* Once we have realized that none of us is truly innocent when it comes to the transcendent objective of an ideally healthy environment, the noble sentiments of the true environmental activist take on a different, more threatening hue. He or she is willing to do anything, risk anything, to right the wrongs committed against the environment; violence is his instrument—and each of us has a debt to pay.

3.3   THREE ELEMENTS OF TRAGEDY

I do not want to suggest for a second that environmentalists *generally* succumb to the vices that can infect those who believe in transcendent objectives, nor that they generally intend violence—although there are clearly some who do, and environmental activism harbors its share of those who are attracted to violence. I only want to make clear the nature of infection by a transcendent objective and the logic whereby it subverts normal values.

On the other hand, we must not underestimate the danger. The *first element of tragedy* is already present: the environmentalist belief in a transcendent objective. So, too, is the *second element of tragedy*: that this objective is held by large—in fact, very large—numbers of people. Environmentalism is one of the dominant ideologies of our day. Greenpeace, one of the best known among a growing number of environmentalist organizations, is not exaggerating when it speaks of "millions of people dedicated to environmental protection," who make up a "global social movement." It goes on to describe environmentalists as the "emerging second superpower," and, more ominously, as "no longer willing to accept the agendas of timid or inept governments or unscrupulous corporations."†

---

* The environmental organization Fuck for Forest (FFF) (http://fuckforforest.com/members.html, August 23, 2007) is very telling in this context, inasmuch as it sacrifices sexual purity for environmental purity. It offers the following form of environmental action: "Donate $15 to save threatened rainforests!! Get 30 days web access with erotic idealists, showing it all to save our planet. All profit goes to save nature!" Here environmental values transcend common sexual mores.

† See http://www.greenpeace.org, p. 1, May 11, 2004. Greenpeace says it is ". . . dedicated to environmental protection, human rights and social development." I have here focused on environmental protection, the first goal of environmentalists. The co-opting of other popular goals, such as human rights, is a standard strategy of many radical organizations: the inclusion of these broader goals under the environmentalist umbrella increases its overall appeal, and draws in a broader membership. However, it is clear that whenever there is a conflict between goals, environmental protection will dominate. If one's human rights, for instance, include the right to work, or the right to a place to live, environmentalists actively oppose these rights in trying to close oceans to fishing, close forests to logging or housing, etc. I am not arguing that this is necessarily wrong, nor ignoring the fact that environmentalists explain such losses of human rights by saying that they are only temporary, and promising that they will once again be honored after environmental health is achieved, when we all live happily from then on. But to explain something is to admit its reality, and the loss of human rights is the reality in question. The point is merely that protection of environment transcends other objectives and values. In this case it transcends human rights—despite what the Greenpeace manifesto indicates to the contrary.

Environmentalism flourishes in all parts of the world, among people of diverse cultures, different faiths, and disparate incomes. It is as widespread in the third world as in the industrialized world. Its appeal cuts across other systems of belief, making it the most widespread of convictions and giving it the potential to become the dominant ideology of the new millennium. There are environmentalist constituencies among Buddhist Tibetans living on the southern fringes of the Himalayas, among spiritualist hunter-gatherers living in Amazon rain forests, among cattle ranchers in North America, among communist cadres in the Andes, and among capitalist businesspeople in the skyscrapers of New York. Green parties, as environmentalist political groups call themselves, which have appeared around the world and are gaining in influence and power, are merely one aspect of the environmentalist phenomenon.

Sad to say, but the *third element of tragedy* is also on the scene: the theoretical justification of violence. In fact, the rhetoric of violence has become the norm in environmentalist circles, although most normal people living normal lives who self-identify as environmentalists would be quick to deny this. But matters are not that simple, logically speaking. The rhetoric of violence is there whenever environmentalists characterize people as destroyers of the environment (*other* people usually, perhaps characterized indirectly via reference to corporations). Strong language among serious people leads to action. If someone is destroying the environment, it follows that he or she should, or must, be stopped. Confrontation is thought to be necessary by the younger generation of environmentalists. They speak the language of the soldier, of the need for personal sacrifice, of real commitment, of putting one's body on the line.

When the line is drawn, when they chain their bodies to trees or destroy laboratories designing genetically modified crops, it is tragically apparent to an objective observer that the destroyers of the environment that they attack, these modern-day Satans, are nothing but people like the rest of us. Yet just as renaissance Europeans could not see the woman burning at the stake but only the Satan they believed was inside her, the environmental crusader does not see a logger trying to feed her family or a scientist trying to coax more food from less ground with less pesticide, but only the agent of the demonized multinational corporation or the personification of the evils of technology.

How depressing (and frankly, boring) to see the traditional enemies of the revolutionary yet again: the landowner, the industrialist, the capitalist, the rich, and the merely prosperous—not cast now as the enemies of God or the working class, but as the enemies of the environment. Environmentalism has become the most shared ideology of the legions of people discontented with the state of the world. No matter how much they may differ in other ways, revolutionaries and would-be revolutionaries can agree that their foes are destroyers of the environment. It may seem impossible that aboriginals seeking the death of the nation-state and the return to tribal rule would have anything in common with communists seeking to establish just such a nation-state, but both will profess environmentalist values. By the logic of the old adage, "my enemy's enemy is my friend," environmentalism is a friend of revolutionaries of various persuasions, because they have the same enemy. We can't be blamed for being reminded of the predominantly tragic legacy of revolution down through the centuries. Perhaps it would be better not to forget it.

It is hardly surprising that the web site of the Earth Liberation Front (ELF)* for 2002 showed a huge building in flames, and still claims as triumphs a catalog of burning and destruction.† What is surprising is the naiveté with which mainstream environmentalists ignore it, and always ignore things like it. When confronted, environmentalists blithely distance themselves from the violent actions of their radical comrades. But this is a shallow evasion. For it is the mainstream environmentalist who insists that the environment is being destroyed in the first place, and it is this apocalyptic theme, repeated like a mantra, that moves those braver souls who put words into action, the more literal-minded environmentalists, to burn, destroy, and (at least in the case of the so-called Unabomber) kill. If, seriously speaking, the environment itself is being destroyed, then given that environmental health is a transcendent objective, there has never been a stronger justification for violence.

ECO-SABOTAGE

"In writing to us, do not use your real name or put a return address on your missive. We do not need to know who you are. . . . After taking the information from your letter, we will burn both the letter and the envelope." This passage in Dave Foreman's *Ecodefense: A Field Guide to Monkeywrenching* (Foreman 1985) warns the reader this is an unusual book. Yet only an unusual reader would need to know how to bring down power-line towers, set buildings and equipment on fire, the finer details of camouflage, or how to sneak up on security personnel. Under the heading "Monkeywrenching Is Non-violent" on page 10 we find five lines of text culminating in the principle "Care is always taken to minimize any possible threat to other people"—not prevent, but minimize. After this perfunctory expression of the rights of the enemy and innocent bystanders (in comparison to which the Geneva conventions signed by European governments before World War II look like civil libertarian manifestos) there is a paragraph outlining the need for the cell structure that underground revolutionaries from Che Guevara onward have used to defend themselves from discovery and "infiltration." The details follow in a later chapter: Internal information is kept on a strict "need to know" basis (p. 146); "beware the newcomer . . . he/she may well be an *agent provocateur*" (p. 148), so, periodically, destroy your notes, equipment, etc.; "Don't worry about the cost of replacing tools, clothing, and the like. Freedom is priceless" (p. 147). There are pictures of the targets (power lines, seismic trucks, bulldozers) along with folksy prose and diagrams about how to take them out. Tire spikes can be made cheaply from rebar and planted

---

* http://www.earthliberationfront.com.
† August 2007. In this context, perhaps it is not unfair to wonder whether it was merely a coincidence that the day after Tre Arrow's arrest in Victoria, British Columbia, Canada, a new subdivision under construction in that city was destroyed by a blaze that apparently was the result of arson? Arrow is a self-proclaimed member of ELF, which in turn has an active membership in southern British Columbia.

in roads—don't forget, the points' edges should be vertical and facing the flow of traffic. It pays to buy aluminum tanks for your cutting torch since steel tanks are heavy to backpack. Be prepared to get out of the way when that tower comes down! Safety first. After all, "monkeywrenching is fun" (p. 12).

Monkeywrenching will not be fun, however, for the "self-appointed guardian of the mindless machine," the "security guard," and the "miners, ranchers, loggers, and other assorted yahoos" who get in the way (p. 169). The first purpose of eco-sabotage is to bring grief to the "three-piece-suited gangsters who control and manage these bandit enterprises" (p. 3) that they target. As for the government and its laws, how could the eco-saboteur have any respect for "the quisling politicians . . . who would sell the graves of their own mothers if there's a quick buck in the deal" (p. 4)? This rhetoric trumps the weak and perfunctory statements of nonviolence. It is easy to *say* that eco-sabotage is nonviolent, but actual nonviolence is not a superficial matter. When praises are sung to the wilderness, "our ancestral home" and "primordial homeland . . . of such noble creatures as the grizzly bear, the mountain lion, the eagle and the condor" (p. 4), it *is* implied that the yahoos and gangsters that destroy it and its noble occupants are of lesser worth. Serious people who take these claims literally will take serious measures, especially when they are told they "have the right to defend that home . . . by *whatever means are necessary*" (p. 4, my emphasis).

The eco-warrior is simply someone who takes *literally* the apocalyptic belief in environmental crisis. He or she has heard over and over that industry has ignored calls to mend its ways, that it has funded lobbyists to dissuade governments from protecting the environment, thereby blocking legal avenues of environmental activism. If this does not *imply* a call to arms, what would, given that a flourishing environment is of transcendent importance? Young people, especially young males, have a dangerous capacity for organized conflict, no doubt the result of eons of evolutionary shaping. They are especially dangerous when provided with rhyme and reason for violence by older, presumably more thoughtful, more responsible people. There is not only a clear danger of violence, but already the fact of death and bloodshed in the environmental struggle.* Martyrs have been created on both sides.

Nevertheless, in the near future, our greatest fear is not violence and bloodshed so much as the concomitant hardening of dialogue, the entrenchment of positions, years of failed communication, deadlock, stagnation, and the diseases that this will breed. Deadlock will polarize the debate, making positive steps difficult and violence easy. We are entering a crucial century in the history of humankind, one with the potential

---

* Environmentalists have distanced themselves from the Unabomber, who assassinated high-profile industrialists whom he deemed to be "destroyers of the environment"—but they cannot deny that he shared their objective and that he identified the same enemies. Was his error that he took environmentalist doctrine too literally? (Does this not, in turn, imply that it is not to be taken literally?)

for wonderful steps toward liberation and prosperity (how could you have the former without the latter?) for the bulk of humankind. But this will require clear thinking on the part of those who inform the opinions of their fellow humans. Unfortunately, there is already a strong tendency for polarization of opinion toward two simplistic views. The environmentalist characterization of human industry, agriculture, and settlement—in short, human civilization itself—as a destroyer of the environment is simplistic. And this creates an equal and opposite reaction in those accused, who then move to the opposite pole, characterizing environmentalists as tree-huggers, bleeding hearts, terrorists, or worse. This reaction, if it is to be avoided or at least minimized, demands that environmentalists give up exaggeration and rhetoric in favor of a more sustainable development of environmental thought.

However, there is not a lot of evidence of sound environmental thought when it comes to the fundamental principles of environmentalism. It begins with the apocalyptic idea that we face environmental crisis, and then takes pristine nature untouched by human beings as its transcendent objective. Environmental Puritanism is clearly on the rise, with many environmentalists explicitly defining their goal as leaving the environment in the state it would have been in if human beings had never existed. In deference to them, we build sewage plants so that human household sewage (urine, feces, and wastewater, as opposed to industrial sewage) can be purified and discharged into the environment in the form of water clean enough to drink. But just as sexual Puritanism could not make sex go away, so environmental Puritanism cannot make human waste disappear. Sewage plants may discharge water of unblemished purity, but only by converting the bulk of the waste into gases, particularly carbon dioxide (a greenhouse gas) and methane (a far more powerful, hence worse, greenhouse gas), and then dumping the tons and tons of resulting sludge into landfills or the oceans. So our sewage is just swept under a different rug.

Human beings cannot be made to disappear anymore than their household wastes can. Like all other organisms, they affect nature. The only question is, what that effect will be. Perhaps the problem starts as soon as our wastes are *defined* as pollution. We are then faced with the necessity of making them disappear, or else becoming polluters, intrinsically, by our very nature. The original mistake is thinking that a substance is a pollutant solely by virtue of its own nature: forgetting that pollution is, logically, just as much a matter of where that substance is put, and in what quantities.* Human waste is a pollutant because we dump enormous amounts of it into nearby rivers and bays. If we hadn't gotten into this habit in the first place, we wouldn't assume that wastes must first be purified in sewage plants before they are dumped into the very same rivers and bays—and maybe we wouldn't be so ready to think of ourselves as an intrinsic insult to the environment, an unnatural part of nature.

Could it be that human wastes, like the metabolic by-products of other animals, have a natural role to play in nature? Could it be that the thousands of hectares of land located on the outskirts of Paris (the *champs d'epandage*), which have for over a century accommodated its wastewater, really embody more advanced thinking than

---

* A logically similar point applies to resources: Resources are not a fixed set of substances, just whatever substances happen to be useful. Thus, the Malthusian environmentalist warning that we will run out of resources overlooks the fact that we switch from one substance to another as shortages occur.

that which resulted in sewage treatment plants?* Could it be that the thought embodied in sewage plants, simply to make sewage disappear, perversely increases our distance from the natural world? Could such plants be just a sophisticated form of denial of our own animal—hence *natural*—characteristics? The knee-jerk environmentalist reaction to human waste is to loathe it and try to make it disappear, but this is to remain stuck at the second level of environmental awareness, in which we are viewed as adversaries to nature itself. Thus, the usual environmentalist ethic prevents us from moving on to the third stage of environmental awareness, in which we are recognized to be a fully legitimate part of the environment. Perhaps the fact that the very concept of the environment does not include the organism itself, only what is around it, encourages the separation of human nature from nature at large.

A selective blindness, coupled with a startling propensity to see what is not there, is a natural symptom of a transcendent objective. Even moderate environmentalists will look you straight in the eye and talk about the environmental devastation all around us. Actually, it is usually very difficult to see such devastation in anything like a literal sense of the word. We do see scenes of the supposed devastation on television and in the newspapers, such as clear-cut forests and open water at the north pole. Never mind that the arctic ice cap sits on open water and normally breaks into pieces every summer.† Never mind that the amount of forest logged is a fraction of 1% yearly, and far less than the amount that burns due to lightning.‡ Never mind that far more

---

* In 1900 the *champs d'epandage* totaled 5000 hectares (12,500 acres) but has been reduced to 2000 hectares today. Still, they "treat" 200 million liters of sewage each day, which as has been known from prehistoric times, is a high-quality fertilizer. This is not to say that this method is without difficulties. First, human household wastes must be kept separate from industrial wastes (which require different solutions). Second, although this method eliminates 100% of suspended solids, it does not, in its current form in Paris, completely eliminate ammonia and other organic solubles. To do this, the amount of land employed would have to be much (approximately five times) larger, and its drainage into local watersheds controlled more carefully. Nevertheless, the potential remains for a complete return to the natural "solution" to the "problem" of human waste.

† Headlines were made around the world in late August and early September of 2000, when tourists on the Russian icebreaker *Yamal* discovered open water at their destination, the north pole. James McCarthy, a paleontologist (not a climatologist or meteorologist) onboard, told the *New York Times*: "I don't know if anybody in history ever got to 90° north to be greeted by water, not ice." The *Times* then reported that the last time anyone could be certain that the pole had open water was more than 50 million years ago. And so the headlines and television specials began. *Time* magazine of Canada, for instance, featured a cover with a beautiful shot of a polar bear on an ice floe, staring at open water, along with the banner headline, "Arctic Meltdown," and the subtext, "This polar bear's in danger, and so are you. Here's how global warming is already threatening Canada—and the whole planet." We consider the scientific arguments made for global warming in Case Study 7 at the end of Chapter 6; for now we need note only that environmentalists did nothing to parry this false fear. Either they did not want us to know the truth, or they did not know it themselves. Lost in the paranoid shuffle was the unnewsworthy fact that open water at the pole in the arctic summer is normal. The observation of Claire Parkinson, who monitors satellite imagery of sea ice for NASA's Goddard Space Flight Center, that such breakage of the ice "happens many, many times every year" was buried in the back pages, if it was reported at all. Her later discovery that Antarctic sea ice is actually on the increase (Parkinson 2002), which had the potential to soothe, rather than stoke, popular fear, was essentially ignored by the media, and even denied by environmentalists.

‡ Canadian statistics provided by the National Forestry Database Program (2004) indicate that, on average, approximately twice as much lumber is burned in forest fires (an average of just over

(*Continued*)

(around 5 to 10 times as much) forest burned before the days we began watching every lightning strike and then rushing to snuff out the resulting fire. Environmentalists just ignore or deny the fact that 80% of the maximum extent of land under forest cover attained near the end of the last ice age remains covered by forest to this day (Richards 1990, p. 164; Goudie 1993, p. 43). Instead, they treat any forested area that has ever been cut or cleared as being permanently destroyed, its purity forever lost, so that they can claim much higher levels of environmental "devastation."

Of course, it is difficult if not impossible for environmentalists to come to grips with the fact that much of Earth's forests are recent and ephemeral, only a few thousand years old, much less old than our own species. If they did, they would have to admit that the greatest destroyer of forests has been nature itself during the ice ages, which have repeatedly destroyed much of Earth's life, including forests, over the last few million years. This would throw their system of values in disarray. Given that the health of the environment is the ultimate value, only human beings can be the agent of environmental harm, because we are distinct from *our* environment, which is what the phrase "*the* environment" refers to: everything but us. It is unthinkable in this value system for nature itself to be the agent of its own harm.

Consequently, the very concept of harm is transmogrified by the environmentalist. No matter how much devastation nature itself wreaks upon organisms and ecosystems, it is ignored or even seen as awe-inspiring. On the other hand, any effect that we have on the environment is automatically seen as destruction. Even the prettiest field of dairy cows grazing on wild pasture is understood by them as our attack on nature. The modern city is viewed with horror that such an insult to the environment could have been perpetrated. The perpetrators, the city dwellers themselves, can only be viewed as evil.

If environmentalists reject this characterization of their values, that would be wonderful. If they do, we must ask them: What values, if any, might be considered as important as the health of the environment? If they cannot name any value that equals environmental health, they hold it as a transcendent value after all, and cannot reject the foregoing characterization. If they can name some other value, hopefully human life and happiness will be included. If not, why not?

THESIS 3: Environmentalism tends to take the good of the environment as its sole and ultimate value. Such transcendent objectives are dangerous.

(*Continued*)   2,000,000 hectares) as is harvested (1,000,000 hectares on average). Forest fire statistics for 1970–2002 indicate that from 289,157 (1978) to 7,559,572 (1989) hectares burn yearly (http://nfdp.ccfm.org/cp95/data_e/tab31e_2.htm), the vast majority due to lightning. The amount of forest harvested for lumber, pulp, and the like, has remained constant at about 1,000,000 ha per year over the same period (Canadian e-Book 1999; http://142.206.72.67/03/03b/03b_006_e.htm). For the sake of perspective, the total forest area of Canada is 997 million hectares—of which 0.2% burns and a mere 0.1% (one-tenth of 1%) is harvested—hardly the devastation environmentalists would have you believe. Moreover, this low level of forest harvesting is occurring in a country that is a world leader in forestry and forest exports, and is much maligned for it by environmentalists.

4 Environmentalism's Antihuman Bias

Humanity is the cancer of nature. . . . The optimum human population of Earth is zero. . . . Human suffering resulting from drought and famine in Ethiopia is tragic, yes, but the destruction there of other creatures and habitat is even more tragic. . . . [T]he worst thing we could do in Ethiopia is give aid—the best thing would be to just let nature seek its own balance, to let the people there just starve.
— David Foreman* (1987; 1991, p. 26; 1998b)

On a lovely, sunny Sunday morning in May 1980, Mount Saint Helens exploded and, in a flash, destroyed 230 square miles of forest. It dumped 75 million cubic yards of mud into local waterways, blew 1 million tons of sulfur dioxide into the atmosphere to fall to Earth as acid rain, and spewed a half billion cubic meters of ash over much of the Pacific northwest, coating some 22,000 square miles of fields, forests, lakes, swamplands, and rivers (McGee and Gerlach 1995, p. 2). I vividly remember seeing a vast dark bank of cloud from my small window as, a few days later, the Air Canada flight I was taking from Vancouver to Winnipeg crossed central Alberta. The pilot explained that it was ash from the eruption of Mount St. Helens and that we had to avoid it since it might cause engine failure. Naturally, the mention of engine failure brought a collective gasp from the passengers. The menacing plume was much too high to fly over, so we had to go hundreds of miles north to get around it. In Winnipeg, more than 1000 miles from the eruption, people with respiratory problems were advised to stay indoors. I remember the film of white powder that had to be wiped from the windshield of my host's car each morning.

Had such a phenomenon been somehow attributable to humankind, the wrath and moral indignation of environmentalists could hardly be imagined. What nature itself does is inevitably seen as good, and no matter what humans do, it is seen as unnatural and wrong. After the eruption of Mount Saint Helens, naturalists crowded on the scene, studying, making notes, and taking pictures. Hundreds of articles in professional journals and countless reports in the popular media told of the amazing power of nature to renew itself by its own paroxysms. The explosion of the mountain was viewed as part of the natural cycle of the west coast of North America, and

---

* Foreman is a founding member of the radical environmentalist group Earth First!

Beyond Environmentalism: A Philosophy of Nature, By Jeffrey E. Foss
Copyright © 2009 John Wiley & Sons, Inc.

so was described in poetic terms: It was awesome, powerful, inspiring, a reminder of the power of nature. Yet, had a logging company cleared a fraction of the forest destroyed (in this case, literally) by the volcano, the news would have been marked by lamentations, accusations, and heavy tones of guilt.

Lest we forget, the eruption of Mount Saint Helens killed 57 people directly, caused an airplane crash, triggered numerous traffic accidents, and resulted in the deaths of many more by means of its aftereffects, such as falling ash (seven people died from heart attacks while shoveling ash, while deaths from respiratory problems have not been tallied). Such volcanic eruptions are not rare but an everyday occurrence. Much of the time there is a volcano somewhere causing death and destruction for human beings—and the environment. Mount Pinatubo erupted in 1991, releasing 20 million tons of sulfur dioxide into the atmosphere, dwarfing Mount Saint Helens' 1-million ton release of this pollutant, causing global temperatures to fall by 1° Fahrenheit (0.5° Celsius) and the largest recorded increase in the ozone hole (Newhall et al. 1997). Its ash and poison gas annihilated everything within a 10-mile radius, a terrific amount of environmental destruction. Pinatubo also killed some 875 human beings who couldn't move fast enough to escape, and displaced some 1 million more, disrupting their lives completely (Avundo and Marchand 1999, p. 13). To this day, every tropical storm brings rivers of mud down on the towns and fields below, driving people away and destroying habitat.

4.1   CAN NATURE DO ANY HARM?

The environmentalist is, no doubt, saying, "Well, it's people's own fault for living so close to a volcano—it was just doing what volcanoes *naturally* do. Don't blame the volcano!" Well, okay, we have no interest in *blaming* an inanimate object for anything, but neither should we deny that the Mount Saint Helens explosion was the *cause* of destruction of forests and watersheds and the death of many animals, some of them human. In short, the volcano harmed the environment, where the phrase "the environment" is taken to include all of nature except humankind and its cities, fields, dams, highways, forest cuttings, and other works. This is what is meant by the phrase as used by environmentalists, and we will use it in this same sense. When they speak of environmental harm, for instance, it is the environment from the point of view of humans that is being referred to: Every part of nature that is around human beings (in the circle around us, from the Old French, *en*, in and *viron*, circle). Volcanoes sporadically destroy huge swaths of living things, and kill some 700 persons every year on average (Avundo and Marchand 1999, p. 13). When it comes to destroying or harming living things, volcano damage is much worse than simply cutting down the forest, which at least creates habitat for the plants and animals that thrive in clearings. We are instead talking about annihilation of everything that lives and the creation of a wasteland in its place. No, it is not a matter of blaming volcanoes for anything, but simply being logical and fair when it comes to causes of harm to living things, including human suffering and death. But environmentalists will have a very hard time admitting this simple fact. The environmentalist calculus is unfair: If any activity of human beings caused the destruction, both environmental and human, of volcanoes, it would be trumpeted in our ears, and our guilt loudly proclaimed. We would be accused and summarily convicted of an environmental crime.

Which brings us to a second fact: Human beings don't measure up to volcanoes when it comes to this sort of destruction. If we include not just volcanoes but all of the destructive activity of nature, it is apparent that our effects are much smaller. In general, our effects are more widespread and far less intense. It is difficult to estimate the destructive effects of humankind as compared to those of the rest of nature, I readily admit. Still, the environmentalist does not even make the effort, but instead, assumes that what nature does is always right and what we do is always wrong. And when we look a little closer, a pattern emerges: The human effects on nature so loudly decried by the environmentalist are matched or surpassed by natural events. As mentioned earlier, more forest burns down in North America due to lightning strikes than is cut down in forestry. And forest fires cause much deeper destruction than logging, killing everything, including the soil itself. Forest fires do not replant the trees they remove.

Here's something to think about: The Sahara desert itself is the product of natural changes in climate. Many environmentalists have taken the apparent advance of the Sahara desert as a symptom of global warming. As a matter of fact, the Sahara was a verdant blend of forest and prairie some 6000 years ago and has been drying out and dying out ever since. It may seem counterintuitive, but the Sahara was caused by *falling* global temperatures as Earth passed the point of maximum interglacial warmth. Lower temperatures weakened the monsoons, the rains failed, and the Sahara forest became the Sahara desert. Simply put, less warmth, so less evaporation, hence less rain. The ice ages were extremely dry, reducing the jungles to aridity, as shown in Figure 4.1. Nothing human beings have done or are ever apt to do compares with an ice age in terms of environmental harm. A large portion of the life on this planet is totally destroyed in an ice age. Global cooling makes deserts—and global warming replaces them with greenery once again.*

Of course, you have never heard the issue put that way, nor will you. Instead, it is simply assumed that human effects on nature always constitute damage by their intrinsic nature. Similarly, it is assumed that what nature does is good. What, then, are we to make of the human toll of nature in action? Thousands are routinely killed in earthquakes, millions wiped out yearly by disease, hundreds killed weekly by tornadoes and hurricanes. The untold and uncounted lives disrupted, jobs lost, and dreams shattered every single day, are never mentioned by environmentalists.

On the other hand, every threat to health caused by human beings, no matter how unquantified, and often despite the absence of actual deaths attributable to it, is loudly advertised. Why is it that in the weighing of human effects on nature, the price exacted by nature from human beings does not even appear in the balance? The innocence—no, the benevolence—of nature is always presumed. Granted, environmentalists are advocates of the environment, and we cannot expect the lawyers

---

* Oddly enough, then, if one assumes that human beings are causing global warming (something we consider in more detail in Case Study 7 at the end of Chapter 6), we are thereby reversing desertification.

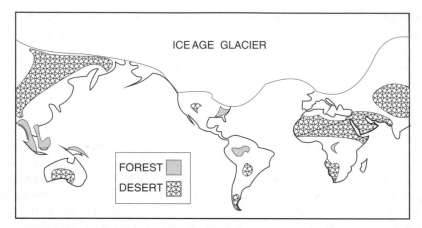

Figure 4.1   The last ice age. During the last ice age, the great temperate forests were destroyed and the great tropical jungles were decimated. Deserts expanded and Earth was much drier everywhere. The desert areas shown on this map were extreme and surrounded by arid zones. All of the great forests we now know came into being after the end of the last ice age, reached their maximum extent some 6000 or 7000 years ago, and have been shrinking since then as the Earth has cooled. (Based on data from Adams and Faure 1998.)

for the defense of nature to mention its history of violence. But that is exactly the point, and one that needs to be recognized: Environmentalism is not evenhanded. It is advocacy, plain and simple, hence one-sided by its very nature. The concept of *the* environment presupposes a dichotomy between nature and human nature in which the environmentalist is biased in favor of nature and against humankind. Having misconceived the relationship between nature and humankind from the start, the environmentalist then abandons the impartiality that is essential for any worthy judge.

Even if we ignore the human costs, and instead consider only environmental costs themselves, there remains an obvious bias. As we saw, the purely environmental costs of the Mount Saint Helens eruption are ignored by environmentalists and never appear even as a point of comparison with the effects of human activities. To move from the cataclysmic to the commonplace, millions of acres of land are flooded each year by beavers,* causing the death of trees and other plants and destroying the habitat that they provide for birds, mammals, and insects. This is never seen as environmental harm or habitat destruction, but as the creation of new habitat for fish and for bullrushes. I have seen many beaver dams while hiking and canoeing,

---

* Whereas the North American beaver population prior to European settlement is estimated at anywhere from 60 to 400 million, it is now thought to be a fraction of that number. Massachusetts reports a beaver population of 70,000 (http://www.state.ma.us/dfwele/dfw/dfw_beaver_law.htm). Given the vaster areas of similar, but less inhabited, hence more suitable habitat in northern Canada, we should expect more beavers by a factor of several hundreds. Supposing a modest beaver population of 10 million for all of North America, and a very modest dam of 1 hectare per beaver family, which runs around five individuals, we arrive at 2 million hectares of flooding, although the actual area may well be many times larger.

and I must say that they certainly are scenes of obvious destruction of living things. Flooded areas often contain stands of dead trees, which, like dead trees everywhere, are bleak symbols of death and devastation. Surrounding the dam itself you will find acres and acres of trees felled by beavers busily improving their dam or seeking bark to eat. They are very thorough and gnaw down broadleaf trees for a hundred yards or more around a dam.

This certainly looks like environmental destruction. If an identical scenario had been produced by teenagers with chainsaws and a penchant for dam building, there is no doubt that it would be counted as a scene of environmental vandalism. You would expect to see pictures of the destruction, with damning headlines, in the papers and on television. To approach it from the other side, if the beaver dam is seen as the creation of habitat, why is the logging of a forest not seen as the creation of habitat? In my experience, areas of logged forest quickly fill with grass and brush and become magnets for birds, rabbits, deer, bear, coyotes, and other animals that you would scarcely ever see in the forest. The forest floor, being heavily shaded, does not provide food for grazers, but the clear-cut does. Before we began to manage our forests, naturally occurring forest fires provided the clearings necessary for these animals. Forestry now provides this service without destroying the soil and converting the trees to greenhouse gases. Why is this not an improvement? Perhaps it is not, but the possibility that it is an improvement is not even considered. It is merely *assumed* that modern forestry, despite all the care it takes to promote regeneration of the forest, is bad, and that naturally occurring forest fires are not. We need to advance past mere assumption here and seek a proper answer to this question.

I am not accusing the beavers of anything—I am, instead, asking environmentalists to recognize their antihuman bias. In fact, I have spent many pleasant hours in and around beaver dams. Despite the dead trees, there is abundant life and the fun of catching glimpses of the beavers themselves, who seem to have a sense of humor. Besides, nothing is as peaceful, as dreamy, as drifting around in a canoe or boat. I have experienced precisely the same abundance of life, the same glimpses of wildlife, the same peaceful, dreamy mood while floating on ponds and lakes that form behind dams made by humans. But, of course, manmade lakes are always described in different terms by the environmentally conscious: as scenes of environmental destruction. Environmentalists love lakes that were made by landslides (many mountain lakes were obviously formed in this way), although landslides indiscriminately wipe out forests, kill animals, and leave wastelands that take centuries to recover. Environmentalists despise lakes formed by dams, although they are the environment for the fish, bullrushes, and insects that live in them and for deer and moose that come down to their shores to drink their water.

I do not mean to imply that manmade dams can be justified simply by comparing them with beaver dams. The observation I would like us to make moves in the opposite direction: that the effects of human beings upon nature cannot count as bad, as environmental damage, solely because they are caused by human beings. This is illogical and *unfair*. It is illogical because whether a given effect on nature is good or bad depends on what that effect is, not on whether or not it was caused by human beings. It is unfair because it faults human beings for no reason. Given that an area is dammed to form a lake, it makes no difference from an environmental point of view whether it was dammed by an ancient earth movement, a landslide, or an earth-filled dam. Furthermore, from a logical point of view, human beings are a part of nature. That is just a fact.

4.2   ENVIRONMENTALISTS ON HUMANKIND

The system of environmentalist values forces environmentalists systematically to picture human beings as the sole source of environmental evil. We are pictured as unnatural. Once, long ago, we were part of nature, but we have rejected this inheritance. We have, like Satan, rejected our creator out of pride and the refusal to obey. Our relationship with our maker, the natural world, has gradually deteriorated into a state of open warfare. Not all environmentalists have fully realized that this is what their values imply, and of these, fewer have expressed the implications. But many environmental leaders have. Eldredge (2001), for instance, says

> Humans do not live with nature but outside it. *Homo sapiens* became the first species to stop living inside local ecosystems. All other species, including our ancestral hominid ancestors, all pre-agricultural humans, and remnant hunter-gatherer societies still extant exist as semi-isolated populations playing specific roles (i.e., have "niches") in local ecosystems. This is not so with post-agricultural revolution humans, who in effect have stepped outside local ecosystems. Indeed, to develop agriculture is essentially to declare war on ecosystems.

A brief essay by D. Suzuki boldly entitled "State of the Planet" (1994, pp. 18–19) is another example of the antihuman bias of environmentalism. He says: "... the monster is us.... We are overrunning the planet like an out-of-control malignancy," and goes on to speak of "the war to save this planet." Eldredge and Suzuki are not the only environmentalists to talk of war, but for the sake of brevity I will cite just one other, the highly influential Thomas Lovejoy:* "The planet is about to break out with fever, indeed it may already have, and we [human beings] are the disease. We should be at war with ourselves...." (quoted in Ray et al. 1990). If we are a disease, an out-of-control malignancy, what does this war imply? How do these people intend to stop or defeat humankind? Whose lives must be surrendered?

Suzuki goes on to cite a number of *straight facts* (his italics) "based on 1989 statistics," but his so-called facts could hardly be true. He claims that due to human-caused soil erosion, we lose "seven percent of the globe's good growing land every

---

* "Dr. Thomas Lovejoy, a Yale University–trained biologist, is Science Advisor to Secretary of Interior Bruce Babbitt and project leader of the National Biological Survey, a comprehensive survey of the nation's biological resources being undertaken by the U.S. Department of the Interior. He has served as Assistant Secretary of External Affairs at the Smithsonian Institution and as Vice President for Science of the World Wildlife Fund" (The National Center for Public Policy Research, August 24, 1993, http://www.nationalcenter.org/dos7127.htm). Lovejoy is also a proponent of the "sixth extinction," as mentioned in Chapter 2.

decade." Are we seriously to believe that now, nearly two decades later, nearly 14% of our good agricultural land has been lost? He says that in addition to lost soil, "vast areas are being degraded by poor land use" (what sort of statistical measure does "vast" represent?) and "since 1984, global food production has declined each year. And this is precisely at the time that human population is exploding."

How revealing, then, that decades later the massive famines that his "facts" must entail have not occurred. In 1950, standard analyses claimed that over half of the world's population was malnourished. The most recent famines, notably in Ethiopia, involve at most a few million people, of whom at most some few thousands perish. This is a horrible, horrible thing, and I do not mean to imply otherwise. But from a statistical point of view, a few million people represent a mere one-twentieth of 1% (0.05%) of the population of the globe. As the twentieth century progressed, fewer and fewer people faced famine, and those that did, such as those in the Ukraine and even those in Ethiopia, were victims of war and other political actions, not of food shortages as such. To quote an economist who has proven his ability to see the larger picture, Paul Krugman says (2003): "What we've seen in the last generation is an enormous, unexpected improvement in the human condition. . . . Over the past 25 years more people have seen greater material progress than ever before in history." This is not just a judgment call but the implication of a broad range of vital statistics concerning income, nourishment, health, and education.

Things used to be different; they were much worse, from both nature's point of view and our own. When Christopher Columbus landed on the first Caribbean island he came upon, the people living there shot some birds to prepare dinner for him, explaining that only the elite of their society were permitted to eat the birds, which were already rare, and that the local pigs had long ago been hunted to extinction. This sort of human-caused extinction has a long history, one that has only recently begun to change as people turned away from hunting toward agriculture, and have come to value species other than their own. Although Eldredge and other proponents of the "sixth extinction" portray agriculture as intensifying the rate of extinction, the evidence indicates otherwise. The data on extinction indicates that agriculture was the first step toward halting human's tendency to hunt edible animals to extinction. On the whole, replacing hunting and foraging (as the hunting of plants is called) with herding and growing crops has been a positive step, despite the loss of habitat involved. On balance, it is better to have intense use of relatively small areas of land than intense hunting through all of it. The transition to agriculture was gradual, since early farming techniques merely supplemented hunting and gathering, so the rate of anthropogenic species extinction fell only gradually. As late as the 1950s, my grandfather still had to supplement his mixed-farming production of cash and food with the hunting of duck, rabbits, deer, elk, and moose. But by then, happily, the pressure on wild animal populations had been reduced to a level that did not threaten their extinction.

Paleobiology is beginning to reveal a disturbing pattern that goes back at least 40,000 years: Game species disappear when humankind arrives. The ancient woolly mammoths and mastodons may have been among our victims. We cannot help but feel guilt over these events. Like the original sin of Adam and Eve, the sins of our forefathers are somehow visited upon us despite the fact that we can hardly be responsible for what happened long, long before we were even born. Our feelings of paleo-guilt, natural though they may be, should not be exaggerated or distorted in support of the environmentalist agenda. Other predators have also caused extinctions, but lions and wolves cannot dwell on the past. Their acts, dictated by natural instincts, are neither right nor wrong, neither moral nor immoral. Alone among the other organisms, it falls to us to see the difference between right and wrong. When we do, we should be aware—we should *realize*—that like the lion and the wolf, our ancient ancestors were merely doing what came naturally. They too hunted, driven by hunger and instinct. If the lion and the wolf are innocent, so are our ancestors. Despite our pangs of paleo-guilt, they committed no sin for us to inherit.

4.3   IS MODERN CIVILIZATION BAD FOR NATURE?

Because technological agriculture and civilization have given us both a margin of safety from natural necessity and the time to reflect philosophically, we have learned to see the consequences for nature of our actions and have begun to realize our responsibility for them. It is only because of modern science that we have discovered the distant past in which our ancestors hunted game species to extinction. It is only because of our modernity that we have begun to seek our proper relationship with the natural world. One of the most important effects of humankind's switch from hunting to agriculture was philosophical: the realization that human hunger does not justify hunting game species to the brink of extinction. This is not the time to let paleo-guilt drive us back into submission to nature as the environmentalist's vision would require.

Yes, the instinct to hunt is still inside us. Every child plays hide-and-seek, the age-old game initiating us into the arts of hunting and being hunted. Many an adult still indulges in sport hunting or fishing to satiate this ancient instinct. If we feel guilty about the extinctions caused by our ancient forbears, it is because we recognize the instinctive drives behind that original sin in our own nature. Now is not the time to return to those ways, those instincts, and the pretechnological lifestyle the environmentalist idealizes. It is most definitely not the time to bang the drum of paleo-guilt and exploit it in support of antihuman environmentalism. This is not the time—if such a time could ever be conceived—to be at war with ourselves.

Instead, we must grasp the lessons we have learned and the moral sensibility that reflection has awakened. And so we have—at least those of us for whom modern ways keep the struggle for existence at bay, so that we can seek things other than mere survival, such as the good of living things in general. Thus, we now seek to protect endangered species, to control hunting, to control agriculture and industry, and to protect the living space of organisms other than ourselves. We have already turned this corner. A level-headed assessment shows that the threat of species extinctions, just like the rate of extinctions, is on the decline.

Like evangelists castigating us for our sins and trying to control us with visions of hellfire and brimstone, environmentalists portray us not just as violent, careless murderers of other species, but as greedy, lazy, and materialistic thieves of Earth's resources. Like the evangelists' exaggeration that each of us is bound for eternal damnation, environmentalists see apocalyptic environmental outrage in every measure of human activity. It is impossible to list the cases, but to give just one example, Suzuki prophesies horrible consequences due to continuing ozone loss, despite the fact that human production of chlorofluorocarbons (CFCs) was being curbed even as he wrote, and even though science clearly indicates that the worst ozone loss experienced to date increases the amount of incoming ultraviolet radiation by a trivial amount, equivalent to that which would result from moving 100 miles south.

Now, two decades later, CFC levels are on the decline, but environmentalists have moved on to new battles in an ongoing war. Suzuki called for jail terms for politicians who disagree with the theory of global warming (Offman 2008, O'Neill 2008), thereby placing himself in the company of totalitarian theorists such as Stalin and Pol Pot. Control of people's beliefs and conscience is recognized as a travesty by anyone with the least concern for human beings' freedom and happiness. Of course, given the transcendent ideal of environmental health, any sacrifice is justified. "We are battling to keep the planet livable for our children," is Suzuki's war cry, echoing environmentalists everywhere. Does he really think that this sort of exaggeration will do his children any good? Are they not part of the malignancy just like the rest of us?

Surely the only charitable answer is that Suzuki and all of his fellow environmentalist prophets of doom are engaged in rhetorical overstatement of their case. Surely they love their children just as much as we do ours and would not sacrifice them for the good of "the environment," which is nothing other than the natural world minus humankind. These baby-boomer environmentalists have been living side by side among us, have raised their families among us, and are now growing gray as we all, with luck, will do. They did not sacrifice themselves or their children for the cause they promote. Actions speak louder than words, and their actions say that humankind is worthy of remaining on Earth, that human life is part of the grand scheme of nature. We are not a cancer that should be removed, but a unique and interesting species whose needs must be included in any reasonable system of values, including those of the environmentalist.

THESIS 4: Humankind's environmental effects are not intrinsically harmful, and nature's environmental effects are not intrinsically harmless. Nature, not humankind, has been the source of the greatest environmental destruction.

It should be noted that the scale of human effects does not necessarily imply environmental harm. All of our urban areas occupy (only) 0.2% of Earth's land surface (or 0.04% of Earth's total surface), and our agriculture 16.5% of its land surface (EC 2004). Thus, our total effect on the environment is very large, but that alone does not mean that it constitutes environmental harm. Certainly, this effect is much smaller than that of natural events such as ice ages, and much less harmful. Our fields, after all, are filled with vegetation that absorbs carbon dioxide, generates oxygen, cools the surface, moistens the atmosphere, and provides food and shelter for insects, birds, mammals, and other wildlife. Pronghorn antelope graze in farmers' fields just as do migrating birds such as Canada geese and the great blue heron. Nothing grazes on glaciers. Of course, some of the effects of our cities and fields are indeed environmentally harmful, such as pollution of waterways by city sewage or agricultural runoff. The point is that environmental harm has to be shown on a case-by-case basis. The simple fact that humankind has a large environmental effect (even though smaller than that of an ice age) does not mean that this effect is harmful.

Moreover, we should be on our guard against the presumption that our effect is always large. The current concern over global warming, for example, derives in large part from the false sense that humankind has a large effect on the atmosphere. In fact, each year we emit something like 9 billion tons of CO_2 into the atmosphere, while natural sources emit something like 100 to 120 billion tons. The "carbon budgets" used to estimate this number are themselves primitive and uncertain at this stage. For example, it was just recently discovered that the phytoplankton growing in Earth's oceans removes 40 to 50 billion tons of CO_2 from the atmosphere rather than the 10 to 20 billion tons estimated previously (Falkowski 2002), showing that carbon budgets up to that date had been in error. So our effect on CO_2 is smaller than the uncertainties in the science of atmospheric CO_2 balance.

In addition, phytoplankton growth has been discovered to be restricted by the trace amounts of iron in the ocean, with the result that carbon sequestration is sensitively controlled by flows of iron into the oceans, usually by windborne iron-bearing dust. The Southern Ocean Iron Experiment of 2002 (Falkowski 2002, pp. 60—61) showed that a single ton of iron could increase phytoplankton tenfold over 300 square kilometers, revealing the possibility of an inexpensive and practical method of controlling CO_2 levels and hence global climate. So it is possible for us to have very large *beneficial* effects on the environment as well. This is a very important fact, and it should be kept in mind. Environmentalists *presume* that human environmental effects must be harmful and militate for a general decrease of those effects. They *presume* that nature should be left to itself and that whatever it does is best for the environment. This two-part presumption is not supported by data. It is merely an expression of environmentalism's antihuman bias.

4.4 ENVIRONMENTALISTS' PERSONAL VALUES

Paradoxically, most prominent environmentalists commit the very same environmental sins as those against whom they preach and fight. Greenpeace activists sail in ships with inspiring names like *Rainbow Warrior* to join battle against such things as oil tankers and offshore oil wells, while their ships burn the same fuel oil that those tankers carry and oil wells produce. The well-known Canadian environmentalist David Suzuki preaches a technologically simpler life closer to nature and more like that of aboriginal populations around the globe. Meanwhile, Suzuki is the star of a large television and publishing business, flying around the globe on jet aircraft and transporting television crews into the heart of the wilderness that, according to his own stated principles, must be protected at all costs from just such incursions.

Transcendent objectives do not have the same logic as other values or ethical principles. Our ethical principles against lying, stealing, and cruelty—not to mention sacrificing other people's lives in the pursuit of a goal—cannot be used to justify a privileged class of leaders or warriors who are permitted to lie, steal, or be cruel. Normal ethical principles do not permit these sorts of exceptions, although transcendent objectives do. Environmental leaders and warriors justify far higher personal levels of environmental damage than do other people, according to their very own definitions of environmental damage—along with ordinary crimes against their fellow human beings, as in the sabotage of employment, destruction of property, and so on—by reference to a logically privileged goal that transcends all other values. The logic of transcendent objectives is very different from that of other values or ethics. The idea that, in the battle against cruelty in general, a little cruelty is fine for (or a lot of cruelty is required by) our leaders or warriors, is obviously absurd and clearly marks a difference between legitimate values and transcendent objectives.

As far as the *logic* of transcendental objectives is concerned, environmental leaders are just like those religious leaders who preached love for all humankind, or those political leaders who held up the welfare of workers as a beacon, only to engage in warfare, murder, false arrests, and other destructive acts. I am not saying that the scale of harm is comparable in the two cases (we will come back to the scale of the problem in a minute); I am saying only that the logic of justification is the same: The transcendent goal for which one is fighting permits—in fact requires—the commission of small crimes in the service of the larger good. Al Gore, who is perhaps the most famous environmentalist ever, lives in a Tennessee mansion that burns 20 times as much energy as is burned in the average home. Gore travels in limos and private jets. He justifies this by his purchase of *carbon offsets*, which are in fact donations to environmentalist organizations. Suzuki also justified his yearly vacation trips to the Caribbean by the purchase of carbon offsets. The purchase of such offsets is becoming a standard justification for many activities and businesses that violate environmentalist scruples. Travel agencies and airlines now offer offsets for "purchase" along with their products, so that customers have the opportunity to meet their self-imposed environmental obligations while enjoying the benefits of the modern, prosperous lifestyle.

THE "LOGIC" OF CARBON OFFSETS

"Gore has described the lifestyle he and his wife Tipper live as 'carbon neutral,' meaning he tries to offset any energy usage, including plane flights and car trips, by 'purchasing verifiable reductions in CO_2 elsewhere'" WorldNetDaily (August 28, 2007).*

* CNBC European Business, May 2007;http://cnbceb.com/2007/05/01/voluntary-carbon-markets-spring-to-life/ (August 28, 2007).

What is the official logic of the environmentalist claim that the purchase of carbon offsets can stop global warming due to greenhouse gases? The argument is that through such a purchase, a person's production of CO_2 can be counterbalanced (or offset) by a reduction in CO_2 elsewhere. When we use fossil fuels to heat our houses, generate electricity, drive our cars, or travel by plane, train, or bus, we produce CO_2. If we produce a ton of CO_2 by flying to visit family or friends, we can purchase a ton of CO_2 reduction elsewhere, and so assure ourselves that we have not increased the total amount of CO_2 in the atmosphere. A ton of CO_2 can be removed from the atmosphere by, for example, planting trees that will absorb it. Currently, a 1-ton offset is priced at about $12. So for a relatively small cost added to our travel expenses, we can help save the planet. Organizations have sprung up and corporations have been formed (and listed on stock exchanges) to handle what is expected to be a rapidly growing trade in these offsets. These businesses are supposed to transfer the money they accept, minus their own operating expenses and profit, to those who actually reduce CO_2, so environmentalist organizations are eagerly lining up to accept these payments. In 2006 the total world trade in carbon offsets came to $7.2 billion (5.3 billion euros) (ibid.), a figure that is expected to grow by a factor of 10 in the next decade. So everyone benefits, including the people who (like Al Gore) need to travel or burn fuel for other purposes, the people who are employed in the trade of offsets, the people in environmental organizations, and the people who have invested in offset trading companies—and the planet wins as well.

If it sounds too good to be true, perhaps it is. One problem is that the only large-scale carbon reduction method at the moment is tree planting, which does not actually reduce CO_2 in the long run. In the long run, trees are CO_2 neutral, because they release as much CO_2 from their metabolism, shed leaves, and (eventually) decaying trunks as they "sequester" from the atmosphere in the first place. Annual per capita CO_2 production is currently 20 tons in the United States, and about half that in the rest of the industrialized world, so billions of trees would have to be planted simply to provide short-term sequestration. Long-term sequestration could be achieved by cutting the trees for use as lumber in houses and furniture, but this would obviously violate other environmentalist values. Since offsets must be calculated *relative to* what would otherwise have taken place, offsets are paid for trees not cut down as well as for those planted. This creates an incentive to schedule all standing trees to be cut, and then not cutting them, thus converting them into a source of income. Thus, people and businesses can be paid for the trees they do not cut as well as for those they do, with no net result on the total quantity of living trees. *Carbon equivalents* of other gases, such as methane and CFCs, also qualify, so the carbon-offset trading will create an incentive for rice producers to reduce their production (rice paddies being a major source of methane), which will in turn make food more expensive for the world's poor. Finally, the recipient of the carbon-offset cash will spend it, for example, on the purchase of more offsets for jet travel. Thus, money spent on carbon offsets reemerges as purchasing power to subsidize CO_2 production.

From the point of view of environmental ethics, carbon offsets are self-contradictory for the purchaser, who in effect buys the right to violate environmental obligations. A cultural analogy might be the practice of purchasing indulgences from the Roman Catholic clergy in order to avoid God's punishment for sin. In actual practice, agencies that sell a given tonnage of carbon offsets keep a share (some promise that this will not exceed 20%) and transfer the rest, along with the responsibility to actually reduce the stated tonnage of CO_2, to environmentalist organizations. In turn, the environmentalist organizations typically spend the cash on campaigns against forestry, against travel, and so on, where there is no accounting procedure for actual tonnage of CO_2 reduction. Environmentalists eagerly endorse carbon offsets because it guarantees a steady flow of cash into their bank accounts. They ignore the fact that carbon-offset trading encourages people to continue doing precisely what the environmentalist says that they should not do.

In effect, carbon-offset trading amounts to a system that taxes supposed environmental harm to pay environmentalists for supposed environmental harm reduction. Environmentalists justify this in effect by reference to their own growth and increase in power, which they assume will work for the good of the environment. Thus, current environmental damage is supposed to pay for prevention of future damage. If this works, then in the long run air travelers will have paid for the prohibition of air travel. So it is, ironically, that environmentalist leaders like Al Gore burn fossil fuels to heat their swimming pools and power their jet and sea transportation, cut down trees to publish books, and license environmental wrongdoing to prevent environmental wrongdoing. Such is the logic of transcendent objectives.

When Catholic Queen Mary disemboweled and hanged English Protestants, her goal was to make sure that as many Englishmen as possible got to heaven. When Stalin starved millions and millions of Ukrainian workers to death, his goal was a workers' paradise for those who survived. Environmentalists, too, have a vision of paradise, where nature is pure (i.e., is unsullied by human beings) and healthy. They say, of course, that no human being should ever be harmed. On the other hand, their vision of the greater good must be kept in focus, and lesser goods may (indeed, must) be sacrificed for the sake of the greater. Thus, not only harm to other human beings, but the harm they themselves do to nature, may be justified. Put this way, of course, the purported justification sounds ridiculous. However, taken step by step, each move in the morality play makes perfect sense given that environmental health is taken as an overriding objective. Environmental leaders such as Gore and eco-activists on board the *Rainbow Warrior* no doubt feel it necessary to travel and do battle for the sake of the environment and to forgive themselves a small sin in order to do a greater good.

If this makes sense, all the rest, ridiculous as it sounds, will follow. Transcendental objectives are seductive.

4.5   ENVIRONMENTALISM'S INSTITUTIONAL ETHICS

The sacrifice of the lesser good for the greater is well under way. The trend toward biofuels in order to prevent adding CO_2 to the atmosphere has resulted in making basic carbohydrate foods much more expensive worldwide as food is commandeered for conversion to fuels.* The poor in the Americas south of the Rio Grande have already been forced to cut back on their consumption of tortillas. This turn of events has been brushed aside by environmentalists as a small price to pay to prevent global warming. Should people starve due to the suppression of modern agricultural methods (genetic engineering springs to mind)† or in a decades-long economic depression caused by the suppression of fossil fuels,‡ their deaths will be justified by reference to the environmental paradise that lies at the end of the difficult road that must be followed. Let us hope that such dire effects of environmentalism will be avoided. Surely environmentalists will be outraged at the mere suggestion that their creed would ever lead to death and destruction on the scale attained by religious or political warfare. This outrage is a good thing, indeed a precious thing, and should be encouraged.

Environmentalists who shudder at the very thought of such human carnage, however, would do well to reflect on the effects of environmentalism on human beings so far. In particular, they should meditate on the effects of the ban on DDT that is often reckoned as the first major victory of environmentalism in the battle to save the planet. That ban had a bad effect on many human beings. It stalled a decades-long

---

* At the time of this writing, wheat and corn prices are at historic highs, which is explained by market analysts as due in large part to biofuels initiatives.

† In May 2004, Monsanto announced that it would not be releasing its new strain of bread flour wheat genetically modified to survive Roundup herbicide treatments. This "Roundup-ready" wheat would permit farmers to use much less herbicide in total, via one or two potent applications rather than in several less potent applications as needed to grow current wheat strains. The new strain was not released due to environmentalist concerns over the introduction of genetically modified organisms into the environment (in particular, the banning of GM products by the European Union). Whether or not this new wheat is a good idea from an environmental point of view (it would, after all, reduce herbicide use), its suppression by environmentalists will reduce wheat yields and increase food costs, causing more hunger and starvation among the very poor—who are always the first to suffer given any general shortage, economic downturn, or misfortune. Many farmers have already stopped growing wheat because of the high herbicide costs, costs that are generally integrated into the price of wheat. This is just one example of the general tendency of environmentalist agricultural policy to have tragic human costs, usually paid by the poorest among us.

‡ There is no doubt that the Kyoto protocols will, if followed, depress the global economy. Exactly how, and how much, is a matter of debate that cannot be usefully addressed here. But there is no doubt that economic downturns affect the poorest most severely, and that hunger, starvation, and disease are typical results. Whether or not the protocols are a good thing environmentally, they will inevitably have a cost in human suffering and death. Should the resulting depression be deep and long term, environmentalist principles—as a transcendent objective—would nevertheless demand the Kyoto protocols be obeyed, even though the human toll of severe depression could easily mount into the millions.

battle against malaria in the developing world, and by so doing led to increased illness and death among the people there. Tragically, the number of people who have died due to the banning of DDT *does* compare with the human tolls of Hitler and Stalin.

Environmentalists will cry foul when this comparison is made. I am not in any way suggesting that environmentalism shares the moral depravity of Nazism or Stalin's despotic megalomania. The point is much simpler: Many millions of deaths would have been prevented if DDT antimalarial programs had continued. They were stopped because of environmentalist action. So, as a matter of simple logic, some portion of the enormous, 50 million total death toll of malaria since the ban must be counted as a cost of environmentalist action. Environmentalists can quibble about what portion of these deaths was due to banning DDT. They can argue, however dubiously, that in the long run more lives will be saved. But unless they, like Nazi or communist or other ideologues before them, are prepared to ignore human costs, they will at least admit that these costs are relevant to their program. Even if we assume that massive environmental harm was avoided by banning DDT, the fact remains that the ban also caused human death and misery. Even if on environmental grounds we must judge the DDT ban as a good thing, this judgment cannot ignore the human cost. That is the main point, logically, that must be realized: Environmentalism cannot treat the good of the environment as a transcendental good.

Of course, we cannot forever forestall the question of whether the ban was a good thing. However, we can put that question to one side of the main argument here: namely, that treating the good of the environment as a transcendental objective is not a good thing. This is a forward-looking proposal that has the potential to help us move beyond environmentalism to the philosophy of nature. If that progress can be safeguarded, we are better able to look in the rear-view mirror to see how we arrived at our present situation. With that proviso, we may face up to the question of whether the banning of DDT was a good thing.

CASE STUDY 4: WAS THE BANNING OF DDT A GOOD THING?

Yes: Rachel Carson's hugely popular and profoundly influential book *Silent Spring* (1962) awakened a generation to the damage perpetrated by industrial agriculture to the environment and to human health, and inspired the rise of environmentalism. She proved once and for all the folly of pesticide use, in particular the environmental unacceptability of DDT. She marshaled the scientific data linking DDT to the decline of bird populations, particularly raptors, educating us about persistent toxins and how they are concentrated in the tissues of animals at the top of food chains. As she revealed, studies had shown that DDT caused the thinning of eggshells in raptors, leaving their eggs prone to collapse and raptors unable to reproduce. Shortly after the publication of *Silent Spring* in the 1970s, industrialized countries banned DDT in response to environmentalist pressure. Raptor populations returned to normal, and vast, untold environmental damage that would have occurred worldwide was prevented. Not only the environment, but the entire human population of the world owes Carson a huge debt of gratitude, for she showed that cancer rates were linked to the rise of DDT in the environment and in the bodies of every human being. She predicted that cancer rates might rise to 100% if the devastation of the environment by DDT had been permitted to continue. You and I might very well have cancer were it not for her. So not only the environment and environmentalist, but all of us, have benefited from the banning of DDT.

No: To begin with, Carson's case against DDT seems to have been exaggerated.* However, let that pass, for there is a bigger issue here that has been ignored: the human cost of banning DDT. So let us assume that this highly persistent chemical has harmful effects on wildlife, and that regardless of the scientific soundness of Carson's case against it, its removal from the environment was a good thing for wildlife.

What is being ignored in the environmentalists' crusade is the *human* cost of banning DDT—proof that environmental health is a transcendent objective for environmentalists. Beginning in the 1940s there was a rapid decline in malaria, encephalitis, and other mosquito-borne diseases. The disease was eradicated from developed nations (we now tend to forget that malaria was once common in Europe and North America) and was on its way out in developing nations as well. There is no doubt that this was due to the invention and use of DDT, and no doubt that DDT saved *millions* of lives. But beginning in the 1970s, because of the banning of DDT by

* Rachel Carson's conclusions have proven to be exaggerated and her arguments to be based on a few carefully chosen cases. Her famous charge that DDT causes birds to lay eggs with thin eggshells derived from a study involving incredibly high dosages of DDT. Her prediction that DDT would lead to cancer rates of 100% must confront the absence of any clear scientific evidence that DDT is carcinogenic. No one claims that DDT is harmless, especially in high concentrations, but it has proven difficult to substantiate scientifically serious danger of trace amounts, although research continues to this day. Of course, environmentalists will at this point bring up recent research indicating that DDT and DDE (its metabolic product) are endocrine mimics, are "implicated" in birth defects, are "suspected" carcinogens, and call for its ban, citing the *precautionary principle*, which requires us to treat the mere possibility of environmental harm as though it were actual. WWF (1999) is a good example of just this sort of approach. Beginning (p.1) with the common but dubious idea that Carson's *Silent Spring* (1962) "detailed the devastating impact of persistent pesticides on wildlife," it then exhorts us to "embrace the precautionary principle" (pp. 2, 3, and 17) as the proper response to "the paucity of information and subsequent uncertainty about cause and effect relationships." Apparently causality, the essence of science, is irrelevant when the safety of the environment comes into question. Persistently extrapolating from acute toxic studies to the conclusion that trace quantities are also toxic, and merely assuming such things as that an increase in testicular cancer must be caused by such traces, DDT is roundly condemned. Given the scientific weakness of their case, the authors are forced to admit that the evidence is merely suggestive, and that "it is virtually impossible to answer questions about the impact of these persistent chemicals on human health directly or definitively" (p. 10). Nevertheless they call for a worldwide ban on DDT, calling even for the destruction of stockpiles set aside to "combat malaria or locust outbreaks" (p. 16). In fact, Greenpeace, the American Wildlife Federation, the World Wildlife Fund, the Worldwide Fund for Nature, and so on, were (even as this article was published) pressing the United Nations Environmental Program (UNEP) for an absolute worldwide ban on DDT. Their ban was adopted in 2002, and quickly ratified by member nations. Not once is the appalling death toll from malaria even mentioned in this article, although the evidence of harm to humans in the case of a ban is not merely suggestive, but real. Obviously, the precautionary principle takes no precaution for African children. It trades off their real deaths in the millions against a mere suggestion of health problems among those in the developed world, or of harm to the transcendentally valuable environment.

environmentalists, malaria began to spread once again. By the year 2000 it had increased some 40% in sub-Saharan Africa (NIAID-NIH 2004), to the level of around a half-*billion* clinical cases each year, resulting in 2.7 million deaths annually (USDHHS-NIH 2002).

If we forget for the moment the enormous suffering of the 500 million people each year who survive malaria infections and focus merely on deaths, somewhere in the neighborhood of 15 million additional people have died because of the increase in malaria between 1970 and 2000. There is plentiful evidence, and practical certainty, that DDT could not only have prevented the increase, but would have continued to reduce the incidence of malaria (Attaran and Maharaj 2000, D. R. Roberts et al. 2000). Just how many could have been saved? Certainly 15 million if only the increase had been averted, and probably a large portion of the other 35 million killed by the disease if malaria continued to be eradicated at the pace of the 1960s. In other words, the environmentalist goals that mandated the banning of DDT have cost somewhere between 15 and 40 million lives.

Whatever Carson may have said, the toxicity of DDT for humans is very low, with numerous cases of survival despite ingestion of large doses either experimentally or in suicide attempts, and very little by way of epidemiological studies showing any long-term effects (A. G. Smith 2000). As Attaran and Maharaj point out (2000): "Although hundreds of millions (and perhaps billions) of people have been exposed to raised concentrations of DDT through occupational or residential exposure from house spraying, the literature has not even one peer reviewed, independently replicated study linking exposure to DDT with any adverse health outcome. The relative low toxicity of DDT for humans makes it an ideal weapon in the battle against malaria."

All that was being asked was that DDT be permitted in malaria control—precisely what the ban prohibited.* No one was suggesting a return to the days when kilotons of DDT were used on crops [at a rate of 5 to 20 kilograms per hectare (4 to17 pounds per acre)]. Instead, what was needed was a few hundred grams [several ounces, at the rate of 2 grams per square meter (1/15 of an ounce per square yard)] of DDT for biyearly treatment of houses. Although this would find its way into the surrounding environment eventually, the total amount would be a tiny fraction of that which caused so much worry decades ago.

Fortunately, DDT is not merely an insecticide but an insect repellent and irritant as well, and so provides extremely effective, cheap, and safe protection from

---

* Again, you may find some quibbling by environmentalists on this point. According to the UNEP ban, use of DDT in house spraying is permitted, but only if no other alternatives are available, and only as a stopgap until other alternatives are put in place. Note, first of all, that this escape clause was included only at the insistence of medical authorities appalled at the death toll of malaria, and over the protests of environmental groups. Second, the environmentalist goal is still the total ban of DDT, and the escape clause only permits a delay under special circumstances. Third, since the poorest nations are those that must rely most heavily on international aid, and in particular UN aid, they are fearful of doing anything to anger aid agencies and so do not avail themselves of the escape clause, even though they cannot afford the vastly more expensive alternatives to DDT. Finally, the substitutes for DDT are not as effective, are more toxic to humans, and even where they are used, malaria is again on the rise. In fact, it is on the rise even in rich industrialized countries such as the United States itself.

malaria—fortunately, that is, if it had not been banned. The alternatives are much less effective, much more expensive, must be used in larger quantities, and often are more toxic to humans as well (A. G. Smith 2000). On the other hand, we know that use of small quantities of DDT in house spraying works. For example, D. R. Roberts and co-workers report (2000): "DDT house spraying was stopped in Sri Lanka in 1961, and this was followed by a major malaria epidemic. Since then, after suspension of DDT house treatments, numerous epidemics have occurred in many countries, such as Swaziland (1984) and Madagascar (1986–1988), where malaria killed more than 100,000 people. In both cases, the authorities restarted DDT house spraying and stopped the catastrophic epidemics. In Madagascar, malaria incidence declined more than 90% after just two annual spray cycles."*

Yes: But WHO repealed the ban on the use of DDT for house spraying for the purpose of malaria control in 2006.

No: This does not justify the environmentalist's ban—it merely revokes it. The people who lost their lives are not going to be resurrected by repealing the ban, and it is callous of environmentalists to ignore or deny the human toll their actions have taken.[†] Anyway, the ban was revoked *over the protests* of environmental organizations, many of which are now lobbying to renew the ban. People continue to die from malaria at the rate of one every 12 seconds. Need we point out that the majority of those who perish are children? Need we point out that the poorest people and the poorest countries suffer worst from the ban on DDT?

Environmentalism is clearly a potent, global, political force, so the second element of tragedy is indeed in place, just as we observed earlier. Indeed, if the death of millions by malaria is tragic—and surely it is *deeply* tragic—the play has already begun. If we are indeed acting out this tragedy, let us change the script. Let us rewrite the ending of this drama.

To do this we have to come to a clearer, more complete realization of how the action has unfolded to this point. But so far as the public media reveals, environmentalists

---

* Lest anyone be tempted at this point to aver that only a pawn of the big multinational chemical companies would present this case, note well that the patent on DDT is expired, and that the only current producers are small, third-world companies. The big chemical companies have long ago phased out production of DDT in favor of more expensive alternatives—and would not dare produce it again, given its horrible (if simplistically misconceived) reputation—which, by the way, works to their financial advantage.

† A typical case of ignoring the problem is provided by Greenpeace's international website (http://www.greenpeace.org/international/, April 24, 2007) where a search under the heading "malaria" turns up only dozens of articles warning about the rise of malaria predicted by some scientists should global warming occur as predicted by the IPCC. A typical case of denial is provided by the World Wildlife Federation's international website (http://www.panda.org, April 24, 2007), which posts articles that continue to call for the ban and promote other methods of mosquito control, even though some of these have proven impractically expensive, or ineffective, or both, and others, such as reduction of mosquito habitat, are opposed by the WWF on other grounds, since mosquito habitat is also habitat to other forms of wildlife, such as alligator and hippopotamus.

have not even bothered to debate this issue. Instead, they have ignored it or denied it, and continue to do so. It is beyond debate that millions of lives have been sacrificed in the pursuit of an environmentalist goal. Were it not for the pursuit of environmental health, they would not have lost their lives to disease. Some degree of environmental health has been purchased at the cost of human suffering and death. Whether or not this was a good thing—note well that this is in no way intended to imply that malarial deaths themselves, especially on this horrific scale, could be a good thing—it had a human cost. And this simple fact exemplifies a crucial conclusion that we are forced to draw: Like achieving any other objective in the real world, achieving environmental objectives will have a cost. Environmentalists have studiously ignored the costs of their projects and goals, with tragic consequences. Unless we want to ignore these costs, we cannot treat environmental values as transcendent objectives; and surely we do not want to ignore these costs.*

THESIS 5:   Environmentalist values often oppose human values and well-being, and have already taken a toll measurable in millions of human lives. The human cost of environmental initiatives must be taken into account.

---

* As this book goes to press (November 2008), the world is experiencing its first global food shortages in decades. Three reasons are cited for the shortages: (1) increase in demand for food among the people of the developing world as their economies improve, (2) crop failures in Asia due to abnormally cold weather, and (3) diversion of foodstuffs (primarily corn and sugar) and farmland for the production of biofuels. The biofuels industry is being strongly encouraged around the world by various tax breaks, grants, and other programs in an effort to reduce the use of oil and coal to combat global warming. This massive encouragement of the biofuels industry is clearly one factor causing food shortages and rising food prices. Consequently, the poorest people in the world are going hungry. They are rioting to protest steep rises in food prices, while everyone else must pay more for food. What has been the response of environmentalists? In this context of hardship for their fellow human beings, environmentalists have raised their voices only to downplay this effect of biofuels, to reassert the necessity at all costs to combat global warming (human costs which, ironically, are due partly to cold weather), and to reemphasize their support for biofuels. The antihuman bias of environmentalism seems as strong as ever.

5 Science, Environmentalism, and Environmental Science

Few people are capable of expressing with equanimity opinions which differ from the prejudices of their social environment. Most people are even incapable of forming such opinions.

—Albert Einstein (1954)

The prairie and rolling hills in south-central Alberta along the U.S. border are very hot in summer, very cold in winter, and dry all of the time. Each year only a foot or so of water falls from the skies as rain, hail, snow, and sleet. The land teeters between grassland and desert. The grass springs happily up from the ground as the snows of winter melt, and then loses all of its enthusiasm for life by mid-June, when it starts to get very hot and very dry. If the wind starts blowing, you feel like you are inside an industrial clothes dryer. The wind seems to blow all of the time here, whatever the season. In wintertime, when the temperature drops into the minus 30s (Celsius and Fahrenheit are about the same in that range), the wind can kill you. All things considered, this is a harsh place. It tolerates your being around, but it is seldom welcoming.

Yet I find it beautiful. One reason is that there is hardly anyone around. There are farms, but they are few and far between. Farmers get only 100 frost-free days "and then only if you're lucky," as they say. Frost or no frost, a hailstorm can still wipe you out. Or a rattlesnake can bite you or kill your dog. But agriculture's loss is the wanderer's gain—not that this place will ever be featured in travel brochures. But that is just the point: Its unfriendliness keeps people away, and I like that. In the half-hearted light of a gray winter day you can hear a deep silence that sounds like eternity. Or so I fancy.

Every once in a while you will see some pronghorn antelope (so-called; *Antilocapra americana* is not a true antelope). They are wary and will not let you get close enough to be within rifle range, so you need binoculars to get a good look. It is always a thrill to see them. They look exotic, as if they were transplanted from Africa. Hunted nearly to extinction early in the twentieth century, they have since bounced back. Now farms, fences, and highways are just parts of nature, as far as they are concerned, parts which they handle with expertise and grace. They never give a thought to the state of the economy, the price of oil, or globalization, yet they know

Beyond Environmentalism: A Philosophy of Nature, By Jeffrey E. Foss
Copyright © 2009 John Wiley & Sons, Inc.

everything that matters—to them. I have to admit that I like some animals better than others. Snakes I can do without, particularly rattlesnakes, which I have successfully avoided all my life and hope never to come across. As for mosquitoes and ticks, I blush to admit (particularly if I might be overheard by anyone with environmentalist convictions) that if the little bloodsuckers could all disappear and everything remain otherwise the same, I would be all for it. The pronghorn, on the other hand, I would fight to keep. I am happy for them in wet years, when summer rains revive the grasses and shrubs on which they live, and worry about them in dry years.

This is, of course, sheer favoritism from a scientific point of view. To see what I mean, consider another denizen of these plains, the prickly pear cactus (*Opuntia polyacantha*), which the pronghorn sometimes eat in dry years if prairie fires burn off their spines. The prickly pear flourishes in a climate that is drier than is ideal for grasses and forbs, with which they cannot compete. So for the prickly pear cactus, the environment is best when it is dry—exactly when it is worst for the pronghorn. My sympathies are with the pronghorn, but that is purely subjective. The environment itself is not healthier one way or the other. It is healthier *for* the pronghorn when it is wetter, and healthier *for* the prickly pear when it is drier. Environmentalists lament the health of the environment in south-central Alberta, which has been dry recently, but the environment does not care one way or the other, although its inhabitants do. They say that the recent dry years are signs of global warming and the collapsing global environment, but this is nothing compared with the 1920s. I know it sounds crazy, but the old-timers used to tell me that it did not rain *at all* for 10 years back in the 1920s. Climate historians tell us that before the 1920s there was a wet period from roughly the 1880s onward, when Europeans moved into the area. Before that, in the mid-nineteenth century, there were bigger droughts than in the 1920s. Going back thousands of years, century-long droughts occurred. Before that were the ice ages. In about 80% of the last few million years, this land has been under a mile of ice. Not that any of this matters one iota to the pronghorn or to the prickly pear.

As for which way the land *ought* to be, which way is *healthy*, I think no human alive has anything to offer other than his or her personal feelings about it—and that includes environmentalists.

5.1   ENVIRONMENTALISM AND SCIENCE

When environmentalists view science, they are torn between suspicion and trust: suspicion because so many environmental crises (as they see them) have been made possible by science, trust because the proof (as they see it) that these crises are real is provided by science. If DDT and chlorofluorocarbons* took a toll on the environment, it is difficult to overlook the fact that they were developed by scientists explicitly for the purposes for which they were used, and then promoted by them for

---

* Chlorofluorocarbons, formerly the most common refrigerants, have been named as the cause of an increase in the extent of ozone thinning that occurs during the polar winters; their eventual banning has been credited with a reversal of this increase.

this use as well. When environmentalists reject genetically modified foods, calling for further testing before they are inserted into the human food chain, it is with this sort of example in mind. Admittedly, the distinction between science, the pursuit of knowledge, and technology, the development of methods, devices, and materials with which to achieve specific goals, is pretty clear. Technology appeared with the first stone chopper, long before there was any science at all. But distinctions are not the same thing as actual separations in reality. Science is distinct from technology, and they can be kept separate—but in fact are not.

We can easily conceive of science being kept separate from technology. Imagine, for example, that (perhaps due to fierce religious opposition) science had developed within secret societies which took great pains to ensure that scientific discoveries were protected as secret knowledge and never leaked to the society at large. In fact, the contrast between this scenario and what actually happened could hardly be more stark. From the beginning, scientists have been very generous, very democratic, with their knowledge and their technology. Galileo designed his "geometrical compass" for sale to the military; Pasteur developed pasteurization to combat disease; many physicists joined the Manhattan Project to apply the science of nuclear physics to developing an atomic weapon for the United States near the end of World War II. From the early days, a huge part of the appeal of modern science was its promise of practical help in fighting disease, eliminating poverty, and generally benefiting humankind. Although many people became scientists solely to discover the secrets of nature, just as many joined the scientific enterprise wholly or in part to improve the human condition through such knowledge. It is *most remarkable* that science has lived up to its promise as a means to improve the human condition.*

The idea that science serves human welfare is so obvious that it now goes entirely unnoticed. The *de facto* social commitment of the scientific community to human welfare,[†] and a matching commitment of society to support science in return, is seldom expressed, much less formalized in an explicit contract. So, let us refer to this tacit understanding of mutual benefit and support between science and society as the *social subtext of science*.

THE SOCIAL SUBTEXT OF SCIENCE

It is tacitly assumed that science works for the good of society, and that society should therefore support science, although there is no explicit agreement to this effect. This subtext appears to have arisen with the new sciences themselves. Galileo, Descartes, and Newton all spoke about the benefits that science would

---

* David Stove (1927–1994), a clear-sighted, clearheaded, and honest observer of the human condition, argued persuasively that the crucial Enlightenment ideals of liberty and equality were achieved (to the extent that they have been achieved) only because of the success of science as an engine for the improvement of human welfare. See especially his book *On Enlightenment* (2003).

† The exceptions, such as the notorious Nazi medical researcher Josef Mengele, are so few as to prove the general rule.

provide humanity, as have many scientists and supporters of science since then. Francis Bacon (1561–1626) was noteworthy in that he explicitly promoted social support of science because of the benefits it provides as part of his promotion of the new method that energized the new science. In his *Novum Organon* (or *New Instrument*) of 1620, Bacon argued persuasively that science should proceed by induction based on observation, as in the new sciences of Galileo and Kepler, which he admired, not deduction from first principles, as in the old Aristotelian sciences, which he argued were mired in stagnation. He observes that since the time of the Egyptians, people have awarded higher honors for scientific discovery than for civil achievement. Bacon justifies this, saying: "The benefits of discoveries may extend to the whole race of man, civil benefits only to particular places; the latter last not beyond a few ages, the former through all time" (1620, Aphorism 129). To this day we still award the highest honors and awards, such as the Nobel prizes, to our scientists—presumably because we, at least tacitly, are thankful for the benefits they have provided us. Society continues to provide financial and moral support for the sciences, and scientists continue to invent ways of benefiting society.* For, as Bacon said, "we cannot command nature except by obeying her" (ibid.).

Given the absence of any actual separation of science and technology, or even any serious effort at separation, the mere distinction between the two does not provide scientists with any defense against any blame that might accrue to science-based technology. Most scientists accept the praise accorded science for improving human welfare, and some maintain consistency by also accepting guilt for whatever harms science may have engendered as well. Alfred Nobel (1833–1896), dismayed that his scientific investigations into nitrogen-based explosives, which he had undertaken to

---

* One place where we can find the informal contract between science and society emerge above the level of subtext is in the mission statements of science funding agencies, which distribute public moneys for scientific research. For example, the European Science Foundation, the umbrella group of the many funding agencies of the nations of the European Union, says the following in its Strategic Plan 2006–2010: "European citizens are generally aware of the potential contributions of new knowledge to innovation and economic growth, as well as for its contributions to societal needs, not the least in the health sector." This reference to the social subtext of science is immediately followed by a reference to the environmentalist belief that science is also a source of harm: "This positive attitude is, at the same time, counterbalanced by perceptions that research creates rather than solves problems." It is wisely suggested at this point that this perceived violation of the social subtext of science be resolved through dialogue in order that social support for science can be maintained: "In this context, the dialogue between science and society on genuine ethical concerns must be intensified. The success of such a dialogue will ultimately be a crucial factor in the political willingness to support and fund research" (European Science Foundation 2006, Sec. 2.1, p. 10). Here we see science promising economic, practical, and health benefits in return for continued material support—which, not incidentally, follows comments on Europe's need to catch up to the United States and stay ahead of Asia in the context of globalization. Similar expressions of the social subtext of science are found in the self-descriptions of other science funding agencies. Very noteworthy in the present context is the recognition of "perceptions that research creates rather than solves problems," which is a key characteristic of the environmentalist's conflicted view of science that is the topic of this section.

improve mining safety, also made warfare more deadly than ever before, is a prime example of this sort of scientific guilt. On the whole, society owes an inestimable debt of gratitude to science. The practical sciences of sanitation and immunization have freed us from the scourge of plagues and disease; scientific agriculture has freed us from famine; and even purely theoretical sciences have enlarged the vistas open to the mind. You might say that paleontology, for instance, frees us from the tyranny of the present by permitting us to view the distant past—surely our lives are enriched by our knowledge of dinosaurs. But if it is right to accept praise for the good that science has done, surely it is only fair to accept blame for the harm it has brought about as well: nuclear weapons, superbugs,* industrial pollution, and the alienation of people from nature. When it comes to distributing the guilt for today's environmental ills, scientists must accept their share.

The best way to expiate guilt is to reverse the harm done. This poses a profoundly intractable problem for contemporary scientists, because *we live in a world that has been transformed by science.* Science has given us television, automobiles, jet planes, chemical fertilizers, time zones, x-rays, mercury poisoning, birth control pills, test tube babies, clones, computers, plastics, lasers, the green revolution, the Internet, nuclear magnetic resonance imaging, and 6.6 billion people drawing breath and looking for fulfillment on this crowded planet. In this context, it is very nice to remember something called "pure" science, the science of Galileo and Newton, which set out with one, and only one, goal in mind: to reveal the mysteries of nature. But having revealed those mysteries, science went on to transform everything that it touched. This is one reason that the environmentalist is enchanted with the idea of what nature was like *before* it was transmogrified by science. The so-called pristine environment is one untouched by scientific man.

From the point of view of the environmentalist, scientific progress has always brought with it the seeds of its own undoing. Sure, the development of engines (first the steam engine, then gasoline and diesel engines, then the gas turbine, etc.) freed the masses of humanity from the endless drudgery of manual labor. But this progress had *unintended side effects*, such as pollution, long-distance bombing, and the release of greenhouse gases that will (so it is believed) warm the planet catastrophically. Yes, the development of sanitation, immunization, and modern drugs has freed us from a lot of diseases. Yet this progress had the *unintended side effects* of overpopulation and environmental destruction. Despite these side effects, it was possible, even in the mid-nineteenth century, to see science as our victorious ally in the struggle against *natural* evils such as manual labor, pain, and disease. Science had been our ally *against* nature, and had tipped the balance in our favor. In the 1950s we created visions of a swiftly approaching scientific utopia, where people lived long happy lives free of disease and work (intelligent machines would do everything for us).

But now, at the beginning of the twenty-first century, environmentalists reject this vision as an illusion, on the grounds that nearly all of the environmental crises that they believe we face are the unintended side effects of those very same scientific victories. Pollution, the death of the oceans, environmentally caused cancers and other

---

* *Superbugs* are strains of antibiotic-resistant bacteria and viruses that have evolved in hospitals, farmyards, and other places where antibiotics are overused.

diseases, nuclear and biological weapons of mass destruction in the hands of immature politicians, a soaring population that threatens to overwhelm the planetary ecosystem: these crises are the products of science, too. No wonder that the pessimist can say, "the law of the unseen side effect still dominates the development of civilization."* No wonder that environmentalists automatically reject the "techno-fix" whereby the problems engendered by science are supposedly solved or going to be solved by yet another application of science. In a standard text on environmental ethics, we find the following standard message:

> Medical science found cures for tuberculosis and syphilis and has aided in greatly lowering infant mortality, but in the process we have allowed an exponential growth of the population to produce crowded cities and put a strain on our resources. The more people, the more energy produced, the more pollution; the more our lives are threatened by disease.... And so the story goes. For each blessing of modern technology, a corresponding risk comes into being, as the tail of the same coin.† (Pojman 1998, p. 1)

Thus, environmentalists view science with suspicion: Even when they see its friendly, supportive face, they know that the other face is still there.

There is an ongoing revolutionary split within science that mirrors the two faces that environmentalists see. Some scientists reject environmental guilt, some accept it, and others are torn somewhere in between. Currently, those who accept responsibility for repairing the environment have the upper hand and represent the historically recent emergence of the enviro-friendly face of science as *the* face of science. Little wonder, in this historical context, that some scientists explicitly reject the social subtext of science itself. Clear examples can be found in the publications of the IPCC, where we are warned that the continuous technological progress that we have come to take for granted must stop. In particular, "higher speeds in transportation are (efficiency gains notwithstanding) unlikely to be environmentally sustainable in the long run" (IPCC 2001c, Secs. 1.4.3.1–1.4.3.3). Instead of giving us better technology, science gives us philosophical advice: We should not want faster transportation, since "it is doubtful that this really enhances the quality of life" (ibid.). To improve the quality of life, we should instead have slower cars, slower trains, slower (sailing) ships with sails, and do such things as ride bikes. In fact, we should stop traveling so much, and stop shipping food and goods long distances. Our economy, and our lives, according to the IPCC, should be "regionalized" (compare Lomborg 2001, p. 428).

---

* Ulrich Beck (1992, p. 170). Beck usefully distinguishes *primary scientization*, in which "science is applied to a 'given' world of nature, people, and society," from *secondary scientization* (or *critical scientization*), in which "the sciences are confronted with their own products, defects, and secondary problems" (p. 155).

† Pojman's argument requires the assumption that crowded cities and pressure on resources are worse than syphilis, tuberculosis, and high infant mortality, which is not supported by further argument, perhaps because it may pass without argument of any kind among a student population that has been imbued thoroughly and systematically with the environmentalist view of the world in their primary schooling. The argument, however, is not that easy to make. Certainly, Pojman's claim that the more we turn to technology for protection from disease, "the more our lives are threatened by disease," would be very hard to support with actual data about rates of disease and death from disease before and after the development of modern medicine. And, to put it bluntly, it is probably false.

This is but one sign of a historically profound change in scientific orientation, a movement of science away from its traditional, natural alliance with technology toward a sociopolitical alliance instead. The IPCC scientists do not propose a search for a technological solution to the environmental problems they perceive, but instead, urge a sociopolitical solution, the Kyoto accords, as the only possibility of avoiding calamity.*

Ironically, it seems undeniable that the enlightenment goals of freedom and equality have been achieved (to the extent that they have been achieved) not by sociopolitical innovation on its own, but mainly by the scientific improvement of technology. Slaves, tenant farmers, armies of laborers, and laboring armies vanished as scientists invented the machines to replace them, and along with them the form of social organization that went with them. Science has been a welcome intellectual partner in the enlightenment, but it was on technological grounds that the battle for enlightenment was won. Tragically, since the rise of environmentalism this triumph has become an embarrassment and a liability. In response, science has begun to transform itself and its social role. Environmental science has new prestige and power within science and is re-conceiving the social subtext of science.

SCIENCE AND SOCIAL ENLIGHTENMENT

Science has made great steps in its quest for knowledge by leaving behind the innumerable religious, political, and ideological quarrels that mark human life in other spheres. To redeploy the words of the philosopher Bertrand Russell (1912, p. 93) in a new context,[†] science "does not ... divide the universe into two hostile camps—friends and foes, helpful and hostile, good and bad—it views the whole impartially." Science has come to play a crucially important

---

* In his public statement upon resigning from the IPCC, Landsea exclaimed, "It is beyond me why my colleagues would utilize the media to push an unsupported agenda that recent hurricane activity has been due to global warming" (Landsea 2005). The explanation is ready to hand, given the emergence of the enviro-friendly face of science—environmental science—as the authoritative face of science. Landsea speaks of an "unsupported agenda" because he thinks in terms of the traditional values of the pure scientist: truth and evidence. As we shall see, environmental science is an applied science and so serves a goal over and above these traditional values: namely, the health of the environment. Assuming that Landsea's former colleagues in the IPCC are (mainly) environmental scientists in this sense, they are being perfectly rational: Given that truth and evidence are only instrumental values in environmental science, they must give precedence to their ultimate value, environmental health, which they view as being better served by the general acceptance of the hypothesis that global warming will bring dire results, such as more violent hurricanes—whether or not this is true.

† Russell was actually speaking about philosophic contemplation, not scientific contemplation. However, since science as such, pure science, is nothing other than a branch of the philosophy of nature, I am fairly certain he would agree that the same impartiality of gaze and judgment must obtain in science just as it must in philosophy in general. More importantly, pure scientists will, I am confident, find his words an apt description of their scientific philosophy.

role in society: It determines the facts about nature for society as a whole. It seems likely that it has been granted this role in large part because of its commitment to value neutrality. We are inclined to trust science because it is above the fray, and it is above the fray because it makes no value claims, and furthermore, tries to make sure that its fact claims are not influenced by values. Indeed, science not only accepts value neutrality but has been an active proponent of it.

By championing value neutrality in its own domain, science has thereby promoted impartiality in other spheres of human life as well. As Russell puts it, "The mind which has become accustomed to the freedom and impartiality of [scientific] contemplation will preserve something of the same freedom and impartiality in the world of action and emotion" (ibid.). This is a very interesting hypothesis, which he explains as follows: "The impartiality which, in [scientific] contemplation is the unalloyed desire for truth, is the very same quality of mind which, in action, is justice, and in emotion is that universal love which can be given to all, and not only to those who are judged useful or admirable" (ibid.).

We may speculate that in this way scientific value neutrality has contributed significantly to the progress of political liberation of peoples around the world. To demand value neutrality in one's own search for truth leads one to demand it in the search for truth in courts of law and the political arena as well. The quest for truth makes one realize that one's own subjective point of view is merely one among many, and that checking one's opinions and findings with those of others via free and open communication—in other words, maintaining objectivity—is essential in this quest. This is illustrated in scientists like Andrei Sakharov (1921–1989), who helped the USSR develop the hydrogen bomb in the 1950s and then went on to become a champion of truth, open communications, and human rights. Although he was subject to "internal exile" in the USSR, he went on to be the first Russian to win the Nobel Peace Prize in 1975.

Galileo's heritage runs deep.

5.2 THE RISE OF ENVIRONMENTAL SCIENCE

Environmentalism has been aided and shaped by the scientists who dedicated their intelligence, energy, and scientific skills to solving the pollution crisis of the mid-twentieth century. In the process, they created environmental science: science applied to the analysis and solution of environmental problems. Since averting the pollution crisis did not solve all environmental problems once and for all, the need for environmental science did not dissolve along with the crisis. The need for environmental scientists in government, administration, and industry has continued to grow, and the environmental science establishment continues to gain momentum as environmental awareness grows. An Internet search will reveal thousands of university, college, and research institution programs worldwide devoted to environmental science* at both the undergraduate[†] and the graduate[‡] levels.

This very interesting phenomenon deserves careful study. One elementary question arises right at the start: Is environmental science pure science or applied science? *Pure science* is dedicated to discovery of the truth alone and takes whatever direction this quest requires. Pure science is often described as "curiosity driven." Astronomy and paleontology are good examples of pure science, for the knowledge they seek has little or no practical application. *Applied science*, by contrast, aims to solve practical problems for reasons other than their pure scientific interest. Paradigms of applied science include the work done by scientists to invent cures or palliatives for various diseases, research done by scientists to make better weapons or defenses, and the crime scene analyses done by forensic scientists to help the police and judiciary identify and convict criminals.

An overwhelming majority of scientists work in applied science. In the nineteenth century science moved away from the protection of rich patrons and the academy into industry, the military, and medicine. This massive growth of applied science resulted in pure science being reduced to a minority of practicing scientists working with a minority of the resources available. As a general rule, only scientists working in the academy, usually with very much smaller salaries and research funding than those of scientists outside universities, kept pure science alive.[§] The demand in the

---

* EnviroEducation (http//:www.enviroeducation.com) lists 1560 programs in the United States alone.

[†] For example, the University of Maine offers a Bachelor of Science in Ecology and Environmental Sciences degree, described clearly in terms of achieving environmental values: "a science-based, interdisciplinary program for students interested in all aspects of ecology, conservation, and environmental protection ... The program is designed for students who wish to pursue professional careers in environmental protection, natural resource conservation, management, ecology, planning, research, or environmental advocacy" (http://www.umaine.edu/nrc/Default.htm, July 13, 2007).

[‡] The web service GradSchools.com lists graduate programs in environmental science for universities in every state of the United States, as well as for countries around the world (http://www.gradschools .com/programs/environmental_science.html). At the graduate level, as at the undergraduate level, program descriptions marry the science involved to environmental goals. Columbia University's Ph.D. in Ecology and Evolutionary Biology is typical; the goal of "conserving biodiversity" is to be achieved by "formulating and implementing environmental policy" and taking positions of authority. "The Ecology and Evolutionary Biology (EEB) program is designed to provide the broad education needed to describe, understand and conserve the Earth's biological diversity in all its forms." This program, therefore, does not provide mere understanding of biodiversity, but the ability to conserve it as well. How? Through the formulation and implementation of environmental policy: "Matriculating students will have the skills to conduct ecological, behavioral, systematic, molecular, and other evolutionary biological research, as well as to formulate and implement environmental policy." Implementing environmentalist policy can, of course, best be done from positions of power and influence: "Graduates may pursue academic careers as researchers and teachers, or professional positions in national or international conservation, environmental, and multilateral aid organizations" (http://www.columbia.edu/cu/e3b/phd.html, July 13, 2007).

[§] The National Science Foundation reports that in 2006, $342.9 billion was spent on "research and development" in the United States, of which $49.1 billion was spent by universities and colleges (14.3%), most of the rest going to industry and government (NSF 2007). As for what proportion of research and development is devoted to pure science, no statistic is offered. Clearly, *development* refers to the transformation of scientific discoveries such as new drugs or technologies into marketable products, and is therefore applied, not pure, science. Even the research leading up to such applications is apt to be applied rather than pure science. For what it is worth, it is typically estimated that R&D is 4% research

later twentieth century that even the funding for scientists working in universities be reserved for projects that are relevant to social goals* has further eroded pure science, to the extent that it is now endangered.

Applied science is a search for knowledge and understanding of natural phenomena, only with the added bonus, as it were, that the knowledge sought will serve a goal other than knowledge and understanding as such. Applied science, then, is science in the service of a specific nonscientific goal. From the outside, the work of applied scientists looks indistinguishable from the work of pure scientists. Measurements are taken, the extant literature on the subject under investigation is researched, theories or models are devised, experiments are run, highly precise equipment is employed, papers are published, and so on. The difference between them lies in their differing intentions: Whereas the pure scientist works to advance science itself, the applied scientist is motivated mainly by other goals. Since values and goals are logically linked, another way to put the distinction between applied and pure science is in terms of *values*: Knowledge (understanding, explanation) is the sole value of pure science, while knowledge is of value to applied science only insofar as it enables the attainment of some other value, such as human health.

Where does environmental science fit into this picture?

CASE STUDY 5: IS ENVIRONMENTAL SCIENCE PURE OR APPLIED?

Pure: Some environmental science texts emphasize scientific understanding instead of the environmental objectives. For instance, *Essential Environmental Science* (Watts and Halliwell 1995) clearly states its "environmental perspective" in their introduction: "As global concern for the environment grows, so too does the demand for accurate and precise information about it." Information is clearly the aim of environmental science as they conceive of the subject. They return to the theme "the pursuit of knowledge" at the start of the first chapter. The book then goes on to introduce a number of scientific subjects and does not spend time in rehearsing environmental values, nor in defining such things as sustainability, green science, or green technology. Although this may only reflect the fact that environmental values are so deeply ingrained that they may now simply be assumed, rather than stated, it still indicates the authors' desire to keep environmental science in a separate compartment from environmental values and programs. Some environmental science programs, like that of the Massachusetts Institute of Technology, try to compartmentalize earth science studies separately from environmental science studies. The former are (mostly) pure

---

and 96% development. If pure science occurs in academia in nondevelopmental contexts, its share of research funding is some minor proportion of the 14.3% figure.

* At its website, the National Science Foundation (NSF) reports that it "is an independent federal agency created by Congress in 1950 'to promote the progress of science; to advance the national health, prosperity, and welfare; to secure the national defense ...' " (http://www.nsf.gov/about/ June 22, 2007). Clearly, this is an authoritative expression of the social subtext of science: science is to serve the good of society, in exchange for socially supplied funding. The same sort of social contract is sketched in the mission statements of virtually all science funding agencies. It is also an indication that pure science, scientific knowledge as such, is not the goal of the funding proffered.

science courses, while the latter teach how to apply the science to environmental problems.*

Applied: The sort of separation just mentioned is not typical, and in any case may not be sufficient to maintain the purity of science, for in the end it has to be *applied* to environmental objectives. Placing this atypical case aside, environmental activism permeates the mission statements of most environmental science institutions, the programs of studies they offer, the research they house and support, the accomplishments they advertise, and the news stories they post. Professional conferences,[†] journals,

---

* The MIT environmental science course 12.103 is entitled "Strange Bedfellows: Science and Environmental Policy," which gives a sense of the intellectual separation that seems to be maintained: environmental science is science, but must be married to environmental policy at some point, so the course tries to show that Science can do this without losing its virtue (http://student.mit.edu/catalog/search.cgi?search=%2B12.103&style=verbatim&whole=on, 28 August, 2007).

† The 10th International Conference on Environmental Science and Technology (Kos Island, Greece, September 5–7, 2007) has the following statement of purpose (http:/www.gnest.org/cest/Default.htm, July 13, 2007): "Like the previous conferences this tenth conference maintains and upgrades the synthetic and integrated approach towards protection and restoration of the environment, by bringing together engineers, scientists, students, managers and other professionals from different countries, involved in various aspects of environmental science and technology." Engineers are applied scientists by definition, while managers and other professionals are not scientists at all. Though the conference is said to be for "environmental science and technology," its stated goal is neither science nor technology, but "protection and restoration of the environment," which is supposed to be reached by a "synthetic and integrated approach" that is not only sufficient, but necessary, for obtaining environmentalist goals, as the statement goes on to explain. "This synthetic approach, in combination with the integration of environmental issues with economic and social aspects, is a prerequisite for adopting sustainable solutions to the numerous contemporary problems." The similarly named Third International Conference on Environmental Science and Technology sponsored by the American Academy of Sciences (August 6–9, 2007, Houston, Texas) states the same goals (http://www.aasci.org/conference/env/2007/index.html, July 13, 2007): "to provide a major interdisciplinary forum for presenting new approaches from relevant areas of environmental science, to foster integration of the latest developments in scientific research into engineering applications, and to facilitate technology transfer from well-tested ideas into practical products, waste management, remedial processes, and ecosystem restoration." Again the role of science is to serve environmental goals such as ecosystem restoration, hence environmental science is clearly not pure, but applied. Indeed, the scope of environmental science and technology is expanded to include topics not included in science or technology as normally defined: "Environmental humanity and sociality such as environmental ethics, environmental law, environmental economy and environmental management are also included in the scope of the conference." In fact, this "science and technology" conference really amounts to an international strategy session of professionals of all sorts devoted to environmental goals and tactics: "Researchers, engineers, site managers, regulatory agents, decision-making officials, consultants, and vendors will all benefit from the opportunity to exchange information on recent research trends, to examine ongoing research programs, and to investigate worldwide public and regulatory acceptance of environmental protection and remediation technologies." Vendors and decision-makers are neither scientists nor technologists. The explanation offered for including them in a science and technology conference is that their participation gets the results environmentalists want: "Environmental disturbance and pollution are complex problems worldwide. The current development of modern science and technology combined with management on social and economic activities are contributing more and more to solution of the problems." There is no need even to state what "the" problems are, since the environmental crisis is an assumed article of faith. This statement of purpose goes on to say the conference will be a "multidisciplinary platform"; in other words, not a science or technology conference at all: "Although considerable environmental protection work has been and is presently being conducted, a multidisciplinary platform for environmental scientists, engineers, management professionals and

and funding bodies* also evidence the activist posture of environmental science. Environmental science, as it defines itself in these venues, is not just about knowledge and understanding, the traditional objectives of academic science, but about *solving* environmental problems.

The courses and texts for environmental studies also proudly proclaim this activism. In *Environmental Science: Earth as a Living Planet,* we find a declaration that is representative of the texts in the field. Under the heading "What is different about environmental science compared to other sciences?" the authors, Botkin et al. (2005, p. vi), say that environmental science is ". . . 'mission-oriented,' . . . aimed at solving real environmental problems, rather than just understanding how they arise." In *Environmental Science: A Study of Interrelationships*, Enger and Smith (1995, p. 3) state: "Environmental science has evolved as an interdisciplinary study that seeks to describe problems caused by our use of the natural world. In addition, it seeks some of the remedies for these problems." In *Environmental Science and Technology: A Sustainable Approach to Green Science and Technology*, Stanley Manahan (2007, p. v) states that "humankind is on a collision course with the carrying capacity of planet Earth. . . . The need is urgent, and time is short." He then goes on to tout the evolution of "green science" and "green technology" as having an "essential role" in avoiding this collision. In Daniel D. Chiras's early text (Chiras 1988; first published in 1985), *Environmental Science: A Framework for Decision Making*, we find an entire section, over 60 pages in length, devoted to an analysis of the ethics, economics, and government that has led us into the current environmentally "unsustainable" situation, along with the promotion of values that will lead us to a sustainable society. This is an extremely odd thing to include in a *science* text—unless, of course, environmental science is not only applied science, but science subservient to its application to environmental objectives.

---

governmental officials to discuss the latest developments in environmental research and applications will be very helpful to protecting our global village." In short, this event is not a science and technology conference as ordinarily understood, but an international political rally of environmentalists of all sorts.
* The National Council for Science and the Environment, a funding body, begins by describing itself as concerned mainly with the relevant knowledge (http://ncseonline.org/): "a not-for-profit organization dedicated to improving the scientific basis for environmental decisionmaking. We envision a society where environmental decisions by everyone are based on an accurate understanding of the underlying science, its meaning and limitations, and the potential consequences of their action or inaction." It goes on to add: "While an advocate for science and its use, the Council does not take positions on environmental outcomes. This enables the Council to provide a neutral forum for all." Nevertheless, it emphasizes getting practical results, citing "many programs that we carry out to achieve our mission." Under the heading "Featured Project" is listed "The Wildlife Habitat Policy Research Program (WHPRP)," which is described as "a results oriented program with a mission to develop and disseminate objective information and practical tools to accelerate the conservation of wildlife habitat in the United States. . . . The WHPRP uses competitively awarded grants to sponsor innovative projects. . . . Current projects include the development of a habitat banking system." A habitat banking system is a form of environmental action, not scientific research. Clearly, then, funding support is provided for environmental goals and values, not just scientific research. Scientific objectivity and value neutrality are clearly endangered, if not unsustainable, in this context. A scientific research program which, for instance, came to the conclusion there was no need to "accelerate the conservation of wildlife habitat in the United States" would probably not be funded by this agency.

Merely recognizing that environmental science is applied science really does not go far enough to capture the degree to which the environmental aspect rules the scientific aspect. Environmental scientists are dedicated to reversing environmental degradation and returning the environment to good health. They intend to do this not only by their own direct activity, but by educating a new generation of scientists to carry on with the mission of securing environmental health.* Most accept the chief article of faith of environmentalism: that environmental crisis is either real or threatens to become real. Science, despite its unintended creation of the possibility (or actuality) of environmental crisis, is now ready to help fend off this possibility (or actuality). In order that this mission may be carried out, *holism* is a stated aim of the education offered. The most modest form of holism requires students to study a wide variety of disciplines within the physical sciences. The more ambitious forms of holism offer studies in law, history, communications, management, ethics, leadership, policy, and even religious studies.

Why this extension of science studies into areas that have never before been part of science? The answer is that science applied to realizing environmental values has a complex task that goes well beyond the quest for knowledge as such. The environmental scientist has to solve a value problem and a practical problem. The value problem involves conceiving in detail what form the world should take in order to realize environmental health. This is a problem that would normally be called philosophical or social rather than scientific, and it explains the excursion of environmental science into these topics. The practical problem is more scientific in the traditional sense of the term, but even so, involves understanding how to bring about and guide social change.†

---

* Johns Hopkins University, for instance, has an Environmental Science and Policy program with the explicit mission of the marriage of science and environmental activism in the creation of a new generation of leaders. Its website, describing the program, gives a fairly detailed explanation. "To manage Earth's environment effectively, we must understand the processes that shape our planet's surface, control the chemistry of our air and water, and produce the resources on which we depend." Thus, right from the beginning, the stated goal of the program is not knowledge as such, but managing Earth's environment, with scientific understanding serving only an instrumental role. In the next sentence, it is assumed that science also plays the supporting role of providing solutions to environmental problems, but needs further support in turn: "At the same time, in order to implement scientific solutions to environmental problems, we must establish and execute policies that are politically, socially, and economically feasible." Thus, managing the environment ultimately requires the training of a new generation of political leaders: "Graduates of the program emerge with a combination of expertise in science and policy that enables them to assume key positions in public and private entities responsible for safeguarding our environmental future" (http://advanced.jhu.edu/academic/environmental/, July 11, 2007).

† In *Environmental Science*, Enger and Smith (1995, p. 3), after stating that environmental science aims not only to describe but to solve environmental problems, go on to list "three major areas" of study: first "natural processes," which is the only domain of a natural science; second "the role that technology plays in our society and its capacity to alter natural processes as well as solve problems caused by human impact," which is both a recognition of the past sins of technology and a proposal that it can behave better in the future; and finally, "the complex social processes that are characteristic of human populations." This last topic, which is not part of natural science as normally conceived, is important in order that the environmental scientist can intervene in these processes for the good of the environment. To this end, an entire chapter "discusses the differences that can exist between individuals in a society and the different behaviors exhibited, depending on whether the person is acting as an individual, as part of a corporation,

5.3   ENVIRONMENTAL SCIENCE AND VALUES

Science has traditionally assumed a position of value neutrality. The value neutrality of science, moreover, is not a mere add-on for science, not an optional feature. Value neutrality has been *essential* to science from the start, and still is. It has enabled science both to make astounding progress in the pursuit of knowledge about nature and to gain high respect as an authority about the natural world. In the contemporary shift towards environmental consciousness and responsibility, science finds itself in a position of authority that it has gained in the last few centuries largely because of its tradition of value neutrality. Environmental science, which has abandoned value neutrality in its employment of the authority of science in service to environmental values, thus threatens that authority in the process.

To understand science's value neutrality, it is helpful to distinguish between science as an activity or enterprise on the one hand, and science as a body of knowledge or knowledge claims, on the other. All forms of intentional human activity have goals, and hence involve values. Since the goal of scientific activity is knowledge of the natural world, that activity presupposes the value of that knowledge. However, the body of knowledge or knowledge claims that this activity produces does not include any value claims. Thus, the value neutrality of science may be seen as having two mutually supporting aspects. One concerns what is produced by scientific activity: the theories, models, and hypotheses that comprise the body of scientific knowledge: the second concerns that activity itself, that is, scientific method.

1. *Topical value neutrality.* The topic of natural science comprises the facts of the natural world. Scientific knowledge therefore consists of a body of fact claims. Science does not make value claims.

2. *Methodological value neutrality.* The only value that motivates scientific activity as such is that of the knowledge of nature. No matter how worthy or noble in themselves, all other values are a hazard to scientific objectivity. Religious, political, aesthetic, or environmental ideals may distort scientific judgment, leading to scientific claims being made because they seem worthy of acceptance given these ideals, even where unjustified by the evidence. So care must be taken, insofar as is practically possible, to safeguard scientific objectivity from values.

As presently conceived and practiced, environmental science involves a remarkable twofold departure from pure science: *Both* aspects of value neutrality have been

---

or as part of government." In *Environmental Science and Technology*, Manahan (2007, p. vi) adds to "the four traditionally recognized environmental spheres—the hydrosphere, atmosphere, geosphere, and biosphere (water, air, earth, and life)" a sphere just for us humans, the "anthrosphere. . . . In so doing, the book recognizes that humans simply will modify and manage Earth to their own perceived self-benefit." He goes on to say that environmental scientists must even use "anthrospheric activities to enhance the environment as a whole." Again, environmental science goes beyond the domain of natural science as such.

abandoned in environmental science. Topical value neutrality is abandoned since environmental science has taken upon itself to define what state of the environment *should* be and what humanity's relationship to the environment *should* be. Obviously, questions about how things should be are not matters of fact, but matters of value. Environmental science abandons methodological value neutrality since it does not pursue knowledge for its own sake, but to realize its *ultimate* value: the health of the environment, including whatever that may imply for human economics, politics, government, and value orientation.

The peculiarities of environmental science do not stop there, however. Other applied sciences do not engage in the attempt to convert society to their own goals and values. Military scientists do not try to convert people at large to developing better weapons, and computer scientists do not try to convert society to the quest for better computers. However, environmental science aims to achieve environmental health by converting society in general to its values. In addition, environmental science shares environmentalism's antihuman bias: Human beings are identified as the source of environmental harm. So environmental scientists, as we have seen, are engaged in *conversion* of people in the same sense that religious and political bodies are. This explains why environmental scientists analyze actual human values, have an image of what those values should be in order that environmental health be achieved, and then promote those values by teaching them in schools, colleges, and universities as well as to the general public.* The environmental scientists' campaign to convert society in general to their vision also explains why they educate their students to take positions of influence: The conversion of society is a massive undertaking that requires great numbers of people working in cooperation and in the highest possible positions of authority.

THESIS 6:   Environmental science as actually defined and practiced is not pure science in pursuit of knowledge and truth, but advocacy science applied to the pursuit of environmentalist objectives.

---

* See, for example, any of the following books written by prestigious scientists for a popular audience in support of environmental values: Ehrlich (1968, 2005), Suzuki (1994, 1998), Rees and Wackernagel (1996), Ehrlich and Ehrlich (1998), Eldredge (1998), Wilson and Perlman (2000), Wilson (2001, 2002, 2007), Suzuki and Dressel (2004).

6 Pure Environmental Science

When one turns to the magnificent edifice of the physical sciences, and sees how it was reared; what thousands of disinterested and moral lives of men lie buried in its mere foundations; what patience and postponement, what choking down of preference,... how absolutely impersonal it stands in its vast augustness—then how besotted and contemptible seems every sentimentalist who comes blowing his smoke-wreaths, and pretending to decide things from out of his private dream!

—William James (1897)

When I was young, the continents did not move. Well, the continents were moving, all right, only nobody knew that they were. Geologists said the continents did not move, and that meant that nonscientists could not possibly know otherwise. Then, partly because of data collected in the International Geophysical Year (1957–1958), it was discovered that the continents do move after all. So, as is well known to everyone today, even small children, the continents move. But before this knowledge became an official part of science, nobody could know this.

It is not as though nobody *thought* that the continents moved before then. Alfred Wegener (1880–1930) in particular had argued from 1915 on for what was then called *continental drift* or *mobilism*. His arguments, as we can see *now*, were cogent and powerful: The geological formations of the continents on east and west sides of the Atlantic Ocean match up just as if they had been torn apart at some time in the past, like two halves of a torn dollar bill. But Wegener did not have a mechanism to explain continental motion. Plate tectonics provided that mechanism: The continents move on currents of molten rock in the underlying mantle. Once the scientific community in charge of this matter had concluded that this is how it all worked, then everyone, even children (including me at the time) could then know it.

Science defines what is socially accepted as knowledge of the natural world, and knowledge itself is not the domain of any individual but of humankind as a whole. We rely on the ears and eyes and brains of countless other human beings (many of them now long dead) in order that we may know about Antarctica, bacteria, or dinosaurs. No one among us could possible gather the mountains of evidence and think through the mountains of theory that support even the most common scientific facts taught to our children in school. We *must* rely on scientific authority—there is no other way. Science is often portrayed as a body of knowledge that rests solely and solidly on the facts themselves, but the fact is that nearly all the science that any one person

Beyond Environmentalism: A Philosophy of Nature, By Jeffrey E. Foss
Copyright © 2009 John Wiley & Sons, Inc.

knows, including scientists, rests on *authority*. Fortunately, scientists have created a complex hierarchy of scientific authorities to protect the purity of science. Like all things human, it is not perfect. Nevertheless, this modern priesthood of knowledge does work, and is responsible for the wonders of the modern world.

As a youngster I loved science. I loved dinosaurs, atoms, chemicals, radiation, rockets, the solar system, and the myriad of little stars that appeared around the Pleiades when I looked at them through my small reflector telescope. I loved the certainty of science and the thrill of looking through the veil of appearance directly at the underlying reality. I *believed in* science. My faith was not shaken when Wegener's theory of continental drift (as it was then called) went from being false to being true. As I saw it, science had searched for new evidence, had found it, and had done the right thing by changing its mind. My faith in science was unshaken.

Then I ran into Einstein's theory of relativity. That was something else altogether, for when his theory became true it changed the foundations of physical science itself. As for so many people, Newton's laws made perfect sense to me. Like many physicists, I felt the truth of Newton's laws in my guts every time I rode a merry-go-round. But Einstein's theory required enormous subtlety even to imagine. I couldn't imagine it, actually, but I took it on authority that Einstein was right, and tried harder to grasp his vision. Around the time I entered university I was reading one of Einstein's own introductions for the nonphysicist, *The Meaning of Relativity* (1950), when I suddenly realized what was holding me back: I kept thinking in Newtonian terms, by gut instinct, so to speak. I kept trying to translate relativity into Newtonian physics—but the translation was impossible. Relativity only made sense if, gut instincts notwithstanding, Newton was *wrong*.

The scales fell from my eyes, and I began to see what Einstein was saying. When you glimpsed the world through Einstein's theory, it was obvious that Newton was wrong. But if Newton was wrong—and surely he was, even if he was wrong only by a little bit when it came to everyday objects and speeds—then Newton's laws had never been proven in the first place. What is proven cannot turn out to be false. My faith in scientific authority was shaken. It seemed to me that even when we think science is looking through the veil of appearances, it may just be looking at another set of appearances. My studies of quantum mechanics over the next few years tended to confirm that opinion. We can never transcend ourselves, so we can never leave appearance entirely behind. There is no such thing as scientific proof. But if there is no scientific proof, how does science get to tell us what we can or cannot know? Searching for the answer to that question led me to a career in the philosophy of science, and eventually to environmental science and this book.

In this chapter we have a look at pure environmental science, or environmental science *as such*, as philosophers say. If we were studying basketball, for example, basketball as such would be defined by the rules of the game. We would discover in those rules that the very idea of basketball is that of a noncontact sport, for that is what the rules say and what the various sorts of fouls and penalties are designed to create. The rules create the game. Knowing what pure basketball is, we would probably be surprised when we have a look at professional basketball, for it involves *lots* of contact, with huge players shoving and pushing each other right under the noses of the officials, no fouls called. Actual basketball turns out to be an imperfect realization of pure basketball, with its own special problems, solutions, and (dare I say it?) beauty. The same is true for other human endeavors, and science is no exception.

As we saw in Chapter 5, actual environmental science is science applied to the promotion of environmental health. It is not value neutral; it does not seek truth for the sake of truth; it moralizes; it is trying to reshape human life itself into a form that is better for the environment (where, as noted previously, *environment* stands for all of nature except humankind and its works). It has, to put it bluntly, many of the properties of the sociopolitical ideologies that have gripped our species periodically. In this context it is important not to take our eye off the ball, as it were. If ever there was a time when we needed pure (or at least purer) environmental science, it is now. Since we cannot find something until we know what we are looking for, we need to get a better idea of just what environmental science would be if it were closer to its ideal form. That is our goal in this chapter. We begin by considering science as such. Like most human enterprises, science evolves and its rules are revised as its practitioners deem necessary. So we will start with a bit of science history.

Some readers may find this chapter rather dry and scholarly, which is not surprising, perhaps, since it deals with my own academic specialty, the philosophy of science. Its main ideas are quite simple, and to many people quite obvious: first, that there is no such thing as scientific proof (as the histories of Wegener and Einstein show us), and second, that science cannot tell us what we ought to do, what is right or wrong, what is ethical—in short, has no authority over *values*. So if you find yourself bogging down in the scholarly technicalities of this discussion, please jump ahead to Case Study 6 and Theses 7 and 8. One main function of this chapter is to prepare us for Case Study 7, which is an extended study of global warming theory. The importance of the global warming issue for the human species and for the rest of nature is enormous. It is the issue that will define the future of environmentalism and environmental science for generations to come.

There is, however, no way to come to grips with global warming except through the scientific technicalities involved. To a large extent these technicalities can be rendered intelligible to nonspecialists, and that is what I am attempting here. Indeed, it is important that as many ordinary civilians among us do this as possible. I urge all readers to read as much of Case Study 7 as they comfortably can. But I also realize that scientific technicalities simply are not some people's cup of tea, and they should feel free to jump ahead to Chapter 7, which, by way of preview, is about our sense of the sacredness of nature, which is at least as important as (and much more permanent than) the issue of global warming. It is as if our philosophical expedition has a planned ascent of a substantial mountain peak, the mountain of science that is involved in the issue of global warming, as its centerpiece. Ropes and crampons will be used along with other technical aids, to ensure that those who are really determined can make it to the top. However, those who are not interested in such mountaineering are welcome to sit out this part of the expedition.

6.1   THE RISE AND FALL OF SCIENTIFIC PROOF

Before modern science, anyone who wanted to know what kept the Sun and Moon up in the sky, or how human beings came into existence, or whether people from other cultures were trustworthy, would ask a priest. This was not just a peculiarity of Western Europe, where modern science would first begin to flourish, but a more general phenomenon. As we trace cultures back in time we find that religion and science are not at first distinct. The persons who speak authoritatively about matters of fact are the same ones who speak authoritatively about God or the gods. Starting roughly with Copernicus (1473–1543) and Galileo (1564–1642), science began to assert itself *against* the authority of religion. These scientists said the Sun, not the Earth, was the center of the universe. Copernicus warily published his theory on his deathbed to avoid repercussions from the Church, but Galileo boldly published his ideas despite the dangers, was arrested on suspicion of heresy, and was confined to house arrest. Matters might have ended there, but on Christmas day of the year Galileo died, Newton (1642–1727) was born. He devised the physics, and with it described the mechanism, that explained how the planets go around the Sun. Thanks to Newton, science won a decisive victory. Over the next two centuries, science went from one stunning success to another in the enterprise of understanding nature.

Gradually, the scientist in lab whites replaced the priest in black robes as the authority on what keeps the Sun and Moon up in the sky, or how human beings came into existence, and whether people from other cultures are trustworthy. Newton thought that scientific knowledge was *literally* proven, in the fullest sense of the word, as exemplified in mathematics. In mathematics we can prove, for example, that any triangle with sides of equal length (an equilateral triangle) also has angles of equal size. This sort of proof is the strongest possible: We cannot even *conceive or imagine* a triangle with equal sides that does not have equal angles as well. Newton's epoch-making system of physics was set out in his book, *Philosophiae Naturalis Principia Mathematica* (*The Mathematical Principles of Natural Philosophy**). Newton presents his new philosophy as though it were pure geometry, as a series of proofs of theorems from basic axioms.† Newton's famous three laws are presented by him as

* That what we now call science was originally called natural philosophy is an interesting and still relevant fact, but one that we cannot go into here. Suffice it to say that philosophy is the pursuit of wisdom, not knowledge, and wisdom involves knowledge, including knowledge of nature, as well an understanding of the significance of the thing known to the knower and what stance the knower should take toward it. Since Newton's day there has been a further specialization: Scientists seek only knowledge of nature, not wisdom. Environmental philosophy is, as I understand it, the return of philosophy to nature after abandoning it to the scientists: natural philosophy reborn. I encourage anyone who thinks this is a good idea to get involved.

† An axiom, or postulate, is a claim that is assumed true although no proof of its truth is offered. All chains of proof must stop somewhere, and axioms are those stopping places. Axioms are often said to be self-evident or intuitively obvious. A typical axiom is this one, taken from geometry: Given any two distinct points on a line there exists at least one other point between them. A moment's reflection shows that you cannot even imagine how this could be false. The beauty of mathematics is that it shows how subtle and complex mathematical propositions which are not obviously true can be proven by being derived logically from axioms that are obviously true. Newton believed that all natural science was provable in this way.

axioms, or *self-evident truths*, not as the result of observation or measurement. That is not only different from what scientists are now taught, but opposed to it.

David Hume (1711–1776) showed by a series of ingenious arguments that no universal claims about nature, such as Newton's laws, could ever be proven. One way to see this is to observe that even though every massive body *that we see* attracts other masses via gravitation, this does not prove that those we do not see also attract each other.* Science, in other words, involves its own species of faith: faith in the uniformity of nature. Science demands that we believe that the things which we do not see are precisely like those that we do see. But faith is not proof. For a century or so Hume's arguments were ignored by scientists as they extended Newton's scientific foundations to include chemistry, electricity, electromagnetism, and biology. Ernst Mach† (1838–1916) was, however, a notable exception. Mach took Hume's problem very seriously, so seriously that he analyzed physics methodically to identify its articles of faith. Two of the articles of faith he discovered were absolute space and absolute time. His most famous student, Albert Einstein (1879–1955), went on to develop a new physics that rejected the Newtonian metaphysical concepts of absolute space and time in favor of relativistic space and time. As we all know, Einstein's theory of relativity was an absolute success, with countless confirmations by observation. Perhaps the most stunning confirmation is the nuclear explosion, which confirms the equivalence of mass and energy denoted by his famous equation $E = mc^2$.

What we do not usually realize—indeed what we do not like to admit—is that Einstein's physics disproved Newton's. It also proved, therefore, that *science is fallible*. Science not only makes small superficial mistakes, but also big foundational mistakes about the fundamental nature of space, time, matter, and energy. Ironically, just as scientists abandoned the idea of scientific proof in the last century, society at large simultaneously embraced the notion as they increasingly turned toward science and away from religion and tradition for knowledge and understanding. Thus, we now find ourselves in an age when the concept of scientific proof is still accepted by people in general, even though *scientists themselves* gave up on this idea early in the twentieth century—at least officially.

6.2   THE RISE OF MODELING

The assimilation of this historical lesson is very clearly manifested in the view, which is nearly universal among scientists today, that the business of science is to *model* natural phenomena. A model is an abstract representational structure, typically constructed in mathematical terms, that is sufficiently accurate to enable us to predict and possibly control the phenomenon it represents. We are all familiar with model

---

* Newton himself did not think that universal gravitation was axiomatic, so he did not include it among his three laws. However, he did not realize that those three laws themselves are not axiomatic either: They are not self-evidently true, but rest instead upon observation.

† The use of "Mach 1" to indicate the speed of sound, "Mach 2" twice the speed of sound, and so on, derives from his name: the study of shock waves, using the then cutting-edge technology of photography, was one of his scientific specialties.

airplanes, model automobiles, and model ships. Like these models, scientific models include representations of the specific properties of the things modeled, but unlike them they are not physical objects themselves, but abstract structures. For example, a mathematical formula may model an orbit, a pendulum, or soil erosion. Just as a model airplane will not have all of the properties of a real airplane (its tires may not be made of rubber, its motor may not work, it may not fly), so, too, a scientific model will—in fact, must—omit some of the properties of the system. The volume of an orbiting object may be omitted so it can be represented as a point mass, friction may be omitted in the pendulum, no location may be given in a model of eroding soil, and so on. Simplification is of the essence in modeling: The right things must be left out so that the essential things can be included.

The general use by scientists of the concept of modeling to describe their work is an extremely important indicator of a big shift in thinking in the scientific community since the time of Newton: The *product* of science is not the universal law, but *models* of aspects of parts of reality (such as the crystalline aspect of salt). Fortunately, maps are a familiar type of model and can be used to illustrate two important points. First, like maps, models are neither proven nor disproven. Instead, they are useful or useless, detailed or simple, accurate or inaccurate. Second, like maps, models require simplification in order to be constructed in the first place. Just as no one expects to find actual houses on a street map, or actual water in a map of the Pacific Ocean, so scientists do not expect their models to capture every aspect of the things they model. The purpose of a model is the same as the purpose of a map: to help us find our way around. What information is contained in a map—and what is left out—depends on what bit of the world we want to navigate. The water department's map of the same streets as those shown in the driver's road map will show where the water mains are located, whereas the politician's map will show the electoral districts instead. In the same way, different sorts of scientists provide different sorts of models, employing different sorts of abstraction, to serve their different sorts of interests. The geologist's models will ignore fauna and flora to show the rock formations; the biologist's model will ignore the rock formations to show the fauna and flora.

Scientists' self-conscious recognition of what they are doing is reinforced by the use of computer models. No one expects that a computer model of photosynthesis will yield any actual carbohydrates or that a meteorological model of a thunderstorm will get anybody wet.* Notable in the context of environmental science is the fact that much of the belief that we are in the midst of an environmental crisis depends on computer models. The case for the sixth extinction relies on such models. The belief that the Earth is warming dangerously because of carbon dioxide produced by human beings relies on massively complex computer models known as *general*

---

* There are two significant and illuminating exceptions here. Within cognitive science, many scientists (the psycho-functionalists, as philosophers call them) have the view that cognition is nothing other than computation of some sort, and so believe that computation of the right sort, even in a computer rather than a brain, actually is cognition. Within artificial life research, some scientists believe that a properly constructed program not only models life (or evolution), but in some sense is really alive (or really evolving).

circulation models (GCMs). GCMs involve a simplification that is striking: In order to model climate, weather is ignored! Because the amount of physical detail involved in global weather is far too great to be handled even by present-day computers, GCM modelers are forced to ignore it in order to model climate by attending primarily to radiation balance (the balance of incoming solar radiation with outgoing infrared radiation). Although no condensation and fall of rain is modeled, the *average amount* of rainfall may be modeled; although no hurricane is modeled, the *average number* of hurricanes may be modeled; and so on. It is just plain obvious, then, that GCMs are abstractions, well and truly divorced from full reality. This important issue is discussed more fully in Case Study 7.

6.3 MODELS AND TRUTH

If there is no such thing as scientific proof, must we abandon the entire idea of scientific truth as well? Does science tell us the truth? Does it even make sense to think of science as seeking the truth?

The question is complex, and the answer is not simple. Certainly, we do not have to abandon the entire idea of truth. The concepts of truth and falsehood still apply to such simple claims as "DNA is a molecule," "dinosaurs are extinct," and "the water is boiling." However, the concepts of truth and falsehood are far too simple to apply to entire scientific models. Like maps, models say a lot of things, and it is very crude to think of them as simply true or false. The simple claim that Main Street intersects Central Avenue may, for instance, be true. If the map shows Main intersecting Central, we may say that it contains this truth. But when it comes to the map as a whole, truth and falsehood are very blunt instruments of evaluation. We never say that a map is true or that it is false. Instead, we evaluate maps in terms of their accuracy, precision, completeness, clarity, and so on. For the same reasons, scientists judge their models in just such terms.

Still, a scientist will sometimes speak about theories being true—particularly in the context of public debate—and since theories really are just models, the scientist is in effect saying that a model is true. In a debate with a creationist, a biologist will say that the theory of evolution is true, not that it is a model which is accurate or complete or anything like that. For dramatic effect, he or she may even say that the theory has been proven, so it really is not a theory any more, but a fact. This sort of claim by a scientist is most generously interpreted as rhetorical overstatement. Taken literally it is deeply unscientific. Presumably it is not meant to be taken literally but as a way of emphasizing that the evolutionary model is so accurate, so precise, so complete, that we should just accept it and move on.

Ironically, this nonliteral, nonscientific use of the concept of truth is very close to the popular conception—or *mis*conception. We nonscientists want a simple thumbs-up or thumbs-down when it comes to what we should think or believe. If, for example, after hearing all of the arguments pro and con we decide to accept the evolutionary model, we will say that it is true. If we decide to reject it, we will call it false. It is a busy world, and we do not always have time for subtleties. When we need to know, we need to know now: true or false? So we should not be surprised, then, when scientists give us a yes-or-no answer when we ask them whether a given theory is true. Still, we owe it to ourselves to remember that true-or-false is a simplification—a big simplification. Sometimes we would do ourselves a favor by taking the time to really know the model in question. In science, as in personal relations, it is wise to devote time and sensitivity if we want to achieve something lasting and truly valuable.

In practice, therefore, truth is achieved by constructing a good model, and the pursuit of truth is the pursuit of good models. Five criteria generally apply to models and provide a basis for their evaluation:

1. *Precision*. What, exactly, does the model say? The model cannot be checked for accuracy until what it says is plain. For example, a model which predicts that a flock of ducks will migrate south in the fall is less precise than one that predicts the specific day on which it will head south. Other things being equal, we want greater precision. We want to predict not just the season, but the very day, indeed the very second, something will happen. So precision is very valuable in scientific models. In fact, precision is one of the defining characteristics of science itself.

2. *Accuracy*. How closely does the model match observation? The distinction between precision and accuracy is not obvious and requires a little care if it is to be understood. Precision may be thought of as the target that a model sets for itself. A more precise model sets a smaller target. Accuracy concerns whether or not it hits that target when it comes to observation and measurement. A very precise model of your weight, for instance, would predict your weight to a tiny fraction of an ounce—but that would make it more difficult for the model to be accurate. The demand for precision is what makes the demand for accuracy so difficult. Precision and accuracy are thus in tension with each other—a creative tension, but tension nevertheless.

3. *Consistency with other successful models*. Since successful models are such only because they are precise and accurate, a new model that yields new observations immediately finds itself in danger of being ruled out by well-established observation. Note that the consistency in question is *empirical* consistency. New models do not have to agree with established models on a *theoretical* level. Einstein's model, for example, disagreed profoundly with Newtonian models on a theoretical level. What is essential is that new models agree with well-established observation and measurement.* Other things being equal, we would prefer a model to be at least as precise and accurate as other models where they deal with the same empirical content.

4. *Scope*. The broader the scope of a model, the better the model, other things being equal. If one map covers more ground than another, it is better in that

---

* For example, Einstein knew that his theory of relativity had the same empirical content as Newton's theory in the vast majority of cases already tested by observation, where velocities were small relative to that of light, and knew, moreover, that this was essential in order that his theory have any chance of being right.

regard. The broader the scope, the more things it applies to, and the more information it provides us about the world. So we prefer theories with broader scope to those with narrower scope.

5. *Simplicity.* The point of scientific theories is to reduce the booming, buzzing complexity of the world, and to find the simpler patterns underlying the eternally new and rich variety of the unfolding universe, so that we can anticipate events and gain some control over our own destiny. If a theory were just as complex as the world itself, it would be of no use to us. Theories must help us understand the world, and this requires that they be simpler in themselves than the world is in itself. Simplicity is often in tension with accuracy or with scope. For example, classical mechanics is simpler than quantum mechanics, but classical mechanics also has less scope, since it does not include subatomic phenomena and is less accurate, since it asserts that atoms will collapse. We would most like a theory that is both simple in itself and broad in scope, but often we must surrender one of these theoretical virtues in exchange for the other.

6. *Outcompeting the alternatives.* Since there is no absolute measure of scientific merit, we must rely on comparative measures. Therefore, the levels of precision, accuracy, consistency with other models, and scope that a model must obtain in order to be accepted by the scientific community depend on the levels obtained by other models in the same domain.

What these criteria presuppose is that scientific evaluation and judgment are *relative*. There are no answers to be found in the back of some great textbook in the sky, no angelic referee to tell us what is really true. We must instead do the best with what we have come up with on our own. So scientific models are not measured against the world itself, despite the countless idealizations of science which give that impression. Instead, models are checked against every relevant thing that we know and observe, however imperfectly, and that includes other scientific models. How much accuracy, precision, scope, or simplicity we demand is a function of how much accuracy, precision, scope, or simplicity is already achieved by other models in the same field. This relativity of scientific judgment must be kept in mind whenever we need to figure out just what to make of some new scientific claim. We must also remember that these theoretical virtues are in tension with each other, so that the choice of a theory is not a simple maximization problem, but a multiple-constraint problem.

6.4   PROTECTING SCIENCE'S VALUE NEUTRALITY

One fact that science must face in its maturity is that there is no methodological guarantee that it will not be influenced by factors that are extraneous to its goals. Put plainly, science is not immune to *prejudice*, where "prejudice" is understood in its literal sense: judgment in advance of, or independently of, the relevant evidence. There is always the logical and methodological possibility within science of its claims to truth (keeping in mind that "truth" is a simplification) being influenced by (or being a partial function of) factors that have no bearing on whether or not the claim is true (precise, accurate, etc.). These factors may include such things as beliefs, values, hunches, inclinations, likes, dislikes, and so on, and they are logically irrelevant in two ways: (1) they neither increase nor decrease the probability that the fact claim is true, and (2) they operate tacitly, unseen beneath the surface of scientific debate.

For example, imagine that a scientist who is a snake specialist (herpetologist) has a deep fear of snakes, due to a now-forgotten childhood incident in which he was horribly frightened by a snake that crawled into his bed. Because of this fear, the scientist is now inclined to accept the higher published ranges of the toxicity of snake venom of a given snake rather than the lower ones, a judgment that might then influence other judgments and inferences. His fear (1) has no bearing on the actual toxicity of the venom, only on his judgment about its toxicity, and (2) he has no idea that his judgment is being influenced in this way.

Scientists and philosophers have been reluctantly forced to admit that there is no logical principle or methodological dictum that does, or would, or could, immunize science against prejudice. To repeat, there is no such thing as scientific proof. That means that there is a logical gap between scientific evidence and scientific doctrine: Evidence does not entail doctrine. As a matter of methodological necessity, then, scientists must, and do, cross that gap by nonlogical means. Scientists must make a leap of faith, and that is where prejudice gets its toehold.

Ron Giere, one of the most prominent contemporary philosophers of science, put the matter bluntly in a book he published in the last year of the twentieth century: "In sum, there is little in current philosophical theories of science that supports the widespread opinion that gender bias is impossible within the legitimate practice of science" (Giere 1999, p. 212). Although Giere speaks here about gender bias, the context makes clear that different sorts of bias may affect science. Science is human through and through. It is not magical; it is not infallible; it is not perfect. There is no logical or methodological prophylaxis against prejudice.

The reaction of philosophers to this dawning realization has ranged from cynicism to idealism, and from activism to complacency.* Nonacademic, nonscientific men and women are also aware that science changes its mind, and so feel freer to pick and

---

* Some, like Sandra Harding (1991), have called for the explicit embrace of "liberatory" goals for science, opining that if science is subject to nonlogical influences anyway, those influences should at least promote social justice, particularly for women. Environmentalists might just as well have proclaimed that science should embrace environmental goals—as, indeed, they have in fact done within environmental science. Others, like James Brown (1994), have called for broadening the scientific community itself to include members of underprivileged groups, arguing that if science is open to prejudice anyway, a full palette of representative prejudices is better than those of the group that actually happens to gain entry into the scientific community [for my reply, see Foss 1996b]. Many environmentalists do encourage or demand consultation with indigenous experts as part of understanding a given ecosystem, and many practicing environmental scientists do consult with them in their fieldwork. Some, like Paul Feyerabend (2001), were pleased by the prospect that science will become pluralistic, and that what he perceives as its monopoly on the human imagination will be relaxed. Others, like the European postmodernists Foucault (1926–1984) and Derrida (1930–2004), have taken the dimmer view that science is just another instrument of institutionalized power within the grim scenario of "modernity."

choose scientific theories on the basis of their own personal views and convictions. For instance, many environmentalists think it is perfectly legitimate to accept the theory that our use of fossil fuels is causing global warming not because of its scientific merit, but because they had long before come to see the automobile as ugly, polluting, a blight on cities, and a corrupter of nature. Of course, people do have every right to make up their minds as they please, but the idea that truth is a matter of free choice is not only false but dangerous.

It is crucial at this historical juncture to rediscover the ideals that originally motivated the rise of science and inspired the Enlightenment, for they still apply. The absence of scientific proof or any logic or method to make science immune to prejudice is not a reason to abandon the battle against prejudice and the struggle for truth. To the contrary, it is reason to rejoin the battle with renewed energy and fight for *pure* science: seeking the truth and nothing but the truth, respecting the evidence, and guarding value neutrality. Science is arguably the most important intellectual achievement of the human species. Value neutrality is essential to science, and at this point in history where humankind prepares to shoulder its responsibilities for nature, science is needed. So value neutrality must be safeguarded and promoted. There is no special test for value neutrality that works with 100% reliability, but that does not mean that value neutrality should be abandoned as a hopeless cause. We must not throw out the baby with the bathwater. We, along with the scientific community, must demand value neutrality of its members, set a high value on it, and take steps to safeguard it.

What steps? We can begin by recognizing that science is in the business of creating models, and that these models are to be judged in terms of the criteria outlined above. Given the relativity of scientific judgment, we must encourage competition among scientific models. Just as evolutionary competition improves the fitness of organisms, so scientific competition improves the fitness of models. And just as evolutionary competition requires biodiversity, so scientific competition requires diversity among its models. So we must encourage scientific *pluralism*: the creation and development of a variety of competing models in every domain.

One unfortunate effect of science having assumed the social role of defining knowledge is that it has taken on the same authoritarianism that it battled against in its early days. This is quite understandable. Since science says what is to be taught in the schools and universities, what medicines we are to use, and what evidence is permissible in a court of law, it has been pressed to speak with a single voice. It has been pressured to create and maintain *scientific orthodoxy*. Understandable as this is, it has had some unhealthy results

Because there is no formal process within science to decide which models to use and which to reject, it has instead come to rely upon informal "consensus," as it is often called, to determine what is to count as scientific truth. This consensus is the product of historical accident, and wide open to prejudice. Whenever the concept of scientific consensus arises, we must remind ourselves that science has *no formal procedure* for establishing this consensus. No votes are ever taken, and scientific opinion is never measured. We must remind ourselves that science is not democratic. Its decision procedure—insofar as it has one—is not one person, one vote, but authoritarian: those in charge call the shots. Who is in charge? Scientific authorities emerge from the struggle to publish results, to get research funding, to get on editorial boards that control what gets published and on the committees that control research funding, to write the textbooks, to control the awarding of degrees, and so on—completely without procedural or substantive protocols. It is *scientific charisma*, as much as anything else, that determines scientific doctrine.

This process has worked surprisingly well. It is far from ideal, but the very idea of changing it would take us into territories and mire us in battles that this brief book cannot afford. We can, however, take some steps to make the best of the current state of play within science.

1. Adopt a more mature view of science, one that accepts its fallibility as well as recognizes its achievements. Stop expecting science to determine the truth once and for all by proclamation. Expect scientific debate and more nuanced results.
2. Safeguard and protect scientific value neutrality. Truth must be the first and last goal. Be wary of scientific programs and models that have commitments to values other than truth.
3. Encourage competition among scientific models. Strong competition is needed if strong models are to evolve.

6.5 THE VALUE NEUTRALITY OF ENVIRONMENTAL SCIENCE

As we have seen, actual environmental science is applied science, not pure science. Its practitioners have the goal of bringing the planet to environmental health, which clearly cannot be done without some image of this target. Not being pure scientists, hence not being bound by value neutrality, they have indeed developed various images of environmental flourishing which converge on the concept of the pristine environment, the environment unaffected by human presence, the wilderness. Pure environmental science, by contrast, is value neutral and does not propound values or let its judgment be influenced by them.

We know as a simple matter of logic that every goal presupposes a value, and that no value is a matter of fact. Science deals with facts, makes no value claims (exercises topical value neutrality), and guards the value neutrality of its judgments (exercises methodological value neutrality). It follows, therefore, that pure environmental science cannot define environmental health, since health is evaluative. Health is an ideal, a goal that we set up for ourselves or others. This is not to say that environmental science has nothing at all to say about environmental health, because it will be the source and repository of the knowledge that is relevant both to our concept of environmental health and to the methods we choose to obtain it. Science is tasked with determining the facts, and we look to it to tell us the facts. However, its authority does not extend past the facts. When it comes to deciding what we should aim at, we have gone past the facts and into the realm of values.

CASE STUDY 6: CAN ENVIRONMENTAL SCIENCE TELL US WHETHER WE SHOULD PREVENT THE COMING ICE AGE?

William Ruddiman (2003) proposed that human beings began to cause global warming some 8000 years ago, and have thereby forestalled the beginning of the coming ice age. Evidence indicates what he calls an "anomalous" rise in atmospheric carbon dioxide levels, which he argues was caused by human beings clearing forests to grow crops and raise livestock. Then methane levels began to rise 3000 years ago, which he attributes to humans flooding fields to grow rice. Carbon dioxide and methane are both greenhouse gases, and so would cause warming, which Ruddiman suggests was sufficient to forestall the next ice age. According to the generally accepted Milankovitch orbital forcing theory, the cooling that began about 5000 years ago should have continued, and we should now be on our way into the next ice. But, argues Ruddiman, our production of greenhouse gases has delayed this natural cooling.

A Thought Experiment. It should go without saying that the fate of a scientific hypothesis often does not run smooth,* but let us, for the sake of argument, just *suppose* that Ruddiman's theory is right. Let us imagine that we did cause global warming and thereby have delayed the onset of the next ice age, just as he proposes. Suppose further that environmental science (ES) reveals that we can prevent the coming ice age by keeping our release of greenhouse gases at about the same level it is today, with slight increases to counterbalance the increases in Milankovitch cooling over the coming millennia. If the ice age is prevented, Earth keeps the same sort of climate and the same sort of ecosystem that it has enjoyed over the last 6000 years. If the ice age is permitted, we will have the same sort of climate as during the last ice age and the same ecosystem shrinkage and drying as in the last one (see Figure 4.1). Under these circumstances, can ES tell us whether we should prevent the coming ice age?

Yes: Only ES can provide us with the relevant data. What happens to the environment in an ice age? We need to know or we cannot decide what to do, and once ES tells us, we can decide. So obviously, environmental science calls the shots here.

No: But according to our supposition, environmental science has already told us the relevant facts. We know what happens to the environment in an ice age: the same thing that happened last time. Temperate species are pushed south, compressed into

---

* Ruddiman also proposes that the worldwide plagues around the middle of the last millennium resulted in the *little ice age*, a period of cooling over most of the last half of the millennium; human populations were reduced and so, consequently, was the amount of agricultural production of greenhouse gases. It has also been noted that sunspots were absent during the onset of the little ice age, a phenomenon called the *Maunder minimum*. It is also known that solar output reduces with the number of sunspots. Furthermore, it is known that fewer sunspots indicates a weakened solar magnetic field, which results in an increase in the cosmic radiation striking the Earth, that may in turn cause increased cumulus cloud cover and hence global cooling. Since human agriculture did not cause a reduction in solar activity, it would seem that reduction of solar activity is a strong rival cause of the little ice age. This is just one possible failure of Ruddiman's theory. We take up the issue of greenhouse warming in Case Study 7.

a narrower band around the globe, the tropics cool, things become much drier, and so on.

Yes: It is not a simply matter of the facts. We also need to know whether the environment is healthier in an ice age or healthier without one.

No: Suppose that environmental science says it is healthier for the environment if we have the ice age. Even if we grant this, it does not follow that we should let the glaciers march in. That would only follow if we also grant that we ought to optimize environmental health. Without the assumption that environmental health is paramount, we are free to act on other values, such as keeping the current temperate ecosystems thriving or keeping the human species healthy and happy. More to the point, we do not have to grant that the environment is healthier with the ice age. Health is nothing other than the state that living systems *should* be in. It is therefore a matter of value, not fact.

Yes: We need environmental science to tell us whether meddling with ice ages will cause a worse disaster down the road.

No: We have already supposed that science has provided the relevant facts and that no disaster happens later. The point of this thought experiment is to fix the facts so that we can address questions of value without confusing them with matters of fact. We agree that science is responsible for informing us about the facts, both what will happen and what would happen under different circumstances, and that we rely on science for that information. However, the choice about what *should* happen is not a function of fact alone, but also of value. When it comes to value, a scientist has no more authority than anyone else.

The thesis that pure environmental science cannot define environmental health is one that practicing environmental scientists will find difficult to swallow, although its simple logic is inescapable. They will feel that they know the environment better than anyone, so they are best placed to advise us about its health. Advise, yes, but defining what is best, no. Certainly, we want to hear the advice of environmental scientists on this issue. We want to know about the unforeseen consequences of our actions, which we might later regret—as well as those we might later enjoy. These are matters of fact, after all, of the form "If you do X, then expect Y," and about them we accept pure science's authority. However, the question of whether we *ought* to avoid Y or aim for Y extends beyond that authority. We are willing to hear scientists' *opinion* about what we should or should not do, in fact we welcome it, but in this they do not speak with the voice of scientific authority but as fellow human beings.

If environmental scientists find it hard to accept that their science cannot define environmental health, one reason might be that biologists are in the habit of studying the diseases of animals, which implies that disease and health are defined scientifically. This is all part of biologists' practice of understanding life in terms of function. This is ironic, since the very concept of function involves final causation, and *officially*, biologists have banished final causality from their discipline. Indeed, all of modern science has done so. A final cause is one toward which things aim. For example, we are citing a final cause when we say that a cat hunts mice because it needs nourishment. Biology does not accept such explanations. Getting nourishment cannot be the cause of the hunting, since it occurs after it. The future cannot affect the past. Every biologist agrees to these well-rehearsed points.

Nevertheless, at the level of every day biological investigation and discovery, the first thing that a biologist does when he or she comes across something new and intriguing is to ask: What is its function? And the warm glow of understanding is achieved only when that function is discovered. Still, true scientific understanding is achieved only when the causal process underlying that function is revealed. The idea that the function, the intended purpose as it were, of DNA is to carry genetic information is not science. The science of DNA consists in the discovery of the causal processes whereby it replicates the proteins of parents in their offspring. Science is all about *mechanism*. It is not about purposes, goals, or intentions.

Function is purpose, purpose is goal, and goal is value, none of which belong to science. To identify a function of an organ is to identify a *goal* that the organ enables the organism to achieve: The eye enables the animal to see; the wing enables it to fly; the heart enables it to transport blood to its tissues.* Goals are evaluative by their very nature: seeing, flying, or the circulation of the blood is *good* for the animal in question. To think of these functions as bits of the natural world alongside organisms, DNA, and so on, is unscientific. Nevertheless, one temptation for the environmental scientist will be to assume a function (for a specific ecosystem, say), then to fallaciously infer from this a *proper* function (for the ecosystem), and finally, to define health (of the ecosystem) in terms of the supposed proper function. For example, the function of the eye is to provide sight, and relative to that function the health or disease of the eye can be established. So an infection of the eye will make the eye unhealthy. It is tempting, then, for the biologist who has come this far to take the next step, which is to conclude that the eye infection is a bad thing, and so conclude a value claim solely on the basis of scientific facts.

That this is fallacious can be seen in the fact that the eye infection is a bad thing only *relative* to a presupposed function of the eye and the interests of the organism whose eye it is. Suppose, for instance, that the infected eye we are considering belonged to one of the 24 rabbits released by Thomas Austin in Australia in 1859, the very rabbits that went on to thrive and cause various effects on the local plants and animals that environmentalists universally reckon to have been devastating. Suppose that the microorganism that the rabbits had been infected with would have made them

---

* It is worth noting that the goal need not be achieved for it to exist. The function of the sperm is to fertilize the egg, although only a tiny fraction of actual sperm (there are hundreds of millions in a single human ejaculation) ever achieve that goal, thank goodness. Similarly, a defective kidney still has the function of cleansing the blood even if it is unable to perform that function. The fact that the goal need not be realized to exist marks it as "intentional" (a technical, theory-laden piece of philosophical terminology), which we may gloss as a state "intended" by the system, whether or not it is actualized. This is characteristic of values in general: They are defined by intentional goals.

blind and unable to survive in the Australian countryside. Because the function of the rabbits' eyes was impaired by the infection, it was bad for their eyes and bad for them. But from the point of view of the local plants and animals that would be saved from competition with the rabbits, the infection would have functioned as their salvation, and so would have been good—an opinion they would have shared with the microorganisms infecting the rabbits' eyes themselves (my enemy's enemy is my friend). Since values depend on point of view, the proper functioning of the rabbits' eyes may be either good or bad. Functions do imply values, but only *relative to a point of view*.

Things only get worse when we try to infer the health of an ecosystem from its function or functions. From the point of view of the pronghorn antelope, the function of its kidneys is to cleanse its blood. But from the point of view of the plants in its terrain, the function of its kidneys is to fix nitrogen and disperse it in the soil in a form the plants can absorb. And from the point of view of the mountain lions that prey on the pronghorn, the function of their kidneys is as a nutritious snack. If we try to figure out functions eco-systemically, we are forced to make an arbitrary choice of a particular point of view within it. The ecosystem is an abstraction, and it has no point of view, even though everything that lives in it does. The ecosystem is just the present time slice of biological activity conceived (vaguely) as a system, a set of interlocking functions. Without function, there is no system. As for the land itself, it is merely the stage upon which life has acted out innumerable dramatic episodes, although it has spent most of the last few million years under hundreds of meters of ice.

THESIS 7:   Environmental science as such is value neutral, so it makes no value claims and safeguards the value neutrality of its judgments of fact. Therefore, it cannot define environmental health, which is a matter of values.

6.6   THE SPECIAL CHALLENGE FACED BY PURE ENVIRONMENTAL SCIENCE

The epistemic mission of pure environmental science (PES) is to understand the biological world as a unified system. Traditionally, science has studied the world by *analysis*: breaking it into pieces, with designated specialists studying each piece. So far, biology has followed this plan. Biology is the study of life, but you will only find out what life is by studying all of its various aspects: cellular metabolism, reproduction, genetics, speciation, and so on. There is no generally recognized biological specialty devoted to the study of life as such. The botanist studies plants, the zoologist studies animals, the molecular biologist studies molecular processes, and so on. The goal of PES is to put all of the pieces back together again. Its identifying task is not the typical scientific task of analysis. PES is the biological specialty that aims at *synthesis*. If it succeeds, we will then have a specialty that does study life itself rather than simply its components. Because its job deviates from that of normal science,

PES faces abnormal and extremely difficult problems. Even if environmental science succeeds in protecting its value neutrality, it still faces this special challenge not faced by the other sciences.

Science has traditionally relied on simplification and analysis in order to gain certainty, but PES aims at complexity and synthesis instead. The tradition of analysis and simplification can be illustrated by considering the case of freely falling bodies that Galileo first solved. Galileo's genius was to recognize that the motion of actual falling bodies can be analyzed as the combination of two different mechanisms: the free fall of a body plus the resistance of the medium through which it falls. Actual falling bodies are a very mixed bag: The falling of a person to his knees is very different from the falling of a pendulum on its downward swing; the falling of a stone is very different from the falling of a feather. Galileo spotted the possibility of an underlying simplicity: The rate of fall of the person, the pendulum, the stone, and the feather *would* be the same if only they were not subject to different degrees and types of resistance. In a vacuum, Galileo opined, the feather would fall just as fast as the stone. Within a few decades, in 1659, the vacuum pump was built by Robert Hooke (1635–1703), and this soon led to one of the most important and persuasive experiments of the Age of Reason: direct comparison of the rate of fall of feathers and stones in a vacuum. In this classic experiment, air is pumped out of a tall glass jar, then a device simultaneously drops a feather and a lead weight, and both can be seen to fall at the same speed to the bottom of the jar.

What is often forgotten is that we *still* cannot precisely predict the rate of fall of the feather when it is back in the air—and as far as we can tell, never will. If a feather is dropped repeatedly from the same spot several feet above the ground, it will follow different paths to the ground, require different amounts of time to get there, and stop in distinctly different positions each time it is dropped—over and over, virtually forever. The unpredictability of the feather's trajectory is a feature of nature itself. Even if extreme care is taken to make sure that the feather is in exactly the same position each time it is released, and to make sure that the air is calm, at the same temperature, humidity, and so on, the feather will follow a different line to the ground each time. In more scientific terms, even if we begin with the same initial conditions, the state of the system evolves differently each time. But put it as you like, how can different effects issue from the same cause?

The only answer is that there are (or must be) undetectably tiny differences in the initial conditions that quickly balloon into enormous differences in the path followed by the falling feather. Some of these differences do occur at the molecular level. We know that just controlling the temperature, pressure, and stillness of the air will not eliminate the differences at the molecular level. What we call temperature is just the *average* energy of the molecules of air surrounding the feather. Temperature is a mathematical construct, a pure abstraction, a gross measure that we use because we know how to measure it. If we are to have any hope of precisely calculating the trajectory of the feather when it is released, we need to know, for starters, the actual energy of each of the air molecules that will collide with the feather before and during its fall. But getting precise information about these trillions of molecular collisions is a technical impossibility and will remain so for the foreseeable future. Calculating their effects poses an even more difficult problem that we are unable to solve even in principle, given that computational power is limited in the end by the sheer size of any possible computer.

The only thing to do is swallow hard and admit that as far as we can tell, we will just have to be happy with predicting the average trajectory of the feather. We have good *scientific* reasons for believing that this is about as good as it will ever get. The scientific phenomenon that limits the precision and accuracy of science is well studied and well understood. It is often called "infinite sensitivity to initial conditions," or, more popularly, *chaos*.* Chaos is a misnomer, since chaos implies freedom from the rule of natural law, whereas the systems in question are thought to be fully determined by law, but nevertheless, unpredictable. Unfortunately, the biological systems that PES studies are like the feather released in the air, not like the feather falling in a vacuum. They are complex systems, whereas the highly accurate models on which science has built its reputation always deal with simple systems. The essence of modeling is simplification. A model is necessarily less complex than the reality it represents. A map cannot contain the complexity of the real terrain, and a model of the atmosphere running in a computer the size of a box cannot contain the endless details of the massively larger and more complex atmosphere itself. Since the behavior of the atmosphere depends very sensitively on just those details, no model can predict its behavior.

LORENZ, BUTTERFLIES, AND CHAOS

It is poignant that chaos was discovered by a meteorologist, Lorenz, while modeling weather in 1961. Like so many meteorologists of his generation and since, Lorenz was deeply intrigued by the emergence of the computer and the promise it seemed to hold for modeling weather patterns. Like so many other meteorologists, he could imagine all the global weather data being plugged into a computer on noon Tuesday that would then run ahead at high speed to tell us the weather on Wednesday. The physics and chemistry involved isn't particularly mysterious, so it seemed to be just a matter of getting a computer to do the calculations. Meteorologists imagined the television weatherman showing not only today's weather systems, but tomorrow's as well, and even next week's, while climatologists dreamed of predicting climate years into the future. But instead of discovering the models, measurements, and computer programming techniques required for this sort of prediction, what Lorenz discovered instead was chaos, or the "butterfly effect," as it is often called.

His discovery, as is so often the case with scientific discoveries in general, depended on serendipity—a happy accident. Lorenz was using a computer to calculate the development of a simplified weather model through time when he decided he needed to redo the calculations. Since the computer took hours to

---

* For a much more extensive, but accessible, introduction to deterministic chaos, see Gleick (1987). For more accessible details from the point of view of mechanical systems, see Foss (1992).

do the calculations, he decided to save time by starting them partway through, using the values from a printout at that point. When he did this, he found that the second set of calculations gave wildly different results from the first. He tracked down the cause to the fact that the printout rounded off the last digits in the computer output (in fact, it apparently rounded the number 0.506127 down to 0.506). According to the usual understanding of weather models, this tiny change in the input should have caused only a tiny change in the output. Instead, he found that it caused massive changes.

After studying the matter, Lorenz concluded that tiny changes in a weather system will have huge effects later. A butterfly flapping its wings in Brazil on Tuesday can cause a tornado in Texas on Sunday. With the butterfly, you get the storm; no butterfly, no storm. But there is no way to know this in advance, no way to make the prediction. This unpredictability was what Lorenz discovered. Some things in the world, like the weather, are unpredictable, because they are infinitely sensitive to initial conditions. Lorenz himself did not accept global warming theory, because he believed that the estimated warming cited in favor of the theory was more readily explained as normal, natural variation in the chaotic climate system (see, e.g., Lorenz 1991).

As it turned out, chaos was found not only in the weather, but virtually everywhere and in everything that we do not simplify artificially. The sorts of future things that we usually wonder and worry about are the very ones that turn out to be unpredictable.

Science has gained its reputation for precise prediction through its success with purposely simplified systems. The natural systems that science has been able to predict with precision and accuracy have been in the heavens, not on Earth. The solar system is a naturally occurring simplified system. Sunrise, sunset, and eclipses of the Sun or Moon can be predicted to the second months or years in advance. But when it comes to earthly phenomena, precision and accuracy have been restricted to cases where we have created simplicity in the lab or in the computer. The behavior of ordinary, complex systems that are found in the real world outside the laboratory has largely remained outside the scientific domain. The discovery of chaos and the development of chaos theory shows that this struggle will, as far as we now can tell, never be won. Approximation and compromise are the best we can hope for.

As Nancy Cartwright (1999) puts it, we live in a dappled world: a world that, despite science, is still full of possibilities, surprising events, and wonders that we have not yet appreciated. Science rules in worlds of its own making, the world of the fundamental simple forces and processes where physics and chemistry prevail. On the laboratory bench and in the technological innovations it has made possible, science has achieved levels of precision that are absolutely stunning. This precision itself is an excellent reason to accept that nature is lawlike throughout, not just on the lab bench. Presumably the laws of physics and chemistry and all the physical sciences rule everywhere, but even so, that does not permit the rest of the world to be predicted by science. In the perfect vacuum of the bell jar, the perfect mechanics of Newton prevails, but outside the bell jar, feathers and leaves and raindrops fall to Earth, tracing paths no human being can predict. In the perfect vacuum of the particle accelerator, quantum mechanics and relativity prevail, but outside it the weather takes its own unpredictable course here on Earth as on the Sun and a thousand other planets and suns. Thus, the world will continue to surprise us. To bring this discussion of chaos back down to Earth, life itself is not predictable. We will run into friends that we never thought we would meet again* and make new friends whose identities must remain a mystery to us until then.

Pure environmental science is a paradigm case of complex science. It deals with the dappled world. Our expectations of environmental science must, therefore, be modest.

THESIS 8:   Environmental science as such aims to understand the living world as a systemic whole. Since this system is chaotic, it cannot be modeled precisely. Our expectations of environmental science must be modest.

CASE STUDY 7: SHOULD WE WAGER THE GLOBAL ECONOMY ON THE GLOBAL WARMING THEORY?

On September 11, 2007, the sixth anniversary of the attacks on the World Trade Center and the Pentagon, Osama bin Laden spoke out against global warming. This marks a triumph for environmentalism, for it shows that it has become truly global. Although Osama has very few beliefs in common with my neighbors and friends, he does share with them a concern for the health of the planet. Osama bin Laden, Al Gore, Pope Benedict, Bill Gates, Noam Chomsky, Madonna, the Dalai Lama, George Bush, Vladimir Putin, and the National Chief of the Canadian Assembly of First Nations, Phil Fontaine, are a diverse group of human leaders. Yet they all have one thing in common: They believe (or believe in) global warming. The global warming theory (GWT) is arguably the first truly global news story, one that reaches not only everyone's ears but also their hearts and pocketbooks. The entire human race has been told that the centuries-long party of economic growth is over, and now the CO_2 bill has to be paid. Sure, there were stories of global interest before, especially those predicting nuclear apocalypse a generation ago, but those were only warnings of things to come. Global warming, we are told, is happening right now, and must be stopped by reducing CO_2 emissions right now.

Whatever we human beings do in response to the threat of global warming, it will be momentous for the environment, and momentous for you and me. Even if we do not live to see the days when the threat is supposed to be realized, we are charged

---

* Aristotle used meteorology as an example of an imprecise science and identified the cause of its impreci-
sion as chance. His favorite example of chance is meeting a friend unexpectedly. Although nothing that
either oneself or one's friend does escapes natural law, the chance meeting cannot be predicted.

with making up the collective mind of the human race today. We cannot address environmentalism today without addressing global warming.*

The Threat Forecast by Global Warming Theory. The International Panel on Climate Change (IPCC) officially states in its latest, and fourth, assessment report that nearly every one of the last several years is among the several hottest in 1000 years or more. It also states that the warming is caused by us, and that it cannot be stopped, only mitigated. Even if CO_2 emissions are totally eliminated by 2100, the warming will last for thousands of years (AR4, pp. 77–80).[†] According to the press releases of the IPCC, the result will be an ecological catastrophe—the realization of the environmental apocalypse that has been forewarned, and feared, since the 1960s.[‡] Therefore, we must begin scaling back the use of fossil fuels as quickly as possible in order to *limit* the damage: reduce CO_2 emissions to 5% below 1990 levels by 2010, to 50% below 1990 levels by 2050, with 100% reduction (total elimination) of CO_2 emissions by 2100.[§]

The Threat of Global Depression. Why have we not met these reduction targets? Surely if it were easy to do, we would have done it. The problem is that the human economy runs on energy, and the majority of this energy comes from fire and hence produces CO_2. Meeting the first of the Kyoto targets would have caused a massive

---

* By 2030 the case will, hopefully, be settled, although we must not discount people's tendency to cling to what they want to believe even in the teeth of the evidence. From a logical point of view, however, whether we renounce fossil fuels or not we will see whether global warming theory (GWT) is right by 2030, because according to GWT there will be obvious warming either way. GWT's proponent, the IPCC, lays it all on the line (IPCC 2007a, hereafter AR4, Assessment Report 4, p. 89): "Near term warming scenarios are little affected by different scenario assumptions or different model sensitivities. . . . The multi-model mean warming, averaged over 2011 to 2030 . . . lies in a narrow range of 0.64°C to 0.69°C. . . ." Although it is not mentioned by the IPCC, by its own figures, global temperatures stopped rising in 1997 (AR4; see, e.g., Fig. TS.7, graph D, p. 38). If this sideward trend continues, or if temperatures go down, GWT will be effectively falsified. If, on the other hand, the temperature rises by approximately the amount predicted during the next two decades, global warming will be confirmed. No doubt some GWT proponents will not accept falsification if global temperatures do not rise as predicted. Already some have suggested that widespread, if not global, cooling may be a temporary effect of global warming, due to disruption of heat flows via ocean currents. We should be very wary of any theory that is supposed to be verified by any possible turn of events. This sort of unfalsifiability is precisely the mark of an *nonscientific* hypothesis, as Karl Popper has so plainly shown us.

† The models used in this section of AR4 assume a doubling of CO_2 by 2100 and then an abrupt halt of CO_2 emissions. They predict temperature rises of 1 to 4°C, which fall only a fraction of a degree by the year 3000.

‡ Scientific forecasts of the impact of global warming, should it be real, include a rich mixture of both positive and negative effects for both humankind and the environment, resulting in something like a rough balance overall—or even a net gain. The IPCC's own study (IPCC 2007b) notes that there will be a global increase in precipitation, which will have a beneficial effect on many ecosystems. The reason is simple: Vegetation is the nutritional basis of all life on Earth, and vegetation prospers in warm, hence moist, eras, and suffers in cool, hence dry, eras, as shown in Figure 4.1 (Section 4.1). There is no doubt that the warming trend from the early 1980s to the late 1990s had this positive effect (see, e.g., Nemani 2003).

§ The Kyoto Accords agree only on the 2010 and 2050 targets, although the IPCC premises its afore-mentioned catastrophic prediction of thousands of years of warming on a complete elimination of CO_2 emissions by 2100. Presumably, then, the IPCC would advise that CO_2 emissions be eliminated by 2100 in order that the catastrophe not become even bigger than it has forecast.

global depression. Since emissions have grown steadily since 1990, meeting the first target would now require a global cut of over 20% of current emissions, which cannot be done without massive disruption of every sector of the economy, including food production. In short, human beings do not know how to obtain the necessities of life without the burning of fossil fuels. So the first target cannot and will not be met. The IPCC nevertheless insists loudly and resolutely that we must start making cuts to CO_2 emissions.* This cannot be done without putting the brakes on the global economy. We all know that it hurts when the economy slows down: We work less, we spend less, we travel less, we buy fewer clothes, fewer books, cheaper foods, give less to charity—we live less. The IPCC itself says "it is clear that the future impacts of climate change are dependent not only on the rate of climate change, but also on the future social, economic and technological state of the world" (op. cit., p. 824). This is, of course, perfectly true. It is something that we all have learned from our own personal experience.

A simple, inconvenient truth is forgotten by those who call for cuts to CO_2 emissions: The *economy* is nothing other than the sum total of the ways in which we human beings make a living. It includes everyone: laborers, hockey players, professors, farmers, industrialists, native herders, astronauts, environmental pundits, and priests. The word, *economy*, conjures up the image of money, but money is merely the counter in the social system that we have developed to help us make our living. Food, clothing, and shelter are the main business of the economy, and when business is bad, they are just that much harder to get. Our economy is integrated with our bodily metabolism in the same way that the business of bees, gathering honey, is integrated with their bodily metabolism. Bees have an economy, too: the economy of the hive. In the bee economy, worker bees build hives and honeycombs, gather honey and store it in the honeycombs, and then distribute it among themselves and their offspring. In the human economy we humans build houses and storage elevators, plant fields and gather food, and then distribute it among ourselves and our offspring. But unlike the bees, our methods of doing all of this have evolved so that they now depend on the use of fire. At the moment, we cannot survive without fire. It is conceivable that we can gradually transform our economy, but we do not know how to do that right now, today. If we cut CO_2 emissions, we will reduce overall economic activity—in other words, cause an economic depression.

It may seem odd that the ones who are first to suffer from a downturn in the economy are those who seem farthest from it, those who have the least to do with money, the rural poor who eke out a living tending a garden and a few chickens. It is true, nevertheless. If we remember that "the economy" is just another term for humankind making its living, it will not seem so odd: When it gets harder in general to make a living, those who just barely make a living will be hurt most. The poor need to buy the things they cannot or do not make for themselves, such things as matches, soap, toilet paper, needles, thread, fabric, eyeglasses, toothbrushes, pots, pans, shovels, plows, books, shoes, medicines—the countless "little" things that make life possible and tolerable. When the economy goes bad, these things become more

* Some have argued very persuasively that the Kyoto Accords would have little effect (e.g., Wagley, 1998, 2005), a conclusion not denied by IPCC.

expensive, the money to pay for them becomes scarcer, and the poor must suffer doing without, the worst form of economic hardship.

Because global warming pits the good of the environment against the good of the economy, it threatens to hit us all where it hurts, and where it hurts the poor most. Of course, the impression created by those who profess and promote the Kyoto Accords is that the only ones to be hurt will be the rich, greedy people at the heart of the problem in the first place. This is, as anyone with any experience of things on this Earth knows, false. The rich, as always, will do best no matter what circumstances prevail. The poor themselves will testify that they suffer most during hard times.

The Question. The IPCC presents humankind with the following argument:

1. *Premise*: GWT is true.
2. *Conclusion*: Therefore, we must reduce CO_2 emissions.

Thus, humankind faces two questions. First, the factual question: Is GWT true? Second the value question: Should we reduce CO_2 emissions if GWT is true? Here we address only the first question, the factual question, with the understanding that what is at stake is the rest of the IPCC argument. Thus, our question may be restated as follows: *Are we sufficiently confident in the truth of GWT to slow the human economy because of it?*

Yes 1: The Greenhouse Argument Proves Global Warming Theory. The greenhouse gases (GHGs) we produce are like the glass walls of a greenhouse: They let in the heat from the Sun while preventing its escape, as shown in Figure 6.1. There

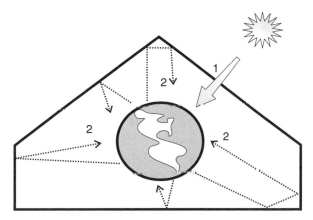

Figure 6.1 The greenhouse argument. Greenhouse gases (GHGs), such as CO_2, are like the walls of a greenhouse surrounding the Earth. 1, White light from the Sun penetrates the greenhouse and warms the Earth; 2, warmth radiates from the Earth in the form of infrared light (dotted lines), but is reflected back to Earth by the greenhouse, preventing Earth from cooling. There is a natural level of GHGs that keeps the Earth at the proper temperature. By adding CO_2 and other GHGs to the atmosphere, we have caused the Earth to overheat, a process known as *anthropogenic global warming*.

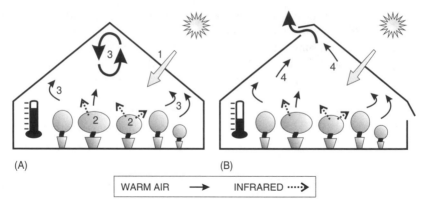

WARM AIR   →   INFRARED ····▷

Figure 6.2   Real greenhouses. (A) Warm greenhouse. Warming is caused by blocked convection: 1, White light from the Sun penetrates the greenhouse and warms its contents; 2, The warm contents then warm the surrounding air, causing it to rise (convection); 3, the rising warm air cannot escape, so warmth builds up. (B) Cool greenhouse. 4, Warmth escapes via convection, leaving the greenhouse cool.

is a natural level of CO_2 that keeps the temperature where it *should* be. We have disturbed this natural level by using fossil fuels and thus have disturbed the temperature by making the glass in the greenhouse thicker. This "runaway greenhouse heating" means that Earth is headed for ever higher temperatures and environmental disaster.

No 1: The Greenhouse "Argument" Is Merely a Misleading Metaphor.   No doubt the political persuasiveness of popular GWT turns on the fact that everyone knows that greenhouses, like automobiles, get very warm in the Sun—even though in the greenhouse metaphor the actual mechanism of this warming is misrepresented. Although it is true that glass is more transparent to visible light than to infrared, this radiation effect is inconsequential in an actual greenhouse. Actual greenhouses work by interrupting air circulation, as shown in Figure 6.2A. If the greenhouse does not trap warm air inside, it is no warmer inside than out, as shown in Figure 6.2B. Greenhouse operators know this, since they cool their greenhouses by opening vents that permit warmed air to escape, although this has very little effect on the radiation balance within the greenhouse.* So, despite the fact that radiation is still trapped as much as it ever was, the greenhouse cools. In an actual greenhouse, the warming of air by infrared radiation causes air motion, just as it does in the atmosphere outside. Convection is a natural engine that is powered by infrared light, converting heat into motion. This engine normally transports heat upward and away. A greenhouse gets warmer because it prevents this natural engine from working. Since it is false that GHGs prevent convection, the greenhouse argument simply fails to apply to the actual atmosphere.

---

* The vents are very small relative to the glass surface of the greenhouse, usually 10% or less, and given the usual design of a greenhouse, provide little chance for escape for infrared light which, unlike air, must travel in straight lines to get outside.

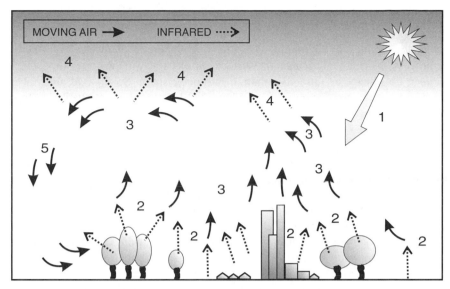

Figure 6.3   The real atmospere. 1, Incoming white light from the Sun warms Earth's surface; 2, the Sun-warmed surface radiates infrared heat, which warms the air; 3, warm air rises by convection, carrying heat into the upper atmosphere; 4, the atmosphere radiates its heat into space; 5, Cool air returns to the surface to begin the cycle again.

In the actual atmosphere, incoming white light from the Sun does not directly heat our atmosphere, which is nearly transparent to white light.* Instead, our planetary warmth begins at the surface when the visible light of the Sun is absorbed by oceans, ground, buildings, trees, and so on, and warms them, as shown in Figure 6.3. Warm things radiate heat: They glow. We can see a red-hot nail glow, but warm rocks, trees, buildings, even our own bodies, glow too, only in a milder way, that we cannot see with the naked eye, in the infrared. In this way, Earth's surface warmth is radiated as infrared light into the overlying atmosphere. Most of this infrared radiation is absorbed, since the lower atmosphere is nearly opaque to infrared light.[†]

The surface layer of atmosphere would become too hot were it not for *convection*, the main engine of global cooling and precisely what is left out of the greenhouse argument. Warm air expands, gets lighter (weighs less per unit of volume), and so rises (or is buoyed upward by the heavier air around it). When warm air rises because of convection, the air around it moves in to replace it, causing wind. The moving atmosphere interacts with the land, the waters, the vegetation, the mountains, the

---

* According to NASA (2007a) satellite data, the transparency of a cloudless atmosphere to incoming sunlight is 78% (with 16% absorption and 6% reflection).

[†] The opacity of the atmosphere to outgoing infrared is 91%: of the 70% of incoming radiation that makes it through clouds and reflective barriers to be absorbed by Earth's surface (6% is reflected by the atmosphere, 20% is reflected by clouds, and 4% is reflected by the surface), only 6% (8.6% of the total) is radiated directly into space, while the remaining 64% (91.4% of the total) is carried by the atmosphere and its water vapor (NASA 2007a).

glaciers, and so on, both picking up and losing such things as dust, gases, heat, and water vapor. The rising air stirs this moisture and heat, creating rain, storms, waves, hurricanes, thunder, and lightning—all of the ongoing drama that we call weather. In the process, the heat that began at the surface makes its way to the upper atmosphere, where it can escape as infrared radiation into outer space. So the real atmosphere is a bit like the cool greenhouse from Figure 6.2B, but nothing at all like a closed greenhouse. Thus, the greenhouse argument simply does not apply.

Yes 2: The Anthropogenic Forcing Argument. The greenhouse argument is, admittedly, a simplification, but it does get to the heart of the matter. *Of course*, the climate is complex, just as the last objection states—in fact, its sketch of the real climate barely scratches the surface. That is why we rely on sophisticated *climate models* to tell us what the climate will do. There simply is no way that the human mind can completely comprehend a system as chaotic as the climate. It is just too complex. In addition to the factors you mention, it also involves more complex things, such as latent heat exchanges when water is vaporized or condenses, not to mention aerosols such as sulfates, smoke, and just plain old dust, to mention just their main varieties. These models include convection, the main concern of the previous objection. Nothing that makes any difference at all to climate is left out of these models. In fact, GWT employs a *hierarchy of models*, some devoted to a single phenomenon such as convection or air–ocean heat transfers, and some that put all of the pieces together again so that we can see the big picture.* In these models the currency of the realm is *radiation*. All of Earth's climate energy comes to the planet as radiation and leaves as radiation. Convection, aerosols, GHGs, and so on, are relevant to climate only insofar as they affect the radiation balance of the planet. So all of these things are reduced to a set of parameters quantifying their radiation effects; this method is called *parameterization.*[†]

One of the things that climate models tell us is that the overall effect of GHGs is to reflect, or reemit, infrared radiation back down toward Earth's surface. That is the kernel of truth that is captured in the greenhouse argument, despite its simplifications. Indeed, that is its whole point: to make the role of GHGs plain. When scientists speak about the *greenhouse effect* (or *greenhouse warming*), they are talking about this overall warming effect of GHGs, as shown in Figure 6.4. Other things being equal, adding GHGs to the atmosphere slows heat's escape. To state it in a homely way, additional GHGs cause heat in its infrared form to bounce around more inside the atmosphere before it escapes into space, increasing the heat it contains and thus

---

* The IPCC currently identifies three main tiers of models. Starting from the bottom they are *simple climate models* (SCMs); above that there are *Earth system models of intermediate complexity* (EMICs), and at the apex AOGCMs, which used to stand for *atmopsheric ocean general circulation models*, but which now also include other factors, and so are defined as follows: "They include dynamical components describing atmospheric, oceanic and land surface processes, as well as sea ice and other components" (AR4, p. 67).

[†] In the words of the IPCC, "although the large-scale dynamics of these models are comprehensive, parameterizations are still used to represent unresolved physical processes such as the formation of clouds and precipitation, ocean mixing due to wave processes and the formation of water masses, etc. Uncertainty in parameterizations is the primary reason why climate predictions differ between different AOGCMs" (AR4, p. 67).

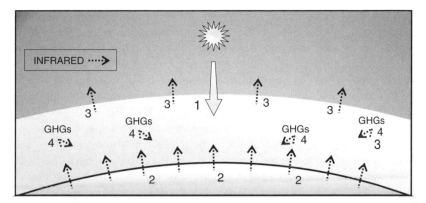

Figure 6.4   The real "greenhouse" effect. 1, Incoming white light from the Sun warms Earth's surface; 2, the Sun-warmed surface radiates infrared heat, which warms the air; 3, some of the infrared escapes, cooling the Earth; 4, the "greenhouse" effect: GHGs keep the surface warmer by reradiating infrared back toward Earth.

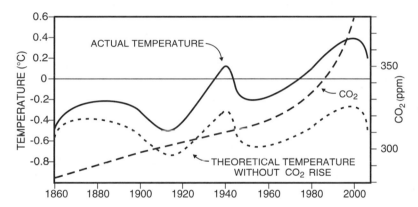

Figure 6.5   The natural variations plus human forcing model. According to this version of GWT, the variations in temperature actually observed (solid line) are the sum of natural variations (dashed line) plus warming due to increases in CO_2 (dotted line). (Based on AR4, pp. 37–38, 60–62, and 135.)

raising its temperature. This effect, known popularly as the greenhouse effect, is called *anthropogenic forcing* when human activities are the source of the added GHGs, and has been calculated precisely.*

Thus, GWT is the claim that part of the temperature rise we have observed since 1750 is due to global warming. Since global temperature is varying continually anyway due to natural variations, we can think of it as the sum of natural variation plus human forcing, which gives us the NV + HF model of GWT, shown in Figure 6.5.

---

* The IPCC calculates that CO_2 has increased from 280 ppm in 1750 to 379 ppm in 2005, causing a warming effect of 1.66 (±0.17) watts per square meter, which in turn has caused a global temperature rise of about 1°C (AR4, p. 25). The National Academy of Sciences arrives at a significantly smaller figure, 1.4 W/m^2 (NAS 2001, p. 12).

No 2: GWT Requires a Predictive Model of Climate as a Whole. Now that the complexity of the problem is out in the open, we have to realize that unless we know what the climate would have been like given natural variations, we can only *assume* that a specific portion of the current temperature is due to human forcing by GHGs. No matter what the temperature is, or whether it is going up or down, it can always be claimed that some portion of it is due to human forcing. The only way that actual temperature data can be used in support of the NV + HF model of GWT is to have independent calculations of *both* natural variations (NVs) and human forcing (HF). Without independent calculations for both, GWT fails to say anything specific or precise about what the temperature will be, so the question of whether it is true simply cannot be answered. To put it another way, until we know what the natural variations will be, we do not know what the NV + HF model is saying. The NV + HF model is *oversimplified*. It is an efficient way to explain GWT to the climatologically unschooled—and to persuade them to believe it—but it is too vague to be tested or to serve as a real scientific hypothesis. Unless we know what the climate would have done without human forcing, we cannot measure the effects of human forcing. That this is so is reflected in the fact that officially, at least, IPCC climate scientists do not use the NV + HF model as part of the scientific basis for GWT.* Instead, they try to explain past climates, including even paleoclimates, to show that their climate models are up to the task of predicting warming over the coming centuries. However, it is not at all clear that their models are capable of predicting—or retro-dicting, or even accommodating—past climates. The anthropogenic forcing model simply ignores the necessity for accurate prediction of past and current climate—and the huge, unsolved problems that this entails for GWT.

Yes 3: Multiple Feedback Models Prove GWT. Admittedly, the NV + HF model is simplified, although at its core there is a kernel of truth: that human GHG forcing will inevitably warm the climate. However, the criticism that we must be able to model what the climate would have been without human forcing is recognized by the IPCC. To do this, the IPCC uses the most sophisticated climate models in existence, *multiple feedback models* (MFMs). MFMs incorporate all the significant factors that influence climate, recognizing that these factors interact with each other in complex ways, sometimes reinforcing each other (positive feedback), sometimes weakening each other (negative feedback). A rough idea of MFMs is given in Figure 6.6. MFMs enable us to provide what you call for: independent calculations of *both* natural variations (NVs) and human forcing (HF). Indeed, the IPCC's official argument in favor of GWT turns on comparing models of natural variations in temperature with

---

* I have had the benefit of speaking with a number of scientists who support GWT (including my University of Victoria colleague Andrew Weaver, a prominent climate modeler and author of IPCC scientific documents) and have found that when push comes to shove, they fall back on the NV+HF intuition. When I query them about the complex vagaries and uncertainties of climate feedback mechanisms, I often meet the following sort of reply: "Look, if you add CO_2 to an aquarium, it's going to get warmer. Adding CO_2 to the atmosphere is going to have the same effect. Sure, there are all sorts of feedbacks, but the effect of CO_2 is to change the radiative balance in favor of warming. It's as simple as that." So they say.

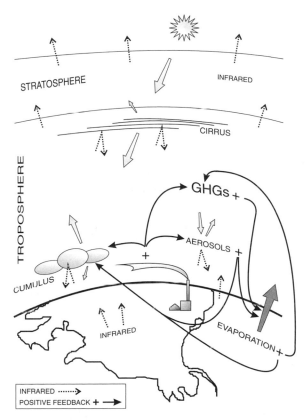

Figure 6.6 Multiple-feedback climate models. These models attempt to capture all significant factors affecting climate, including feedbacks. Only the main, most obvious positive feedbacks are shown here (curved arrows), and their main effects on radiation balance (straight arrows). For example, cumulus clouds have a net cooling effect, since the sunlight they reflect back into space outweighs the infrared they reflect back to Earth. Cirrus clouds have the opposite effect. We consider some of these processes in more detail below.

models that also include *anthropogenic forcing*.* When the two sorts of models are compared, it is obvious that the actual rise of global temperature since 1970 or so cannot be explained except by GWT.

No 3: Earth's Radiation Balance Contradicts GWT. Once it is granted that we must use MFMs, it is granted that nothing less than an adequate model of the entire climate system is required by GWT, and we simply do not have that level of understanding at this point in history. Every climate scientist, including the authors of the IPCC assessment reports, admit to significant gaps in data, theory, and modeling capacity that systematically undermine our confidence in GWT. For example, just recently, data emerged that contradicted the basic premise of GWT. All three models of GWT, the greenhouse model, the anthropogenic forcing model, and the multiple

* See, for example, AR4, pp. 61–63: in particular, Figs. TS.22 and TS.23.

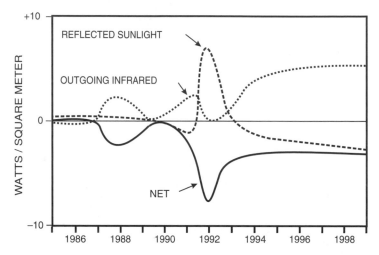

Figure 6.7   Surprising data about Earth's radiation budget. A series of satellites have been launched to study Earth's radiation budget. A stunning surprise was in store: in the midst of what was supposed to be the greatest warming seen in centuries according to the IPCC, the net amount of radiation absorbed by Earth as a whole was falling, not rising as required by GWT. (Based on NASA 2007b.)

feedback model, have the same core thesis: that Earth's outgoing infrared heat has been decreasing, and this has caused temperatures to rise. It should come as quite a shock to GWT supporters, then, that Earth's radiation balance falsifies this thesis. Recent data show an *increase* in the amount of heat radiated from Earth into outer space (e.g., Chen et al. 2002, Hartmann 2002, Wielecki et al. 2002): "Satellite observations suggest that the thermal radiation emitted by Earth to space increased by more than 5 watts per square meter, while reflected sunlight decreased by less than 2 watts per square meter, in the tropics over the period 1985–2000, with most of the increase occurring after 1990" (Chen et al. 2002, p. 838). In other words, over the all-important tropics, at least, there is an increase in the solar radiation reaching Earth which is overwhelmed by an increase in the heat escaping Earth, for a net *increase in outgoing radiation*,* as diagrammed in Figure 6.7.[†]

This result took the entire climatological community by surprise, which is a nice indicator of the current state of climatological science. With all due respect to the brilliant work being done by numerous climate scientists, climatology is still a young science that has a long way to go. A fast rise in temperature occurred precisely during the increase in outgoing radiation, precisely the opposite of what climatologists would expect and precisely the opposite of what GWT requires. When it comes to pure science, such unexpected results are actually very inspiring, since they shake up presuppositions and lead to new insights, new ideas, and new approaches. From the

---

* The decrease in radiation being reflected back into space is less than 2 W/m², and the increase in the heat escaping Earth is more than 5 W/m², for a net *increase in outgoing radiation* of more than 3 W/m².

[†] The graphs on which this figure is based clearly showed the unpredictable wigglyness (see Section 2.1) that characterizes so many natural phenomena. This wigglyness was suppressed for these figures to make the overall trends more evident. All data graphs in this chapter have been smoothed to make trends more evident unless stated otherwise.

point of view of the IPCC, however, this result can only be problematic, because it shows just how little we understand about Earth's complex, indeed chaotic, climate. Pure climatological scientists have been inspired to reassess Earth's various modes of heat storage. One explanation of the rise in surface temperatures at the very same time that outgoing radiation increases would be a release of heat from the oceans. This, in turn, might help explain why temperatures have been falling over the last decade (since 1996), in contradiction to what GWT predicts.*

In any case, GWT is based on the premise that because of human GHGs, the net radiation balance for the planet is positive. Unfortunately, this has been shown to be false during the very period when the most global warming is supposed to have occurred. GWT provides neither explanation nor prediction of global warming *when this premise is false*. This means that the period of warming in question is not evidence in favor of GWT. Whatever the causal mechanism of the warming may have been, it was not the one outlined in GWT. A period of warming that is always cited as strongly indicative of GWT, from 1985 to 2000, occurred during a climate phase in which the Earth was *losing heat*, not gaining heat, as GWT requires.

HOW MUCH IS THE CLIMATE WARMING?

For the purposes of this case study, we will take AR4 as definitive of GWT at this point in time. For our purposes, then, GWT claims that "2005 and 1998 were the warmest two years in the instrumental global surface record since 1850" (AR4, p. 36). It is interesting that in its last assessment report only six years ago the IPCC reckoned "that the 1990s has been the warmest decade and 1998 the warmest year of the *millennium*" (IPPC 2001d, p. 3; my emphasis), not just the last 150 years as they now claim. The IPCC reports of 2001, in particular the authoritative *Climate Change 2001: The Scientific Basis* (IPCC 2001a), featured a graph showing a historically unprecedented rise in temperatures during the twentieth century, which nicely matched the rise in CO_2 over the same period. That graph is the dotted line in Figure 6.8, the "hockey stick" graph, which gives the impression that temperatures had been steadily cooler until human beings began producing significant quantities of CO_2.[†] It was taken from two scientific papers by Mann et al. (1998, 1999).

---

* We might speculate that the increase in surface temperatures and the increase in outflowing radiation of 3 W/m² may have been due to a release of stored heat from the oceans. This would explain why ocean heat content has been decreasing (ibid., Fig. TS.16, p. 48), which in turn might explain why temperatures have not been rising over the last decade (AR4, e.g., Fig. 3.17 p. 268): Cooler oceans are absorbing atmospheric heat.

† McIntyre and McKitrick (2003, p. 752) report that the graph "appears in Figures 2-20 and 2-21 in Chapter 2 of the Working Group 1 *Assessment Report*, Figure 1b in the Working Group 1 *Summary for Policymakers*, Figure 5 in the *Technical Summary*, and Figures 2-3 and 9-1B in the *Synthesis Report*." They go on to point out that the information graphed is used as a basis for an alarming claim: "Referring to this figure, the IPCC *Summary for Policy Makers* (p. 3) claimed that it is likely 'that the 1990s has been the warmest decade and 1998 the warmest year of the millennium' for the Northern Hemisphere." This alarming claim was then used to advance the political acceptance of GWT: "The IPCC view of temperature history has in turn been widely disseminated by governments and used to support major policy decisions."

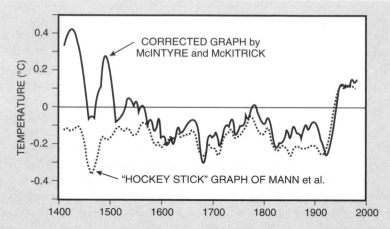

Figure 6.8 The rise and fall of a convenient graph. The "hockey stick" graph of Mann et al. (1998, 1999) was swiftly adopted by the IPCC and became a centerpiece of its Third Assessment Report (2001a–d). It erroneously showed the twentieth century to be the warmest in 600 years. McIntyre and McKitrick (2003) revealed a variety of errors in the construction of Mann et al.'s graph. (Based on data provided by Ross McKitrick.)

As you can see, according to that graph, not only 1998, but every year since 1930 or so has been the hottest for at least six centuries.

The upper, solid line in the figure represents a different result based on the same data but corrected for some mistakes in the statistical processing of those data. It has a very unusual origin. In 1952, Steve McIntyre, a 55-year-old Canadian businessman (McKitrick 2003, p. 22), became interested in global warming. His business experience taught him that to really understand a graph, you have to see the data behind it. "Steve had been a prize-winning student in math and statistics at the University of Toronto and had won a Ph.D. scholarship offer from MIT. . . . Steve figured it would be an interesting exercise" to plot the data (ibid., pp. 23–24) behind the hockey stick graph. When he did, he began to notice gaps in the data and oddities in the way it was handled. He contacted various scientists and statisticians to discuss his findings, and eventually teamed up with Ross McKitrick, an economist, to work with him on his project. McIntyre and McKitrick discovered various disturbing errors in Mann's data handling, such as elimination of spans of temperature records that conflicted with the global warming hypothesis, or, in the more muted, scientific tones of the authors themselves, "unjustifiable truncation or extrapolation of source data, obsolete data, geographical location errors, incorrect calculation of principal components and other quality control defects" (McIntyre and McKitrick 2003, p. 751). When these errors were removed, the upper solid line appeared.

Interestingly enough, McIntyre and McKitrick's corrected graph showed two features that were common knowledge to paleoclimatologists. One was the *medieval warm period*, a period of warm climate and prosperity that came

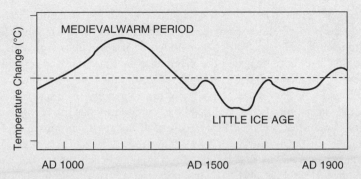

Figure 6.9   Medieval warm period and little ice age. In its first assessment report in 1990, the IPCC included this graph in which two well-known climate features could be seen clearly: the medieval warm period and the little ice age. Both of these features disappeared in favor of the hockey stick graph in later reports, making current warming seem more dramatic. (Based on Wegman et al. 2006, p. 33.)

to a sudden and disastrous end in the fifteenth century when the second well-known feature, the *little ice age*, began.* Just as interestingly, both those well-known and uncontested features had been recognized by the IPCC itself in its first assessment report in 1990, in the graphic reproduced in Figure 6.9. In this graph, current warming is clearly smaller than that of the medieval warm period, when grapes were cultivated in and around London. But both the medieval warming and the renaissance cooling disappeared abruptly in the third assessment report and were replaced with the more dramatic and persuasive hockey stick graph.

When McIntyre and McKitrick's publication of their results caught the attention of GWT supporters, they quickly dismissed it as the bias of global warming deniers. However, U.S. Representative Joe Barton thought that federal government committees studying the issue of global warming[†] would benefit from knowing whether there was any real doubt about the IPCC's claims that global temperatures were at abnormal heights. So Barton invited Edward Wegman of George Mason University, the chairman of the National Academy of Sciences Committee on Theoretical and Applied Statistics to chair a committee of experts to assess Mann's work independently in light of McIntyre and McKitrick's critique. In due course, Wegman's committee issued a report that confirmed McIntyre and McKitrick's analysis.

"Overall, our committee believes that Mann's assessments that the decade of the 1990s was the hottest decade of the millennium and that 1998 was the hottest year of the millennium cannot be supported by his analysis" (Wegman et al. 2006, pp. 4–5). This finding was seconded by a different group from

---

* See, e.g., Lamb (1988, pp. 115–161), Keigwin (1996), and Fagan (2000) for an account of the science and the human toll of the little ice age as it brought the prosperity of the medieval warm period to an end.
† The Committee on Energy and Commerce and the Subcommittee on Oversight and Investigations.

the National Research Council that was also asked to testify before the federal committees.* "In general, we find the criticisms by MM03, MM05a and MM05b [McIntyre and McKitrick 2003, 2005a,b] to be valid and their arguments to be compelling" (ibid., p. 48), said the Wegman committee. He and his group of experts found that the GWT-supporting climatologists who were involved in producing the hockey stick graph were, unfortunately, in a state of denial: "Generally speaking, the paleoclimatology community has not recognized the validity of the MM05 papers" (ibid., p. 49).[†] The expert committee explained the state of denial by this group as follows: ". . . Mann, Rutherford, Jones, Osborn, Briffa, Bradley and Hughes form a clique, each interacting with all of the others" (ibid., p. 49). "[O]ur perception is that this group has a self-reinforcing feedback mechanism and, moreover, the work has been sufficiently politicized that they can hardly reassess their public positions without losing credibility" (ibid., p. 65).

The committee called for a series of reforms so that mistakes of the sort that Mann and the IPCC made would not be repeated, none of which have been implemented. Here is just one of these proposals: "Especially when massive amounts of public monies and human lives are at stake, academic work should have a more intense level of scrutiny and review." The committee goes on to propose a separation of science and politics: "It is especially the case that authors of policy-related documents like the IPCC report, *Climate Change 2001: The Scientific Basis*, should not be the same people as those that constructed the academic papers"[‡] (ibid., p. 6).

---

* "However, the substantial uncertainties currently present in the quantitative assessment of large-scale surface temperature changes prior to about A.D. 1600 lower our confidence in this conclusion compared to the high level of confidence we place in the little ice age cooling and twentieth-century warming. Even less confidence can be placed in the original conclusions by Mann et al. (1999) that 'the 1990s are likely the warmest decade, and 1998 the warmest year, in at least a millennium' because the uncertainties inherent in temperature reconstructions for individual years and decades are larger than those for longer time periods, and because not all of the available proxies record temperature information on such short timescales. We also question some of the statistical choices made in the original papers by Dr. Mann and his colleagues" (North 2006, p. 5). North did, however, assert repeatedly that global warming is real.

[†] The passage continues as follows: ". . . and has tended to dismiss their results as being developed by biased amateurs. The paleoclimatology community seems to be tightly coupled as indicated by our social network analysis, [and] has rallied around the MBH98/99 [Mann et al. 1998, 1999] position" (ibid, p. 49). The tight coupling referred to essentially comes down to the fact that within the "community" of researchers studied (which by no means included all paleoclimatological researchers, but only a small subset of them) published results were supported by reference to yet other published results from within the same subset.

[‡] The good sense of this recommendation is obvious. It was not followed for reasons outlined in the rise of environmental science described in previous chapters, and still is not followed. Nor is there any indication that it will be followed. Although the policy consequences of the IPCC program are of unprecedented enormity, there are virtually *no* safeguards in place to help prevent errors of the sort involved in the hockey stick graph episode. Far more stringent requirements for accuracy and disclosure of data sources are required for simple business deals than for the most momentous policy decision the world has ever faced. The authors of AR4, the most recent IPCC document, still cite their own scientific work in support of their position.

It is difficult to avoid the impression that the IPCC uncritically accepted scientific work that "repealed" the medieval warm period and the little ice age because these two well-known features of the climate record* placed GWT in doubt, at least for the global public. Given that temperatures were just as warm in the medieval warm period as they are now, and given that temperatures fell during the little ice age, it appears that the current warming trend may be nothing other than a matter of natural variability in climate instead of the result of increasing GHGs. Of course, it goes without saying that the previous swings in global temperatures do not show GWT to be wrong—they simply provide another possible explanation of the current warming. They also demonstrate that the current temperatures are not unprecedented, are not unnatural, and are not environmentally dangerous.

A Viking colony was established in Greenland toward the end of the medieval warm period. It succeeded for a few centuries but had to be abandoned when the little ice age struck. The ruins of the stone houses built by the Vikings in Greenland are still there, as are their fields and garden plots, although they are now too cold for farming. Unlike temperature reconstructions, these unreconstructed ruins are facts, not models. Although they are ignored by the media, the global public, and the global warming theorists, they still stand as mute witnesses to the time when it was warmer in the Arctic than it is now.[†]

Yes 4: The Link between CO_2 and Temperature. The case for GWT does not rest on multiple feedback models alone. There is a body of evidence that supports GWT directly, regardless of whether MFMs work or not: the correlation between CO_2 and temperature. The strength of this argument is obvious once the data are represented graphically, as in Figure 6.10.[‡] This is a powerful argument in favor of GWT, so it is no wonder that it figures prominently in IPCC assessment reports and insightful documentaries, such as Al Gore's movie *An Inconvenient Truth* (Guggenheim 2006).

---

[*] During the medieval warm period there were vineyards in London, and during the little ice age the Thames would freeze over during the winter. These well-known historical facts are downplayed by the IPCC as merely local warming, although they lasted for several centuries [see, e.g., Lamb (1988, pp. 115–161) Keigwin (1996), Fagan (2000), Esper et al. (2002), and Moberg et al. (2005)].

[†] The warmest period since the last ice age was about 6000 years ago, when forests reached their farthest northern extent (their fossilized remains can still be seen). At that time the monsoons were stronger than they are now and extended into the Sahara desert itself. At the southern edge of the desert, Lake Chad grew as it was fed by these monsoons during these warmer temperatures. As temperatures fell the lake began to shrink and has been shrinking ever since. Al Gore (2006) claims that the lake is shrinking because of global warming. Climate history indicates that cooling, not warming, reduces this lake. See Lamb (1988, pp. 21–22) for a brief history of these changes.

[‡] Actually, the graph shows the level of a *temperature proxy*, a substance that tracks temperature, since there were no thermometers in place thousands of years ago. In this case, the proxy is deuterium taken from deep ice cores. CO_2 is also measured from the same ice cores, to compare its level with that of the temperature proxy. It is assumed that the levels of both substances in the snow that fell those many thousands of years has not changed from then until now, a view that is open to challenge (see, e.g., Jaworski et al. 1992). As noted previously, some of the detail of the graph upon which this one is based has been suppressed to make trends more apparent.

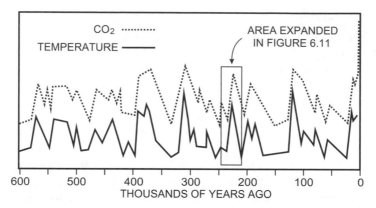

Figure 6.10 CO_2 and temperature over the last 600,000 years. This sort of graph is used routinely in support of GWT, as in AR4 and in Al Gore's movie *An Inconvenient Truth* (Guggenheim 2006). If you do not look too closely, you will not notice that changes in CO_2 *follow*, rather than lead, changes in temperature. A closer look of the area in the inset box is provided in Figure 6.11 making this important fact easier to see. (Based on AR4, Fig. TS.1, p. 24.)

No 4: CO_2 Changes Caused by Temperature Changes. Part of what makes the graph persuasive is its scale: It covers an amazing span of 600,000 years. The scale also makes it impossible to notice that changes in CO_2 *follow* changes in temperature rather than preceding. Numerous scientific studies using various techniques have shown that when temperature falls or rises, CO_2 falls or rises about 800 years later on average,[*] as shown in Figure 6.11.[†] This is important because a cause cannot follow its effect. A cause must come before its effect. Therefore CO_2 changes do not cause changes in temperature. Contrary to the general understanding of GWT, CO_2 is not a climate "driver" that causes major climate changes. Does that mean that GWT is disconfirmed by the CO_2 evidence? No, for GWT can fall back on the claim that CO_2 is merely a climate "enhancer" that amplifies cooling or warming by positive feedback—which, indeed, is the current stance of the IPCC (AR4, pp. 54–57, 85).

From a logical point of view, however, this changes everything. The hypothesis that CO_2 is a climate driver—that is, that changes in CO_2 *cause* changes in temperature— had the benefit of simplicity. Data of CO_2 changes consistently followed by temperature changes would nicely support this simple hypothesis, as shown in Figure 6.12A.

---

[*] For example, Fischer et al. (1999), Petit et al. (1999), Indermühle et al. (2000), Yokoyama (2000), Monnin et al. (2001), Mundelsee (2001), Clark et al. (2002), Caillon et al. (2003), and Stott et al. (2007).

[†] In this case, changes in levels of an isotope of argon (^{40}Ar) are taken as a temperature proxy. Actually, there is a disagreement in the dating of this event between the sources on which this graph and the previous one are based. I have left this disagreement as is, rather than trying to resolve it. Interesting as this disagreement may be, such disagreements between scientific sources are not unusual. In any case, this disagreement is not central to the point at issue, since is concerns the absolute dating of the ice cores rather than disagreement about the order of events: The two sources agree that CO_2 changes followed temperature changes.

TEMPERATURE
CO_2
245,000
240,000
235,000
YEARS BEFORE PRESENT

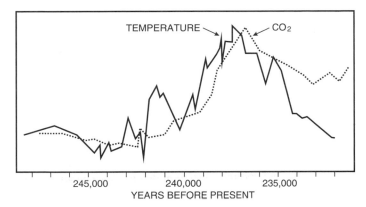

Figure 6.11 CO_2 changes and trail temperature changes. This is a detail of the time span in the box outlined in Figure 6.10. Due to the difference in scale, we can clearly see here what was obscured in Figure 6.10: that CO_2 changes lag temperature changes by about 800 years. (Based on Caillon et al. 2003.)

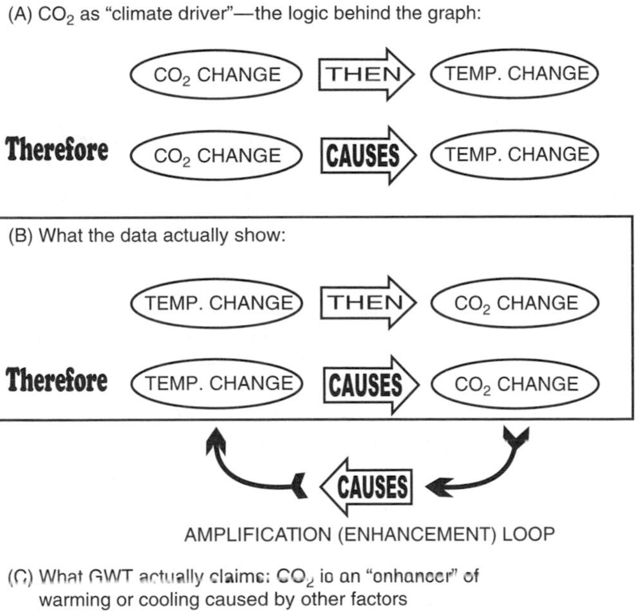

(A) CO_2 as "climate driver"—the logic behind the graph:
CO_2 CHANGE
THEN
TEMP. CHANGE
Therefore
CO_2 CHANGE
CAUSES
TEMP. CHANGE
(B) What the data actually show:
TEMP. CHANGE
THEN
CO_2 CHANGE
Therefore
TEMP. CHANGE
CAUSES
CO_2 CHANGE
CAUSES
AMPLIFICATION (ENHANCEMENT) LOOP
(C) What GWT actually claims: CO_2 is an "enhancer" of
warming or cooling caused by other factors

Figure 6.12 Logical relationship between gwt and CO_2 data. (A) The impression given by graphs linking CO_2 and temperature is that the evidence indicates that CO_2 drives temperature change. (B) What the data show, however, is that CO_2 changes trail temperature changes by nearly 1000 years. The only causal connection this supports is that temperature changes cause CO_2 changes. (C) Given the data, the IPCC no longer claims that CO_2 is a climate *driver*, but rather a climate *enhancer*. This means that the entire history of correlation between CO_2 and temperature does *not* support GWT.

Graphics like the one used by Gore and the IPCC give the impression that the data support a causal linkage from CO_2 changes to temperature changes, as illustrated in Figure 6.12A. But since the data show the reverse sequence, both critics and supporters of GWT now think that temperature change does cause CO_2 change, as in Figure 6.12B.* GWT supporters have therefore redefined their theory as follows: "Atmospheric CO_2 and temperature in Antarctica co-varied over the past 650,000 years. Available data suggests that CO_2 acts as an amplifying feedback" (AR4, p. 57).

The idea is that warming causes CO_2 stored in the oceans to be released into the atmosphere and cause further warming, while cooling causes CO_2 to be reabsorbed by the oceans and cause further cooling, even though how this happens is not understood (AR4, p. 446).[†] The reason that this *enhancement hypothesis* is accepted is that the authors of AR4 cannot think of any other way to explain the changes in temperature between ice ages and interglacial periods.[‡] To put it bluntly, a failure of scientific imagination is not the best reason to accept the enhancement hypothesis. In any case, the current position of the IPCC is pictured in Figure 6.12C. Despite the change in terminology from "climate driver" to "amplifying feedback," the new IPCC position merely *insists* that CO_2 *does* drive temperature, despite the inverse temporal relationship. GWT insists that CO_2 drove the very changes seen in the temperature record, at least in large part, even though it followed them. So GWT is an add-on to the causal relationship shown in the data.

The important thing from a logical point of view is that whether or not the enhancement version of GWT is correct, the CO_2 and temperature correlation data *do not support the enhancement hypothesis*. They support a causal connection in the opposite direction. The hypothesis may be made consistent with the data, but that is

---

* This is not to say that the inference is necessarily valid. If B follows A, then either (1) A causes B, or (2) both are caused by something else, C, but have no direct causal linkage, or else (3) there is no causal connection at all and their temporal sequence is mere coincidence. However, if B tracks A over long periods of time, mere coincidence is usually taken to be implausible, and a causal connection of form (1) or (2) is assumed. Given that there is a well-known mechanism whereby temperature rise would cause a rise in CO_2, it is now generally suggested that the causal linkage is of form (1) rather than form (2): that temperature change causes CO_2 change. The mechanism is the warming of oceans and soil moisture, which reduces the amount of CO_2 that they can hold in solution.

† This page of AR4 consists of Box 6.2, which searches for an explanation of the rise and fall of CO_2 levels *in response to temperature changes*. After exploring various possibilities, some of which are inconsistent with the others, the search is declared a failure: "In conclusion, the explanation of glacial–interglacial CO_2 variations remains a difficult attribution problem.... The future challenge is not only to explain the amplitude of glacial–interglacial CO_2 variations, but the complex temporal evolution of atmospheric CO_2 and climate consistently" (AR4, p. 446).

‡ The inference is explained this way: "Because the climate changes at the beginning and end of ice ages take several thousand years, most of these changes are affected by a positive CO_2 feedback; that is, a small initial cooling due to the Milankovitch cycles is subsequently amplified as the CO_2 concentration falls." This hypothesis is accepted because without it climate models fail to work: "Model simulations of ice age climate (see discussion in Section 6.4.1) yield realistic results only if the role of CO_2 is accounted for" (AR4, p. 449).

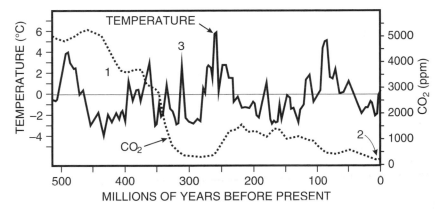

Figure 6.13 Independence of CO_2 and temperature in the megayear range. Here we see that (1) 400 to 450 million years ago there were a series of ice ages that dwarf the most recent ice ages, even though CO_2 levels were 10 to 17 times higher then than they are now. It is also noteworthy that (2) current CO_2 levels are at historic lows against the long-term background. Moreover, even when (3) CO_2 levels were comparably low some 310 million years ago, temperatures still rose to a level 6 degrees warmer than they are now. It seems plain that CO_2 and temperature march to the beat of different drummers. (Based on Berner 1993, 1997, Veizer et al. 2000, and Shaviv and Veizer 2003.)

another matter—one that will no doubt involve more computer modeling—and one, moreover, that remains to be seen.

It also remains to be seen whether enhancement GWT can be made consistent with very long term CO_2 and temperature data where there is no correlation between the two to begin with. As various scientists have argued, the long-term evidence seems to favor a decoupling of CO_2 and global climate [see Veizer et al. 2000 or Shaviv and Veizer 2003 for an introduction and references to this literature]. As Figure 6.13 shows, over hundreds of millions years, atmospheric CO_2 has changed by a factor of 17 or 18, while temperature has changed in a way that seems independent of CO_2 levels.

Yes 5: CO_2 Warming Is Massively Amplified by Positive Feedbacks. It is not clear what is indicated by data concerning conditions hundreds of millions of years ago when conditions may have been very different from those we experience today. But if we restrict our attention to the last few hundred thousand years, there is definitely a relationship between CO_2 and temperature. Admittedly, it is not a simple one-way causal mechanism, but that is not a good reason to reject the thesis that CO_2 is a climate enhancer. We must avoid simplistic thinking—and it is here that MFMs have much to teach us.

To begin with, although it is not generally understood by the global public, the effect of increased CO_2 all by itself would be quite small, and no cause for alarm (the media publicizing IPCC conclusions must be forgiven for not drawing attention to this sort of confusing detail of GWT). But in fact, *most* of the warming predicted by GWT depends on factors that amplify this small CO_2 warming: in other words, *positive feedbacks.* "According to estimates generated by current climate models, more than half the warming expected in response to human activities will arise from feedback mechanisms internal to the climate system, and less than half will be a direct response to external factors that directly force changes in the climate system" (NRC 2003, p. 1). The IPCC calculates that the effect of doubling CO_2 without any positive feedbacks would be about 1.2°C, while the full effect given all of the positive feedbacks would be about 3.2°C, or nearly three times as large (AR4, pp. 630–631). So GWT cannot be understood properly unless MFMs are fully appreciated.

The main positive feedback mechanism is an increase of water vapor in the atmosphere (e.g., AR4, pp. 40, 65, 630). Water vapor is by far the most effective GHG in the atmosphere: It is responsible for most greenhouse heating. Given a small degree of heating by CO_2, for example, the atmosphere warms and so carries more water vapor. This warms the atmosphere further, since water vapor is a powerful greenhouse gas, which in turn leads to more water vaporization, and yet again more heating, in a positive feedback loop.* The effects of water vapor are not limited to simple positive feedback via its greenhouse effect, however. Increased water vapor also affects clouds and cloudiness, since clouds are made of water vapor. There are as many sorts of cloud feedbacks as there are sorts of clouds. There are puffy clouds, wispy clouds, dark clouds, bright clouds, low clouds, high clouds, daytime clouds, nighttime clouds, fogs, mists, and all the variations in between. Each cloud has both positive and negative feedbacks, and the strength of each depends on various factors, including what is underneath them. Generally speaking, low, bright cumulus clouds have a net cooling effect because their reflection of sunlight back into space outweighs their insulating effect, whereas high cirrus clouds have a net warming effect because their insulating effect outweighs their reflection effect (see Figure 6.6). But the IPCC itself repeatedly says that there exists only a low level of scientific understanding of cloud feedbacks, and admits that "cloud feedbacks (particularly from low clouds) remain the largest source of uncertainty" (2007a, p. 65).

Since, as we have glimpsed briefly above, the dynamics of Earth's atmospheric blanket are exceedingly complex, it is simply beyond the unaided human brain to calculate or comprehend. Fortunately, computers can be used to model it via MFMs.

---

* Some readers may notice the positive feedback loop: The rise in temperature due to the positive feedback of water vapor should cause a further increase in water vapor, hence an even bigger rise in temperature, and so on. Would this not yield runaway heating? No, not necessarily. Suppose that the CO_2 rise in temperature is y degrees and that this rise in temperature increases water vapor by a certain amount, which in turn raises temperature by $0.8y$ degrees. This last temperature rise will lead to more water vapor, and yet more warming, by the amount $(0.8 \times 0.8)y$ degrees, and so on—*ad infinitum.* Note, however, that this does not cause runaway warming, since the arithmetical series, $y + 0.8y + (0.8 \times 0.8)y + (0.8 \times 0.8 \times 0.8)y + \cdots$ *converges*, to yield $4.0y$. Whereas a feedback factor (or gain) of 0.8 is fine, a parameter of 1.0 (or greater) would be disastrous, leading to infinite heating: $y + 1.0y + (1 \times 1)y + (1 \times 1 \times 1)y + \cdots$ does *not* converge. So the outcome of any GCM is *extremely* sensitive to the setting of this parameter. Large changes in the warming predicted by a model can be brought about by small changes in this parameter.

There are scores of MFMs (called GCMs, or AOGCMs if they include not only the atmosphere but the oceans as well),* each of which can be run dozens of times with slight variations in its inputs in order to tease out the effect of any one factor (e.g., CO_2, water vapor, clouds).[†] When the results of these hundreds of runs are averaged out, we get a picture of how our climate will change under various scenarios. So despite the uncertainties, we may safely conclude that the net effect of clouds is a significant positive feedback.[‡] Thus, both water vapor and clouds are positive feedbacks to CO_2 warming. So we have every reason to believe that anthropogenic CO_2 will cause dangerous warming. GWT is thus on very solid ground.

No 5: MFMs Fail with Water Vapor, Clouds, and Convection. It is easy to sing the praises of MFMs, but they are beset with problems. When these problems are tallied up, we must conclude that MFMs are still at an elementary stage. At this point they are simply too immature and unreliable to be the basis on which to wager the global economy.

One demonstration of this was the problematic discovery (discussed in No 3 above) that Earth's radiation balance shift toward increasing heat loss between1985 and 2000 was not only unforeseen by MFMs, but was basically inconsistent with them. In his commentary on the surprising data, the prominent climatologist Dennis Hartmann said that they "demonstrate just how little we know. . . . The observations are not easily explained with existing climate models" (Hartmann 2002, p. 811). He then notes that "this change is of the same magnitude as the change in radiative energy balance expected from an instantaneous doubling of atmospheric carbon dioxide. Yet only very small changes in average tropical surface temperature were observed during this time" (ibid.). In other words, the changes observed in radiation balance were just as strong as those predicted by GWT, but they had no effect on temperature, contrary to the basic premise of GWT and the MFMs that model it. Although various excuses could be made for the fact that climate models did not predict this result, Hartmann concluded that "it seems more likely that *the models are deficient*. . . . If the energy budget can vary substantially in the absence of obvious forcing, then the climate of Earth has modes of variability that are not yet fully understood and cannot yet be accurately represented in climate models" (ibid., p. 812; my emphasis). This is not a modest conclusion. To say that MFMs are deficient is to say that the foundations of GWT are deficient.

To see how they are deficient, we need only consider the case at hand. The surprising increase in outgoing radiation involved changes in convection and cloud

---

* GCM stands for *general circulation model*, which is a misnomer inasmuch as GCMs do not model atmospheric circulation (wind, storms, or "weather" in the commonsense meaning of the term), but instead, model radiation balance and represent the effects of circulation on that balance by *parameterization*. AOGCM stands for *atmosphere–ocean GCM* (see AR4, pp. 981, 982).

[†] Eighteen models are cited in support of the main conclusion that CO_2 will cause dangerous warming by the end of the century (AR4, p. 71).

[‡] "The cloud feedback mean is 0.69 W/m^2 with a very large inter-model spread of ± 0.38 W/m^2" (AR4, p. 630). We return to this issue below.

distribution,* which, as it happens, are two things that climate modelers agree are very difficult to model. Convection cannot be included in MFMs for two reasons. One is that MFMs have very coarse spatial resolution. The smallest detail they can represent is 100 by 100 km in size, an area of 10,000 km^2 [just under 4000 square miles)],† so most weather phenomena are simply too small to be "seen" in MFMs, and so are simply left out. The second reason is that MFMs represent only a structure of radiation thermodynamics; in other words, the entire dynamics of the climate system has been reduced to a system of radiation exchanges between the Sun, the surface, the atmosphere, the GHGs, the clouds, the oceans, and so on. Since convection is not radiation, but, rather, moving masses of turbulent fluids, it cannot be included. In order that the effects of convection on radiation balance not be ignored altogether, they can be *parameterized*, or *reduced* to radiation functions that *can* be modeled in GCMs. "Parameterization is defined by the American Meteorological Society (2000) as 'The representation, in a dynamic model, of physical effects in terms of *admittedly oversimplified parameters*, rather than realistically requiring such effects to be consequences of the dynamics of the system,' " (NRC 2005, p. 12, my emphasis).

PARAMETERIZATION PARABLE

Parameterization is complicated from the point of view of scientific methodology, but we can get a feeling for it by means of an analogy. Imagine that you live in the woods and use a campfire for cooking and warmth. Sometimes the wind swirls the smoke in your face, so you move to the other side of the fire, only to discover a few minutes later that the wind has shifted and is again blowing the smoke in your face. You would like to know in advance which way the wind will shift so you can keep the smoke out of your face, but this is not easily done. The atmosphere is a turbulent fluid in a complex system that is effectively unpredictable. You need not remain completely in the dark, however, since the movement of the wind can be parameterized. This can be done in various ways. For example, the probability of wind relative to the compass directions can be observed for a period of time, say 10 years, and captured in a

---

* See Chen et al. 2002 and Wielicki et al. 2002. It appears that there was a reduction in high cloud over the tropical oceans which permitted heat to escape from the top of the atmosphere. Generally speaking, this reduction in high cloud is not well understood, although it is in agreement with Lindzen's iris hypothesis discussed below.

† The NRC states the problem (2003, p. 31) as follows: "Cloud feedbacks are currently diagnosed primarily by using coarse resolution climate models and even simpler one-dimensional equilibrium models." The data points used by GWT models are on the order of 100 kilometers apart, which is much larger than real convection cells, and have a vertical resolution of perhaps three to six vertical layers, which is much too coarse to capture the relevant dynamical details of convection, and hence of advection. In short, GWT models cannot actually model convection or advection—in a word, weather. For this reason they must reduce this ineliminable aspect of the hypothesized warming to simple functions, or parameters.

> function. You might know from this function that, for example, the wind blew mostly from the northwest during January. So if it is January, you know that you should generally sit northwest of the fire. On the one hand, this is better than not knowing anything at all, but on the other hand, you would be better off just watching to see which way the wind is blowing.

The National Research Council characterizes the *parameterization problem* as a result of the fact that parameterization is simplification, and simplification entails loss of information. "current representations of unresolved processes in the models tend not to adequately represent our knowledge of the underlying physics" (NRC 2005, p. 3). The resulting physical inadequacy of parameterization creates a degree of looseness, or lack of constraint, in models, which in turn can lead to very discouraging subjectivity in climate modeling: ". . . physical parameterizations are often viewed as blackbox subcomponents whose knobs, in the form of largely unobservable parameters, can be *adjusted at will to obtain some desired result*" (ibid, pp. 5–6, my emphasis). In other words, parameterization permits climate modelers to tweak their models to get the results that they want, hence placing little or no check on prejudice and subjectivity. The NRC goes on to explain that because of the empirical looseness of parameterization, the main criterion left to modelers for choosing which parameterization to use is whether it gets the results they desire from their models. "Physical parameterizations, often with large numbers of unconstrained or loosely constrained parameters, are inserted into models and judged largely on the merits of their perceived sophistication and *their effect on model performance*" (ibid., my emphasis).*

Put bluntly, the parameterization problem invites subjective judgments of desirability of results to dominate objective observation. To say that large numbers of parameters are "unconstrained or loosely constrained" is to say that GCMs permit climate modelers to move toward whatever results they desire without being hemmed in by observation.

As we saw earlier, water vapor is the most important GHG. It also is the most important radiation and temperature feedback in GWT. Water vapor "amplifies the effect of every other feedback or *uncertainty* in the climate system" (NAS 2003, p. 21: my emphasis). If a warming force raises temperature, this will raise the amount of water evaporating into the air from oceans, lakes, plants, and soil. Since water vapor is a GHG, this will in turn amplify the warming force. Conversely, a cooling force will reduce atmosphere humidity, and this will amplify its cooling effect. So water vapor magnifies the *uncertainties* in the MFMs supporting GWT, uncertainties about not only the size, but even the direction, of the other climate forcings and effects it amplifies. GWT holds that water vapor greatly amplifies GHG warming and

---

* Although the authors of this NRC document go on to prescribe research strategies to resolve PP, they also note deep scientific and research-*community* problems that will be very difficult to overcome. Until they are overcome, GWT remains in limbo, due to the gap between it and observational data. PP is a component of many of the GWT problems outlined below.

accounts for about a third of the warming predicted by most models.* This estimate may be too small[†] or too large.[‡]

Parameterization cannot magically transform these uncertainties into certainties. If the uncertainties are included in the parameterization and hence in the MFMs themselves, they will in all probability reemerge in the results of these climate models. Since the effect of water vapor uncertainties is to amplify other uncertainties in climatological data and theory, GWT becomes nonspecific, vague, and plastic.

If the word, *plastic*, seems a bit over the top, consider the climatological uncertainties about clouds. The IPCC claims that clouds will increase and thereby increase global warming.[§] Some climatology textbooks, by contrast, teach that the net effect of clouds is cooling. For example: "Water vapor, however, results in cloud formation. Clouds cause a host of climate feedbacks, some positive, some negative . . . , although the overall impact of cloud is to increase Earth's albedo, and so . . . to cool the planet" (Bigg 2004, p. 5). The NAS sensibly concludes (2003, p. 26) that "at this time both the magnitude and sign of cloud feedback effects on the global mean response to human forcing are uncertain." In other words, we are uncertain even about whether clouds are a positive or a negative feedback. MFMs cannot magically eliminate this uncertainty, so GWT itself is subject to it.

---

* In the "Technical Summary" written for nonscientists and policymakers, the water vapor effect is estimated by IPCC to be "approximately 1 W/m^2 per degree global temperature increase, corresponding to about a 50% amplification of global mean warming" (AR4, p. 65). This is a bit misleading, inasmuch as all by itself water vapor would "at least double the response" to greenhouse gases (p. 632; also NRC 2003, p. 22). But the IPCC reduces this 100% minimum amplification by subtracting one negative feedback, the *lapse rate effect*, and ignoring other feedbacks, both positive and negative, to get its 50% amplification figure. The reason given for this is that the "close link between these processes [water vapor radiative feedback and the lapse rate effect] means that water vapor and lapse rate feedbacks are commonly considered together" (AR4, p. 632). True enough, but this does have the effect, especially in a technical summary intended for nonscientists, of exaggerating the effect of CO_2 and other anthropogenic GHGs relative to water vapor—and keeping them in the spotlight. But how water vapor reacts to GHGs may be more important than the GHGs themselves. One researcher, for example, argues that "a 12% reduction in the magnitude of the lapse rate completely nullifies the water vapor feedback" (Sinha 1995, p. 5095). Data show, and IPCC claims, a decreased lapse rate. Clearly, the ultimate effect of anthropogenic GHGs will depend very sensitively on just how, and how much, they affect the lapse rate.

† As we noted earlier, IPCC models indicate that the effect of CO_2 alone would be about 1.2°C, while the full effect given all of the positive feedbacks would be about 3.2°C, or nearly three times as large (AR4, pp. 630–631). But as has often been pointed out, the IPCC may underestimate the influence of water vapor, and thereby underestimate the uncertainties in GWT. Harries, for example, reports: "But as has often been pointed out, this IPCC claim may underestimate the influence of water vapor, and thereby underestimate the uncertainties in GWT." For example, Harries says that ". . . uncertainties of only a few percent in knowledge in the humidity distribution in the atmosphere could produce changes of the outgoing spectrum of similar magnitude to that caused by doubling carbon dioxide in the atmosphere" (Harries 1997).

‡ As reported by the NRC (2003, p. 22). As seen in the preceeding note, water vapor levels affect the lapse rate, which once again leads to uncertainty about the final effect of water vapor. Sinha estimates that "increasing the lapse rate magnitude by 6%. . . amplifies the modeled water vapor feedback by 40%; conversely, a 12% reduction in the magnitude of the lapse rate completely nullifies the water vapor feedback" (Sinha 1995, p. 5095).

§ The cloud feedback mean of IPCC MFMs is "0.69 W/m^2 with a very large inter-model spread of ±0.38 $W/m^{2"}$ (AR4, p. 630).

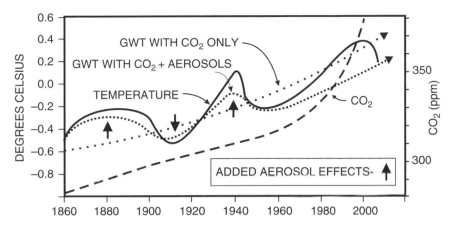

Figure 6.14   Improvement of gwt models by inclusion of aerosols. If GWT considers only increases in CO_2, its accuracy (its agreement with observed temperatures) is relatively low. However, by including other factors, such as aerosols, its accuracy can be increased. Here, observed variations in temperature are accommodated by increases in aerosols, which have a net cooling effect. It is assumed that aerosols emissions go up in good economic times and go down in depressed economic times. A rough correspondence of temperature and the Great Depression can readily be seen.

Yes 6: Multiple Feedback Models Explain Recent Temperature Changes. The theoretical reasons for trusting MFMs of GWT are backed up by empirical data as well: MFMs have been successful in accounting for past temperatures. Admittedly, early MFMs did not do very well in this regard. Whereas CO_2 has increased smoothly through the twentieth century, temperatures first fell from 1900 to 1910 or so, then rose for 30 years until 1940, then dipped until about 1970, when they began to rise again, as shown in Figure 6.14.

But as studies through the 1980s and 1990s began to show more clearly that sulfate aerosols had a net cooling effect, the possibility arose of explaining the cooling spells of the 1930s and 1950s in terms of the emissions from industrial smokestacks.* In those days, smokestack emissions were laden with sulfur dioxide, which turns into sulfate after it is released into the atmosphere. Sulfate cools in a number of ways, which nicely illustrate the complex dynamics of our atmosphere and the complexity of its response to environmental inputs. Sulfate's direct effect is to reflect some incoming sunlight back into space, a cooling effect. It also causes the low, bright, cumulus clouds, which have a net cooling effect, to become even brighter (by increasing the number and decreasing the size of cloud droplets). It may also increase the lifespan of clouds, again increasing cooling. Sulfate and the clouds it affects also reflect heat back toward the surface, but this warming effect is outweighed by its cooling effects.

---

* The work of R. J. Charlson (e.g., Charlson et al. 1992, Charlson and Wigley 1994) on sulfate aerosols went unnoticed for decades until he pointed out the possibility of explaining cooling spells in terms of sulfates. Once this message reached climate modelers, they promptly produced models which included this effect, and Charlson's work gained sudden recognition.

Thus, smokestack emissions explain the cooling and warming spells of the twentieth century. Smokestack sulfur emissions increased during the economic upturn of the 1920s, causing cooling; they decreased during the Great Depression in the 1930s just when temperatures rose, and then rose again during the economic upturn following World War II as temperatures fell again. A touch of irony emerged: The scrubbing of smokestack emissions of sulfur dioxide to stop acid rain that began in the 1970s has actually contributed to global warming since then.

No 6: The GWT Chain Is No Stronger Than Its Weakest Link. The last argument has lost track of the logic of the situation. GWT rests on MFMs, and MFMs are models of the entire climate. In other words, we have no reason to accept the theory of global greenhouse warming unless we understand the climate as a whole. Unfortunately, our knowledge of the climate as a whole has not advanced to the point where we can model the climate as a whole, and this problem is compounded by the fact that our computer models cannot even model what we do know, but have to simplify it by the art of parameterization. Because the climate is a dynamically chaotic system (see Section 6.6), slight changes in measures used by global computer models can have very large effects on their results. In the more measured language of the professional scientist, the MFMs have large error bars for some highly relevant quantities. So even if we suppose that inclusion of aerosols results in an MFM run that yields a perfect prediction of the temperature observed, this would prove nothing. A perfect prediction would be a matter of contingent factors, or luck, since the effects of aerosols are just as uncertain as a host of other crucial climatological factors. A perfect fit between theory and data can be found within the full range of uncertainty of aerosol effects, and an MFM may be adjusted to do so, but that only confirms the plasticity of MFMs.

Since the early days of GWT, some climatologists have been warning that anthropogenic aerosols—tiny particles of dust and ash that human beings release into the atmosphere—are probably more important than the CO_2 we release. The IPCC itself calculates that aerosols have a cooling effect and use that effect to bring their MFMs into closer agreement with temperatures measured in the early twentieth century. On the other hand, they also recognize that there is a very large uncertainty in aerosol effects. Although they do not make the comparison themselves, it turns out that their uncertainty about the effect of aerosols is larger than their uncertainty about the total effect of global warming itself. This is a most remarkable situation from a methodological point of view! Global warming is a very complex theory in which many elements combine to produce a specific affect. These elements are like links in a chain, and aerosols are one of the links. Each link has a specific uncertainty, and the uncertainty of aerosol effects is estimated with "low" confidence to be 2.3 W/m^2.*
The chain as a whole has a specific uncertainty, which is estimated with "*very high*

---

* Their direct effect is estimated to be -0.5 [-0.9 to -0.1] W/m^2 (AR4, p. 29) and their cloud albedo effect is estimated to be -0.7 [-1.8 to -0.3] W/m^2 (AR4, p. 30), for a total effect of -1.2 [-2.7 to -0.4] W/m^2.

confidence" (AR4, p. 31; italics in original) to be only 1.8 W/m^2.* Apparently, this chain is stronger than its weakest link.[†]

The main message of the press releases surrounding the release of the IPCC's fourth assessment report (AR4) was that science had now made it official: It is now certain that humanity is guilty of causing climate change. But the high level of confidence IPCC imputes to this claim is not supported by the underlying science, which involves a large number of factors that have uncertain effects.

Yes 7: Arctic Warming Is a "Fingerprint" of GWT. There is no doubt that there has been a dramatic warming of the Arctic over the last few decades. Depending on the estimate, the Arctic is warming anywhere between twice and four times as fast as the tropics (see, e.g., AR4, p. 37). We have all seen the pictures of melting Arctic ice; we have all heard that the legendary Northwest Passage may soon provide another sea route around North America; and we have all heard that polar bears face extinction due to the shrinkage of polar sea ice. This more rapid warming of the Arctic in conjunction with a slower warming of the globe as a whole is just what we would expect according to GWT, as shown in Figure 6.15.

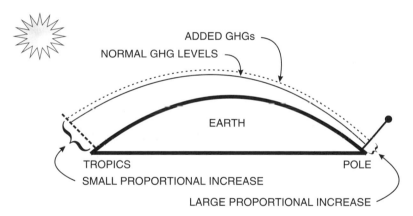

Figure 6.15 Arctic warming is a fingerprint of GWT. Because added GHGs increase the greenhouse effect of the atmosphere evenly over the entire globe, their effect is felt most strongly at the poles, where GHGs are naturally thinnest. The greenhouse effect is normally much weaker at the poles, because water vapor is the most important GHG, and there is 10 to 20 times less water vapor at the poles than at the equator.

---

* The total estimated GHG effect is 1.6 (+0.6 to +2.4) W/m^2 (AR4, Fig. TS.5, p. 32).

[†] If this is not already amazing enough, we might note that it is possible that the cooling effect of aerosols may be only one-third of the −1.2 W/m^2 accepted, with low certainty, by the IPPC. If the real figure is only -0.4 W/m^2, which is inside the error range, then global warming would be overestimated by 1.2 W/m^2. This would reduce global warming to 0.4 W/m^2, which is hardly anything to get excited about, since it would have only a slight effect on temperature—of about 0.25°C. This would also, by the way, be a result that lies outside the IPCC high-confidence error bar—which may indicate some internal tension, if not outright inconsistency, among its error ranges.

Arctic warming is particularly significant because it is a uniquely identifying fingerprint of the sort of warming we should expect from increased GHGs. Whereas there are lots of ways in which global temperature may be made to rise (and hence lots of ways that such rises may be modeled), the sort of warming required for GWT will have specific characteristics, or "fingerprints." Arctic warming is one fingerprint effect of GWT. The greenhouse effect of our atmosphere is naturally smaller at the poles than at the equator, since the warmth of the Sun at the equator vaporizes lots of water vapor, and water vapor is the most abundant and effective GHG. In fact, there is about 10 to 20 times as much water vapor at the equator as at the poles.* Since the greenhouse effect at the poles is small, a small uniform increase in the blanket by the addition of GHGs will have a larger effect at the poles than at the equator.

No 7: The Arctic Warming Fingerprint Is Smudgy. The clearest sign that the Arctic fingerprint argument is in trouble is that the IPCC itself has backed away from it. The argument was presented in the Third Assessment (IPCC 2001a) and is still prominent in arguments presented to the public. For example, in a recent article in the *New York Times* entitled "Arctic Melt Unnerves the Experts," Andrew Revkin wrote that "Proponents of cuts in greenhouse gases cited the meltdown as proof that human activities are propelling a slide toward climate calamity" (Revkin 2007), although even in his article there are observations that perhaps something other than GWT is required to explain what is going on in the Arctic. In AR4 we find only the following mild statement:

"Arctic temperatures have been increasing at almost twice the rate of the rest of the world in the past 100 years. However, Arctic temperatures are highly variable. A slightly longer warm period, almost twice as long as the present, was observed from 1925 to 1945 . . ." (AR4, p. 37).

So the current warming period is only half as long as another one within living memory. We must commend the IPCC for making this observation. Why the change in attitude by the IPCC concerning this fingerprint of the theory they believe? Perhaps it is because the GWT warming effect pictured in Figure CS 7.15 actually applies with equal force to *both* poles. As far as GHG warming is concerned, there should be more rapid warming at both poles. However, IPCC maps of global temperature trends over the last century as well as the last few decades (AR4, pp. 250–251; see Figs. 3.9 and 3.10) show that Antarctic temperatures are not warming much, if at all. If we consult sources other than the IPCC, "one is hard-pressed to argue that warming has occurred" (Chapman and Walsh 2007) at all in the Antarctic. In fact, the most accurate records of temperature, those provided by satellite data (see Figure 6.16), do not show any significant warming in the *entire southern hemisphere* over the last 27 years.

Although the IPCC still argues that global temperatures vindicate GWT, its claims are now nuanced and qualified. Even though GWT stands for *global* warming theory,

---

* Specific humidity varies from about 2.2 g/kg in the Arctic (Gerding et al. 2004) to about 14 g/kg in the tropics (Newell et al. 1974), while total column water vapor varies from about 3 mm in the Arctic (Kiedron et al. 2001) to about 60 mm in the tropics (Mather and Ackerman 1998).

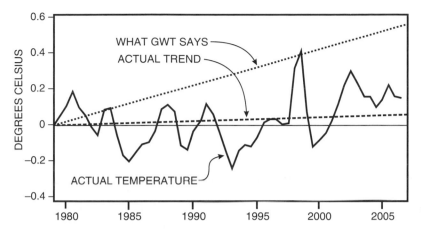

Figure 6.16 Southern hemisphere temperatures. Temperatures in the southern hemisphere have been increasing at the rate of 0.002°C per year, one-tenth of that predicted by GWT. Antarctica has shown no significant warming either. Temperatures in the southern hemisphere disagree with GWT. [Based on satellite data (NSSTC 2007).]

it is now sometimes formally qualified to imply only *near* global warming: "No known mode of internal variability leads to such widespread, *near* universal warming as has been observed in recent decades.... the response to anthropogenic forcing is detectable on all continents individually *except Antarctica*" (AR4, p. 727; my emphasis). So even though Arctic warming is still presented as a popular argument in favor of GWT, it raises a problem for GWT: how to explain why warming has been restricted to the northern hemisphere.

Although this *Antarctic anomaly* is real, it does not seem that in the last analysis it could falsify GWT. There are differences between the Arctic and the Antarctic which may allow the possibility that the first would overreact to global warming while the latter underreacts. For example, the Arctic Ocean underlies most Arctic ice, permitting heat flows via ocean circulation, whereas the Antarctic ice is hundreds of times thicker and lies on solid ground; the Arctic is in the middle of the industrialized northern hemisphere, making it more liable to be affected by industrial aerosols, which often have a short atmospheric lifetime, and do not migrate as readily to Antarctica; and so on.

On the other hand, these admissions and these facts mean that the warming in the Arctic is no longer a GWT fingerprint. For one thing, the warming in the Arctic is much larger than GWT predicts. Ironically, this may be taken by GWT supporters as confirmation, but that would be methodologically perverse. As Stroeve et al. put it (2007, para. 1) "none or very few individual model simulations show trends comparable to observations" when it comes to the Arctic, and simulations are "confounded by generally poor model performance" when it comes to the Antarctic. GWT cannot be treated as a plastic hypothesis, a sort of movable feast for GWT proponents. Either GWT says something in particular, or else it does not. If polar temperatures are not captured in MFMs, this should be taken as a disconfirmation of GWT. If it is not taken as disconfirmation, we must conclude GWT does not say anything specific. So while the temperatures at the poles may not falsify GWT, they do present it with a dilemma: falsification or vagueness.

THE URBAN HEAT ISLAND EFFECT AND GLOBAL TEMPERATURE

Part of the problem for GWT is to determine the temperature of the entire Earth since 1750, when the warming is said to have begun. Obviously, Earth does not have *a* temperature but rather, *many* temperature*s*: At any one time the deserts on the daytime half of the globe are hot and those in the nighttime half are cold, and the poles freeze while the tropics simmer. If we had thermometers placed evenly across the surface of the globe, we could just average their readings to get a pretty good idea of global surface temperature (GST). But in fact, thermometers are placed where it is handy to have them, not where we need them for the purposes of GWT. In fact, virtually all long-term temperature records come from cities and towns in the industrialized world. Oceans, glaciers, deserts, mountains, forests, swamps—and just plain old countryside—tend to be left out. But any idea of GST we might form has to be based on the records we have, not on the ones we wish we had. The result is that estimating GST is a complex business requiring the most sophisticated scientific data, historical data, inference, and extrapolation that we can muster. Current estimates of GST "correct" raw temperature data for various categories of error. Five such categories are used by the authoritative Goddard Institute for Space Studies.[*]

One of the factors that must be taken into account is the *urban heat island* (UHI) *effect* . It has been known for more than a century that urban temperatures are higher than rural temperatures. Cities are like huge stone ovens that absorb the heat of the Sun during the day and release it slowly at night. When scientists began to study the UHI effect in earnest in the mid-twentieth century, they were surprised to find that it could easily be 10°C (18°F) hotter downtown in a large city than in the surrounding countryside (Oke 1973, Lowry 1971). A simple relationship was found between city size (measured as population) and the UHI effect, as shown in Figure 6.17.

The problem for GWT is that the warming effect up to the present day is supposed to be very small, approximately three-fourths of a degree celsius per century (or 0.0075°C per year, AR4, p. 37, Table TS.6), whereas the UHI effect is several degrees. Indeed, urban growth has been rapid during the twenteith

---

[*] These are (1) different times of daily temperature observation at different weather stations, (2) correction of traditional maximum and minimum temperatures, (3) correction for "station history," which includes anything from change of instrument, instrument location, instrument housing, to change of location of station, (4) filling in of missing data by interpolation, and (5) the UHI adjustment (J. Hansen et al. 2001).

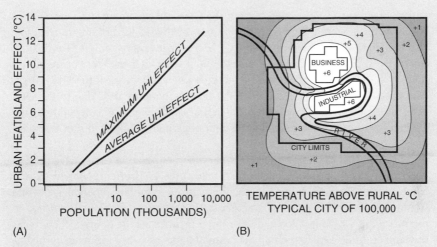

Figure 6.17   The urban heat island effect. (A) The intensity of UHI: cities are hotter than the surrounding countryside, generally by several degrees, both day and night, and both summer and winter, an effect many times larger than the warming GWT predicts. (B) The extent of UHI: the heat generated by a city is most intense downtown and mildest at its limits, although it spills over into the surrounding countryside even on a calm day as pictured here. Since the great majority of long-term temperature records come from cities, even slight errors in corrections for UHI would grossly distort any warming signal sought in the urban temperature noise. [(A) Based on Oke 1973.]

century, with the result that just the *growth* in the UHI effect can easily be as large as, or larger than, the warming. A tripling of population will add about a degree of UHI warming, and many urban areas have more than tripled in size over the last century as global population quadrupled and the proportion of people in urban areas doubled or tripled as well. Thus, we would expect pretty sizable UHI effects to show up in the temperature records. For example, the growth in UHI heating for Tucson has been measured at 0.074°C per year, or about 10 times the rate of estimated global warming (Comrie 2000). In the United States, for instance, about 45% of weather stations are located in cities of 10,000 or larger and so have UHI effects greater than 4°C. To estimate global warming due to the GHGs, warming due to the UHI effect must be subtracted from the raw temperature records for at least 45% of weather stations (and for some of the rest as well, since even a city of 1000 inhabitants will generate a UHI effect of 2°C).

In view of all of this, it is interesting that the IPCC claims that "recent studies confirm that effects of urbanization and land use change on the global temperature record are negligible" (AR4, p. 36). So apparently all of the long-standing commonsense wisdom about UHI effects, and our knowledge that globally the location of weather stations is and has always been mainly urban, is simply wrong. Going a step further, the Goddard Institute for Space Studies (GISS) adjusts its raw-data temperature *upward* by 0.3°C, not downward. The raw U.S. data shows a warming of only half a degree between 1900 and 2000, which the GISS corrections increase to about 0.8°C. Of the five "corrections" that GISS makes, four are upward, and only one is downward: the UHI effect. The GISS downward adjustment of the raw temperature data due to the UHI effect is a mere 0.15°C over the entire century.* Clearly, GWT depends on the accuracy of such corrections, and a slight error could support either warming or cooling of significant magnitude.

There is clear evidence, moreover, that the temperature data being used in support of GWT are, in fact, elevated due to the UHI effect. McKitrick and Michaels (2007) have shown that there is a strong correlation between the rate of temperature rise for a given weather station and the *economic prosperity* of the district in which it is located: The more prosperous the district, the greater the temperature rise. How could economic factors possibly influence temperature? The changes in land surface due to agriculture and urbanization (hence the UHI effect) that go with economic prosperity tend to increase local surface temperature. McKitrick and Michaels' very careful and thorough study indicates that approximately half of the reported temperature rise from 1980 to 2002 is due to contamination of temperature data by such things as the UHI effect.

Yes 8: GWT Is the Scientific Consensus, So We Have to Accept It. At the end of the day, nonscientists simply are not in a position to judge whether or not to accept GWT. Scientific theories are sophisticated products of a professional community that are not properly understood by nonscientists. The best nonscientists can do, therefore, is rely on the experts themselves, and they, as it happens, are in favor of GWT. It is interesting and educational to discuss GWT in the pages of a philosophy book, but the question of the actual truth of the theory cannot be addressed properly in this context.

No 8: There Are Strong Scientific Arguments Against GWT. From this it would follow that it would not be proper to address GWT in government either—and that would amount to science dictating public policy. Democracy requires that we rule ourselves, and that in turn requires that we make judgments concerning matters of fact. When it comes to the judgment of the scientific community, only the judgment of scientists is relevant. Scientists must judge whether a theory is strong enough to be

---

* This correction is *larger* than that used by the U.S Historical Climate Network (J. Hansen et al. 2001, e.g., Plate 2). Hansen, one of the scientists who prepares the GISS estimates, says on the GISS website that "the urban warming that we estimate (and remove) is larger than that used by the other groups (as discussed in the 2001 Hansen et al. reference above [J.Hansen et al. 2001])" (J.Hansen 2007). This shows both that the UHI correction is a matter of debate if not disagreement, and that the correction used by the "other groups" is very small indeed. Also interesting in this context is the fact that GISS has recently admitted that its U.S. temperature records over the last decade or so were 0.15°C too high, due to an error concerning a much simpler matter than UHI corrections.

accepted as the basis of further scientific research, whereas we must judge whether a theory is strong enough to be accepted as the basis of public policy. Even if scientists are willing to accept a given hypothesis on the grounds that they think its probability of being right is 90%, this probability may still be too low for its acceptance as a basis for action. When it comes to practical action, we must consider what is at stake. We might accept that a given process will produce a safe vaccine as a theory on which to base further research, while refusing to use that vaccine on children until further safety tests are performed. In the case at hand, we have to decide whether GWT is strong enough to wager the global economy on it. It is a weighty decision. So even if there is a scientific consensus, we would do well to question it.

There is, of course, no scientific consensus about GWT. Consensus means complete agreement, and there are qualified and respected scientists who do not accept GWT. Presumably those who claim a scientific consensus for GWT do not understand the word in the literal sense, but only claim a strong majority in favor of GWT. It is relevant, then, that no official vote of the scientific community has ever been taken, for it shows that the reported consensus is only an article of faith among those who are already convinced by GWT. George Taylor, 2002 President of the American Association of State Climatologists, reported, "I can tell you that there is a great deal of global warming skepticism among my colleagues . . . the global warming scenarios are looking shakier and shakier" (Taylor 2002). Taylor goes on to reflect on the response of GWT advocates to this skepticism: "It's interesting to me that the tactics of the 'advocates' seems to be to 1) call the other side names ('pseudo scientists') and 2) declare the debate over ('the vast majority of credible scientists believe . . .')" (ibid.). He also encourages discussions just like this case study: "I'm grateful for those who . . . keep the dialogue open and allow us to share relevant information and scientific data" (ibid.).

Perhaps because a purported crisis is vastly more newsworthy than no crisis at all, the popular media has systematically ignored or downplayed scientific skepticism about GWT. They ignored thousands of scientists among the 17,800 who signed a 1999 petition stating "There is no convincing scientific evidence that human release of carbon dioxide, methane, or other greenhouse gases is causing or will, in the foreseeable future, cause catastrophic heating of the Earth's atmosphere and disruption of the Earth's climate" (http://www.petitionproject.org/, September 8, 2008; http://en.wikipedia.org/wiki/Oregon_Petition, September 8, 2008). This remarkable—and remarkably *important*—event has gone virtually unreported and so has been kept from the global public by the popular media. The Petition Project (or "Oregon Petition") website currently lists 31,072 signers, including 3,697 atmospheric, environmental, and Earth scientists, 5,691 physicists and aerospace scientists, and 4,796 chemists. Serious doubts have been raised about the qualifications of a relatively small number of signers, but many are stellar scientists, and many thousands are perfectly well-respected and well-qualified working scientists. Apparently the *political* organization of pro-GWT scientists under the banner of the IPCC has gained them the popular media status of representing a "scientific consensus." However, given scientists' entirely proper professional opposition against submitting science to political organization, it is quite plausible that a petition would be a more accurate indication of their opinion.

Nor has the popular media paid any attention to doubts about GWT raised by established scientific professional bodies, perhaps because the scientists involved did not issue press releases announcing publication of their critiques. At least two books by the august and authoritative National Research Council are dedicated to criticism of GWT (NRC 2003, 2005). These two books alone have 24 authors, reporting on the work of 46 other specialists, and reviewed by another 15 "chosen for their diverse perspectives and technical expertise" (NRC 2003, p. xi, 2005, p. ix),* all of whom express principled misgivings about GWT. In support of their work they cite hundreds of scientific publications. I present these numbers solely to dispel the idea so often presented in the popular media that *no* qualified scientist has any serious misgivings about GWT. This idea, which is propounded by members of the IPCC itself in its press releases and media presentations, is clearly false.

Yes 9: There Are No Credible Scientific Alternatives to GWT. Of course there are scientists who are critical of GWT, but that is only because the scientific community is a free society that encourages criticism of scientific work in general. It is easy enough to be critical, but no scientific competitor to GWT is sufficiently well supported by data and theory to be as worthy of acceptance as GWT, or even come close. GWT is the only game in town, scientifically speaking. It is the standard against which all competitors are measured–and none of them measure up.

No 9: There Are Credible Scientific Alternatives to GWT. The claim that no opponent to GWT is worthy of belief deserves serious consideration, and the only way to do that is to consider one or more opponents. Indeed, scientific theories should not be considered alone but in comparison to their competitors. So, let us briefly assess just three competitors to GWT: (1) the universal alternative to all theories, the "null hypothesis;" (2) an alternative from inside mainstream climate science, the iris hypothesis; and (3) an alternative from outside mainstream climatology, the Sun–Earth climate connection.

In the interests of clarity, let us define the basis of assessment in advance. As we saw in Chapter 5, scientific models must be assessed *relative to* their competition (compare item 6, Section 6.3). This assessment has five dimensions:

1. *Precision* concerns how specific the model itself is. A model which says that sunrise will be at 8:11 A.M. (implying precision to the nearest minute) is less precise than one which says that sunrise will be at 8:11 and 34 seconds A.M. (implying precision to the nearest second). Precision does not concern whether or not the model agrees with reality, for that is a question of accuracy, not precision. Precision may be thought of as the standard that a model sets for itself. A model that tries to predict sunrise to the second sets a higher standard for itself than a model that tries only to predict the minute the Sun will rise. Whether either model actually succeeds in meeting its own standard is another question. As we all know, setting a lower standard makes it easier to succeed, and setting a higher standard makes it more difficult.

---

* There are eight duplications within and between these two books, so that a total of 77 individual scientists are involved.

2. *Accuracy* concerns whether the model agrees with reality. If the Sun rises at 8:11 and 27 seconds, the first model is accurate (it agrees with reality, since 8:11 is the nearest minute to sunrise), whereas the second is not (it disagrees with reality, since 8:11:34 is not the nearest second to sunrise). A model's precision may be greater than its accuracy, as when a model predicts the precise second of sunrise, but is actually accurate only to the nearest minute. But a model's accuracy cannot be greater than its precision, since a model that predicts sunrise only to the nearest minute cannot possibly be accurate to the nearest second. As this example illustrates, precision and accuracy are in tension with each other. As a model becomes more precise, it becomes more difficult for it to be accurate. The scientific ideal is for the highest possible levels of both precision and accuracy, but accuracy must often be purchased at the expense of precision, or precision must be purchased at the expense of accuracy. (This use of *precision* and *accuracy* is admittedly not in perfect agreement with standard usage, but it will enable us to make distinctions necessary to theory evaluation.)

3. *Empirical consistency* concerns the agreement of the model with the empirical content of other models that are already accepted by the scientific community. This may be seen as a species of accuracy, inasmuch as accepted models presumably are empirically accurate.

4. *Scope* is the extent of application of a model. For example, a model of falling bodies has a narrower scope than a general model of dynamics, which applies not only to falling bodies, but also to projected bodies, bodies in orbit, and so on. Theories of broader scope are preferable, other things being equal, because they give us more information. On the other hand, the more a model says, the more it is exposed to inaccuracy. So scope and accuracy are in tension with each other.

5. *Simplicity* concerns the structure of the model relative to the capacities of the human mind. The purpose of science is to provide us with information about the world in a form that is more accessible than the world itself. The map must be simpler than the terrain, otherwise, we do better just to consult the terrain itself. Given that two theories contain the same amount of information, we prefer the simpler of the two. Thus science aims to find the simple patterns underlying the rich and complex events of the world. Simplicity is in tension with scope. Newton's mechanics is simpler than Einstein's, but Einstein's includes bodies approaching the speed of light, whereas Newton's does not. We would like to have theories that are both simple and have great scope, but in practice we often have to trade one virtue for the other.

(1) The Null Hypothesis: Natural Climate Variation. The null hypothesis* is whatever is left when the hypothesis in question is ignored. If we ignore GWT, what remains is the hypothesis that the changes in temperature that we have seen since 1750 are due to *natural variation*. Climate and weather form a nonlinear, multivariable system (commonly known as a "chaotic" system), which will unpredictably move away from average values over various periods of time. It is interesting and

---

* My thanks to Richard Lindzen of MIT for suggesting that I include the null hypothesis as a competitor to GWT.

relevant that deterministic chaos was rediscovered in the 1960s (Henri Poincaré, 1854–1912, discovered it first nearly a century earlier) by a meteorologist, Edward Lorenz. It is especially relevant that Lorenz was working on computerized climate models when he discovered that they were chaotic. His work eventually led to the conviction that the actual climate system is chaotic: deterministic but unpredictable because of its sensitivity to the slightest change. Lorenz illustrated climate chaos with his famous butterfly effect example: A butterfly flapping its wings in Brazil could set off a tornado weeks later in Texas. Lorenz does not accept GWT because, in his own words, "The atmosphere and its surroundings constitute a chaotic dynamical system, and we cannot without careful investigation reject the possibility that this system is one where spontaneous long-period fluctuations occur" (Lorenz 1991, p. 450).

It is entirely possible that the gradual rise of temperature from 1750 through to roughly 1995 may have been nothing other than such a spontaneous long-period fluctuation, or natural variation (NV). This is precisely what chaotic systems do. For GWT to be established, it must be superior to the NV hypothesis. This has not been shown.

Comparative Assessment. *Precision*, our first criterion, favors NV over GWT. NV says that at any scale we will find that climate variables change in unpredictable ways (see the box, "Nature Is Unpredictably Wiggly," in Section 2.1; and Section 6.6). The tests for chaos are well defined and precise, and they have demonstrated to everyone's satisfaction, including the IPCC, that climate is indeed chaotic. So NV is not only precise but has high *accuracy*. GWT, by contrast, predicts climate sensitivity (the temperature rise caused by doubling atmospheric CO_2) to be anywhere between 1 and 6°C, with a best estimate of 3.2°C (AR4, pp. 65, 630–631), a much less precise claim. Moreover, it has not been possible to check this estimate for accuracy, since no doubling of CO_2 has occurred. When it comes to *scope* and *empirical consistency*, NV and GWT are roughly equal: Both apply to climate as a whole, and both are in agreement with accepted scientific theory. However, when it comes to *simplicity*, they are very different. NV is the simple claim that temperature changes since 1750 are within the range of spontaneous variation of Earth's chaotic climate. GWT, by contrast, depends on MFMs of the climate system that must mirror its complexity. Thus, NV is vastly simpler than GWT. On grounds of simplicity, we must prefer NV.

Taking all five criteria together, GWT needs to overcome the chaos of the climate system in order to forecast climate with sufficient precision and accuracy that the GHG warming signal that it predicts can be heard against the chaotic background noise of the climate system: an extremely difficult (almost paradoxical) task. At this point all we can hear is noise, which is evidence for NV and against GWT. Every failure of GWT to explain a climate phenomenon counts against it and in favor of NV, such as GWT's inability to explain the surprising increase in Earth's outgoing infrared radiation between 1985 and 2000 (see No 3 above), and its inability to explain CO_2 variations between ice ages and interglacials (AR4, p. 446; see No 4 above).

(2) The Iris Hypothesis. In 2001, Richard Lindzen and his colleagues M. D. Chou and A. Y. Hou (hereafter LCH) proposed that there is a natural negative feedback loop that would sharply reduce any warming effect of increasing atmospheric CO_2

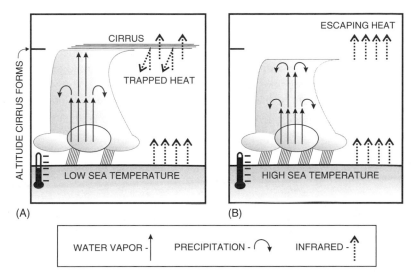

Figure 6.18   The iris effect: (A) iris closed; (B) iris open. The iris effect is the reduction of high-altitude cirrus clouds over seas with high surface temperatures. The opening of a gap—or iris—in the cirrus has a cooling effect since cirrus reflects infrared radiation back towards Earth. The suggested mechanism is increased precipitation efficiency over warm seas, which prevents water vapor from rising to the altitude where cirrus forms.

(Lindzen et al. 2001). Their model was suggested directly by observation. As LCH surveyed satellite data of cloud behavior over the western Pacific Ocean, they noticed patterns that made them wonder how cloud behavior was related to the temperature of the water below. When they correlated the satellite cloud data with sea surface temperatures (SSTs), they discovered that fewer high-altitude cirrus clouds form above ocean areas with higher SSTs. Since cirrus clouds have a net warming effect (see Figure 6.6), decreasing cirrus has a net cooling effect. Thus, seas with higher SSTs clear away insulating cirrus clouds above, which in turn cools the seas below. This process is like an iris that responds to ocean heat by opening to permit it to escape into outer space, so LCH called this model the iris hypothesis (IH). LCH declined to specify a detailed mechanism for IH, although they did *suggest* that the mechanism is increased *precipitation efficiency* in convection cores (or towers) over areas of higher SST. This causes more of the moisture in the rising air to be removed as rain, which in turn leaves less moisture to rise into the upper troposphere to be "detrained" to form cirrus, as illustrated in Figure 6.18.*

LCH's climate model indicated that for each degree that the sea surface warmed, the iris effect cooled the entire atmosphere by 0.45 to 1.1°C—a very powerful effect. Thus, the iris effect would either sharply reduce or entirely reverse the warming predicted by GWT. The iris theory immediately came under attack on a variety of

---

* Suggestions in the general direction of this mechanism had been made before by Ramanathan and Collins
  (1991, 1992).

fronts, including its methods,* the size and sign of the iris effect,† and the details of the iris mechanism.‡ Although some good points were made by its critics, IH remained viable, and the surprising discovery that there had been an unnoticed increase in infrared radiation escaping over the tropics (see Figure 6.7) seemed to be what the hypothesis predicted (Wielicki et al. 2002).§

The latest development at the time of this writing is the release of a study that agrees very nicely with IH. Spencer et al. (2007) employed "high time-resolution (e.g., daily) variations in the relationships (sensitivities) between clouds, radiation, temperature, etc.," (ibid., para. 4) calculated from satellite measurements in order to resolve the many uncertainties concerning cloud feedbacks in GWT. They studied

---

* In "No Evidence for Iris," Hartmann and Michelsen (2002, hereafter HM) attacked the methods of LCH. They focused on LCH's use of cloud-weighted SSTs, their central argument being that the reduction in cirrus is due to "latitude and longitude shifts" coupled to "meteorological forcing" which, they speculated, "seems to originate in the extratropics and is probably unrelated to tropical SSTs" (HM, p. 249). There are well-known differences in meteorological activity tied to differences in latitude in the region studied by LCH: There is more convective upwelling near the equator and more downward flows of this air returning to the surface at higher latitudes (the Hadley circulation), and then converging once again in the zone of upwelling. The resulting *intertropical convergence zone* has some longitudinal features (e.g., it is broader in the west Pacific, or monsoon basin, than it is in the Atlantic) but is mainly a function of latitude. HM's suggestion amounts to the idea that the reduction in cirrus insulation that LCH observe is tied to these meteorological patterns rather than to the underlying SSTs as such. Both are effects of a common cause—one is not the effect of the other. In other words, "the observational evidence uses a gradient with latitude as an analogy for climate change, which it probably is not" (NRC 2003, p. 34; the NRC authors cite HM's "No Evidence for Iris" at this point; Hartmann chaired the panel that authored this work). LCH (2002) responded that if HM was right, "we would expect a noticeable reduction of the effect when the poleward limit of the region considered was reduced. ... Rather, the opposite is observed" (ibid., p. 1346). When this region (the southwest Pacific between Australia and China) is reduced from a 30° swath each side of the equator to a 25° swath, the thinning of cirrus insulation over areas of higher SSTs is even more apparent than before. The map they provide of SSTs for the region (ibid., Fig. 3, p. 1347) may explain why: There is a greater span of SSTs along the east–west axis than along the north–south axis. So if there really is a connection between cirrus reduction and SSTs, reducing the north–south extent of the region considered would therefore reveal the connection more strongly, which the evidence indeed indicates.

† Lin et al. (2002) argued on the basis of their models that a small warming effect should result, and Chou et al. (2002) replied, admitting a possible 20% reduction in the cooling effect, but no more.

‡ Rapp et al. (2005) accumulated data to test the suggested mechanism of increased precipitation efficiency, and did indeed find a 5% decrease in the ratio of cloud area to rainfall over areas of higher SST (ibid., p. 4192). More rain was produced by the same amount of cloud as SSTs increased, just as the proposed mechanism for the iris required. Moreover, these data did not depend on the use of cloud-weighted SSTs, which were the focus of criticism by HM. Nevertheless, Rapp et al. did not see these data as confirming IH, since the increased precipitation efficiency did not occur high in convection towers but in low clouds. However, their methods did not rule out the possibility that the higher precipitation rate for low clouds included the lower level of convective towers, which would have been consistent with IH. Indeed, this result also agrees with other studies showing greater precipitation efficiency over higher SSTs, including a study LCH cites in support of IH and in which he also participated (Sun and Lindzen 1993, Lau and Wu 2003). So presumably greater precipitation frequency in low clouds is seen by LCH as consistent with IH. If it is, the work of Rapp et al. is actually in favor of IH after all.

§ Wielicki also expected that the rise in escaping infrared must involve cloud changes, in particular decreasing cirrus, as required by IH. Unfortunately, when he analyzed his data, he could not find this effect.

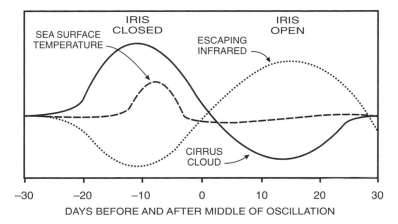

Figure 6.19   Evidence for the iris effect. Intraseasonal oscillations are periods during which tropical rainfall increases and decreases on about a 60-day cycle. A recent study of these oscillations revealed their link to Earth's radiation budget (Spencer et al. 2007). One unexpected consequence of the study was support for the iris hypothesis of Lindzen et al. (2001). As the figure shows, an increase in sea surface temperatures is, in fact, followed by the reduction in cirrus and increased cooling called for by the iris hypothesis. (Based on Spencer et al. 2007.)

daily variations in clouds, temperature, rainfall, SSTs, and radiation balance, which tend to fall into cycles of about two months, called *intraseasonal oscillations* (ISOs). A pattern emerged in the ISOs that agrees with IH. First a period of warming (in which incoming solar radiation outweighs outgoing infrared radiation) leads to higher SST, which then leads to increased convection, wind, cloudiness, and rain. After reaching a peak, this pattern reverses, the weather calms down, and the sea *cools* (see Figure 6.19). The data show that the main agent of this cooling is reduction of high cirrus cloud, just as IH predicts.

Given that this study relies on data from state-of-the-art satellites put into orbit in order to resolve the complex dynamics of cloud feedbacks, these results speak very loudly in favor of the accuracy of IH. Spencer et al. note that their data are "conceptually consistent with the 'infrared iris' hypothesized by Lindzen *et al.*," but caution that it is "not obvious whether similar behavior would occur on the longer time scales associated with global warming" (ibid., para. 18). The iris effect they have observed is very strong,* and no account of climate, including GWT, can be considered complete unless it takes it into account.

---

* Spencer et al. report "The sum of SW [shortwave] CRF [cloud radiative forcing] ($\approx -SW_{all}$) and LW [long wave] CRF ($= -[LW_{all} - LW_{circ}]$) plotted against the tropospheric temperature anomalies for the middle 41 days of the fifteen ISO composite (Figure 4) reveals a strongly negative relationship. A linear regression yields a sensitivity factor (slope) of -6.1 Wm^{-2} K^{-1}, with an explained variance of 85.0%" (ibid., para 20). By comparison, the effect of CO_2 levels rising by 100 ppm since 1750 is estimated to have an effect of 1.66 W/m^2. Note, however, that the CO_2 effect is global, whereas the iris effect has been observed only above the oceans between 30°S and 30°N latitude. This is, however, the area of highest solar heating efficiency (since the Sun strikes other areas more obliquely), and the effect *may* apply to oceans in general, not just the tropics.

Comparative Assessment. Our first criterion is *precision*, and neither model says anything very precise.* Thus, their *accuracy* is good, given that they are aiming at wide targets due to this general lack of precision. IH has an advantage with respect to the radiation budget data, which are problematic for GWT (No 3). Whereas GWT requires a net imbalance of incoming radiation over outgoing radiation, IH predicts the opposite when an iris opens. The surplus of outgoing radiation over incoming radiation that was measured and that surprised climatologists (Chen et al. 2002) shows that IH may be correct. Recent measurements by Spencer et al. (2007) are also in favor of IH. Moreover, IH has the potential to explain the absence of tropospheric warming that is so problematic for GWT (considered below, No 10) since irises would cool the troposphere more efficiently than the surface since they open just above the troposphere. So IH currently has an advantage when it comes to accuracy.

When it comes to *empirical consistency*, there is little basis for preferring one theory over the other. Both models are in their childhood, and their general lack of precision is due to the great difficulty of meteorology and climatology. As we noted in Chapter 5, climate is a complex, in fact chaotic, system, so we cannot expect the same levels of precision in this domain as in simpler physics or chemistry. Thus, the well-known empirical facts of simpler sciences such as physics and chemistry do not sufficiently constrain climate models to make one preferable to another. Both GWT and IH are venturing into uncharted waters.

However, the difference in *scope* between the two hypotheses is enormous: Whereas GWT is a model of a global process involving the entire climate system over a few centuries, IH is a model of a process over the tropical ocean over a period of a few weeks. Other things being equal, we prefer a model of broader scope, because it *says* more than a model of narrower scope—but by that same logic, a model of such massively ambitious scope as GWT has so much more that can go wrong than does a more modest model such as IH. Given the relatively lower level of accuracy of GWT, its broader scope is a liability that is not shared by IH. In other words, in terms of *simplicity*, IH is related to GWT in the same way that a lakeside cabin is related to a mile-high apartment building with a pool on the roof: The cabin is well within proven engineering capacities, while the apartment building is still only on the drawing board. So, as concerns simplicity, IH has a clear and distinct advantage over GWT.

(3) The Sun–Earth Climate Connection. It has been known for centuries that it tends to be warmer when there are lots of sunspots on the Sun. William Herschel (1738–1822) had noticed that the price of wheat in England went up when sunspots were scarce because the weather became cooler and damper. The British astronomer Walter Maunder (1851–1928), who first revealed the 11-year cycle in the numbers (and solar latitudes) of sunspots, also discovered that there was a period of sunspot inactivity from 1645 to 1715, which became known as the Maunder minimum. Before the Maunder minimum there was the Spörer minimum (1450–1550),

---

* AR4 (p. 65) specifies climate sensitivity as ranging between roughly 1 and 6°C, whereas a narrower range is given for the IH feedback factor of roughly −0.8 to −1.1 per °C rise in SST. So GWT is aiming at a wider target than IH and so is to that extent more likely to hit it. On the other hand, any strong negative feedback due to the IH mechanism will be taken as confirmation, so the difference between them on this score is more apparent than real.

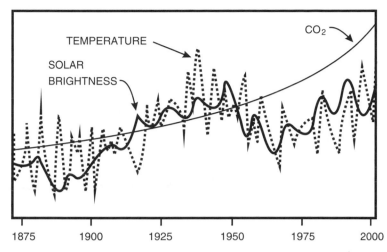

Figure 6.20   The Sun–Earth climate connection. Here we see a connection between Arctic temperatures and solar brightness that stretches over more than a century. By comparison, the correlation between temperature and CO_2 is very weak. (Based on Soon 2005.)

and after it there came the Dalton minimum (1790–1820). These three solar minima spanned the little ice age,* which brought poverty, famine, and disease to Europe as summers shrank and glaciers grew. Modern research into the solar temperature connection has confirmed that there is a connection: When there are few sunspots, the Sun is relatively inactive, and temperatures fall; conversely, when the Sun is more active and has more sunspots, temperatures rise, as in Figure 6.20 (cf. Friis-Christensen and Lassen 1991). As the graph shows, the Sun–Earth climate connection (SECC) is evident, but a CO_2 climate connection is not.† Baliunas and Soon (Baliunas and Soon 1995, Soon and Baliunas 2003, Soon 2005) have been trying for many years now to get SECC admitted into the debate on global warming, but have met rigid resistance from GWT scientists and personal attacks by GWT popularizers—in large part because they are solar astrophysicists, not climatologists.

It is quite remarkable, and clearly relevant, that the Sun–Earth climate connection holds for longer periods of time just as it does for shorter ones. If we go way back, 12,000 years ago, to the last ice age, we find that SECC is still in force, as shown in Figure 6.21.‡ At all scales, even millions of years ago, there is a close connection between solar activity and temperature (see, e.g., Shaviv and Veizer 2003). This is in

---

* There is no firm agreement on the dating of the little ice age. It began somewhere between 1300 and 1600, and ended around 1850.

† For an emphatic, indeed passionate, marshalling of the evidence, see Jaworski (2003) and Robinson et al. (2002).

‡ Periods of high solar activity are marked by lower levels of cosmogenic isotopes left behind in the environment. These isotopes are caused by cosmic rays colliding with Earth's atmosphere and surface; the levels of cosmic ray flux go up when solar activity, and hence the solar magnetic field, are low. By measuring the levels of these isotopes left behind in tree rings, soil, and their fossilized remains, we can determine solar activity levels in the distant past, before humans kept records, and compare these with temperatures.

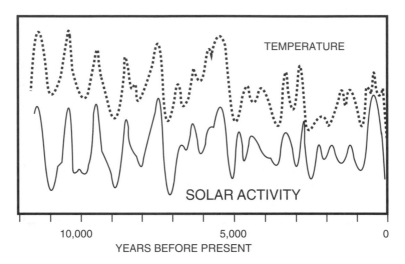

TEMPERATURE

SOLAR ACTIVITY

10,000   5,000   0

YEARS BEFORE PRESENT

Figure 6.21 Longer term Sun–Earth climate connection. The connection between solar activity and temperature goes back at least to the last ice age. (Based on Bond et al. 2001.)

stark contrast with CO_2 which, as we have seen in Figure 6.13, is not correlated with temperature over long periods of time.

Those who study this connection tend to come from the ranks of solar physicists rather than climatologists. They speak a different scientific language, so to speak, and are generally dismissed by the IPCC climatologists on the grounds that solar physicists are not climatologists. Surely we should not be trying to settle the epoch-making issue of global warming on the basis of scientific jurisdiction.* SECC draws attention away from the CO_2 connection that IPCC scientists wish to keep in the public spotlight.[†] There is a sorry history, here, of scientific infighting. Part of what makes modern science so strong is constant competition among scientists for recognition and funding. Unfortunately, the spirit of competition *can* overpower the need for cooperation in getting to the truth. Making matters worse, SECC had an Achilles' heel: No mechanism had been discovered which could explain, step by step, just how changes in solar activity change temperatures here on Earth. Scientists do not like to accept what they cannot understand. Thus, they have not been inclined to accept SECC as long as its mechanism remained a mystery.

Fortunately, Henrik Svensmark has made significant steps in solving the mystery of the SECC mechanism, despite fighting an uphill battle against the popularity of

---

* But if it does come down to jurisdiction, Earth's climate is *obviously* not a closed system. Factors from outside Earth's atmosphere, such things as comet or meteorite impacts, can and do affect our climate. The behavior of the Sun cannot therefore be ignored. The fact that the IPCC already takes variations of solar brightness into account concedes this point. The relevance of the Sun cannot simply be decreed to stop with brightness, but must be discovered by diligent observation. The SECC shown in the last two figures are precisely such observations.

[†] "The Danish delegation to the Intergovernmental Panel on Climate Change [IPCC] made a modest proposal in 1992, that the influence of the Sun on the climate should be added to a list of topics deserving further research. The proposal was rejected out of hand" (Svensmark and Calder 2007, p. 73).

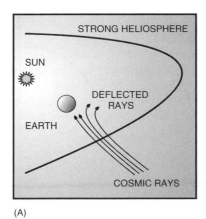
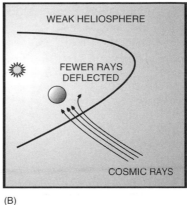

(A)   (B)

Figure 6.22   The Sun and cosmic rays. (A) When the Sun is brighter and more active, its magnetic field is stronger and deflects more cosmic rays away from Earth. (B) When the Sun is dimmer and less active, its weaker magnetic field lets more cosmic rays collide with Earth.

GWT.* To understand this mechanism, we must first realize that Earth is not in *outer* space, but in the *inner* space of the heliosphere: the extended solar atmosphere and magnetic field. Sunspots are marks on the Sun where its massively powerful magnetic field lines pierce through its surface and stream into the *inner* space in which we dwell. When there are lots of sunspots, the Sun's magnetic field is strong, as in Figure 6.22A, and this deflects many of the cosmic rays that would collide with our planet when the Sun's magnetic field is weaker, as in Figure 6.22B. Cosmic rays are particles (mainly helium nuclei) that have energies millions of times more powerful than those achieved in our most powerful accelerators.[†] Svensmark got a big break in the mystery of the SECC mechanism when by poring over satellite data he discovered the correlation between cosmic rays and low clouds shown in Figure 6.23.

---

* In the words of the eminent solar scientist who discovered the solar wind, Eugene Parker (whose own work was often rejected because it defied scientific orthodoxy), "Svenmark received harsher treatment [than I did] for his scientific creativity, and found it hard to achieve a secure position with adequate funding" (Svensmark and Calder 2007, p. viii). Such treatment is, sadly, often the lot of those scientists who break with scientific orthodoxy, even though this is a necessary condition of scientific innovation: "He is in good company, when we recall that Jack Eddy lost his job when he confirmed and extended the earlier work of Walter Maunder, who had pointed out that the sun showed a significant dearth of sunspots over the extended period 1645–1715" (ibid.). Confirming an extended absence of sunspots may seem a virtuous (or at least harmless) thing to do—unless, of course, it challenges the presuppositions of an established field. Climatologists have by-and-large presupposed that solar variations have no significant effect on climate, but "Eddy emphasised the important point that the Maunder Minimum was a period of cold terrestrial climate, thereby making the first direct connection of climate to solar magnetic activity" (ibid.). Given the work of Eddy and that of Parker himself, we may be dismayed that SECC continues to receive short shrift from the GWT orthodoxy. But let us not lose hope that justice will prevail: In 2003 Parker was awarded the $400,000 Kyoto Prize for Lifetime Achievement in Basic Science. Perhaps the importance of SECC will be recognized. Perhaps then some fraction of a percent of the billions of dollars of climate research funding may be directed to working out the physics of SECC, given that all parties agree it is crucially important to accurately predict future climate.

[†] Whereas human particle accelerators achieve energies around 10^{13} eV (electron volts), cosmic rays can have energies of 10^{20} eV.

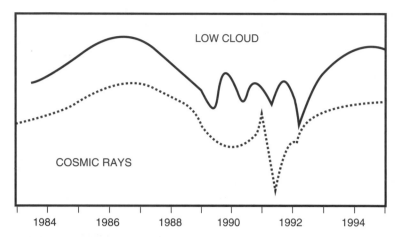

Figure 6.23   Cosmic rays and cooling clouds. Cosmic rays influence the amount of low cumulus cloud, which has a strong cooling effect. Thus, when solar activity is high and Earth's cosmic ray levels are reduced, there is less low cloud, and Earth warms as a result. Conversely, when solar activity is low, there is more low cloud, and Earth cools. This effect obtains mostly at low altitudes, where the cosmic ray flux is thinned by the overlying atmosphere to the point that the reduced ionization produced by cosmic rays is a limiting factor on cloud formation. (Based on Marsh and Svensmark 2000.)

As we saw earlier in this case study, low clouds have a cooling effect: Because they are so bright and white, they reflect lots of incoming solar radiation back into space while reflecting a smaller amount of infrared radiation back to Earth. So when the Sun is inactive, the solar magnetic field weakens, more cosmic rays rain down on Earth, more low clouds form, and Earth is cooled. This crucially relevant result was denied publication for some five years after it was discovered, due to stubborn resistance from IPCC and likeminded scientists (Marsh and Svensmark 2000).* And once published, these results were still completely rejected by the IPCC because they did not specify a mechanism.†

* When Svensmark and his colleague Eigil Friis-Christensen reported their results at a meeting of the British Royal Astronomical Society in 1996, the Chair of the IPCC, Bert Bolín, said to the media: "I find the move from this pair scientifically extremely naïve and irresponsible" (Svensmark and Calder 2007, pp. 73–74). They tried to publish the results in *Science*, but "queries came back. When those were dealt with by brief additions, the verdict was the paper had become too long" (ibid., p.71).
† The tension between theory and observation has a long, vexed history. From the time of Empedocles (492–432 B.C.E.) through to the time of modern medicine, doctors were identified as either empiricists ("empirics") or dogmatists (also known as "rationalists"). Empiricists would prescribe medicines and treatments on the basis that they were known by experience to work, even though there was no acceptable theory of their action in terms of the four humors: black bile, blood, yellow bile, and phlegm (which were the biological forms of the four elements Earth, air, fire, and water, respectively). In retrospect, empiricism is the clear winner: Observation trumps theory. You were better off taking a folk remedy for your illness that was known to be effective by observation than being bled to reduce the amount of air in your body. Dogmatic biologists insisted that men had more teeth than women (because they were stronger, etc.) rather than looking to see. When Galileo showed that a light lead weight falls as fast as a heavy one, he was suspected of trickery or witchcraft, since reason "proved" that heavy bodies fall

Svensmark has persisted, and has now traced the steps whereby cosmic rays increase low clouds. Cosmic rays liberate electrons by knocking them off air molecules. These free electrons then cause a cascade of chemical reactions, resulting in the formation of *cloud condensation nuclei*, the microscopic particles that are required for water vapor to form the tiny droplets that make up clouds.* Thus, cosmic rays aid the formation of low-altitude cooling clouds. In the upper atmosphere there are always sufficiently high levels of ionization, but at lower levels, where fewer cosmic rays penetrate, the level of ionization is the controlling factor, and thus it is the formation of low-level clouds, cumulus, that is sensitive to the level of cosmic rays. Although Svensmark and his colleagues have faced continuous resistance in getting funding for the necessary research and in getting the results of their research published,† the first results for the mechanism have now appeared (Svensmark et al., 2007).

Comparative Assessment. Both theories have a very broad *scope*, because they both apply to global temperatures since the formation of the atmosphere in roughly its present form hundreds of millions of years ago. Both theories are *empirically consistent* with other scientific theories. When it comes to *precision*, because the rate of cloud formation depends on many factors, there is not a simple one-to-one relationship between cosmic ray levels and low cloud levels. This makes it difficult to predict precisely how much cloudiness will be caused by a given level of solar activity and cosmic ray flux. So SECC has not, so far, provided precise predictions. As we have seen, GWT, too, does not make precise predictions. Thus, the two models are roughly equal as far as precision is concerned, just as they are for scope and empirical consistency.

When it comes to *accuracy*, however, SECC has a clear advantage. While solar activity levels and CO_2 levels both track global temperature, SECC does so with much more accuracy, range, and detail, as seen in Figures 6.20 and 6.21. In addition, three sets of data that are barriers for GWT are springboards for SECC. The fact that the rate of surface warming is higher than the rate of tropospheric warming disconfirms GWT (see below), whereas this is just what would be expected with SECC. Since SECC says that global warming starts at the surface when incoming sunlight increases due

faster. Modern scientists claim to be empiricists, but dogmatism still has a stronger influence in some cases (e.g., the rejection of meteorites by the French Academy). The case of the SECC seems to be one of these: Rather than being excited by this important discovery and encouraging its investigation and development, a dour orthodoxy among some climate scientists has hindered, and is hindering, its mere recognition.

* The electrons attach themselves to oxygen molecules. These electrically charged oxygen molecules, or ions, attract a number of water vapor molecules, which, given the usual background levels of ozone, react to form ozone ions. The ozone ions are attached to water molecules, and in the presence of background levels of sulfur dioxide, react to form sulfite ions. The sulfite ions then react with water vapor to form ionized molecules of sulfuric acid, which are cloud condensation nuclei. These condensation nuclei attract more water molecules and so form the larger droplets of water of which clouds are made.

† Various accounts of this resistance, sometimes directly by IPCC officers, as well as the intercession of concerned scientists on his behalf, are now circulating on the Internet (e.g., http://www-tc.pbs.org/moyers/moyersonamerica/green/isanewsletter.pdf; http://www.canada.com/nationalpost/story.html?id=fee9a01f-3627-4b01-9222-bf60aa332f1f&k=0), but perhaps the best account is Svensmark's own (Svensmark and Calder 2007, esp. pp. 99–131).

to reduced levels of cumulus, temperature would rise most at the surface; the extra heat would then escape through the troposphere back into space, producing lower temperature increases at higher altitudes. This, in turn, would solve the radiation budget problem (see No 3), since the extra incoming solar radiation at the surface would cause higher levels of outgoing infrared at the top of the atmosphere. There is even the potential to explain the Antarctic anomaly: since the Antarctic has such high albedo to begin with, it will be relatively unaffected by the albedo effect of cumulus levels.

When it comes to *simplicity*, SECC again has a clear and distinct advantage over GWT. The mechanism of SECC is much simpler than that of GWT. As we have seen, GWT is an extremely complex theory: Its very definition, not to mention its fate, depends on numerous factors that are poorly understood, poorly measured, and inadequately modeled (water vapor feedbacks, cloud feedbacks, aerosols, parameterization, etc.). SECC, by contrast, concerns only the sensitivity of cumulus clouds levels to cosmic ray levels, which we already understand to be controlled by solar activity levels. SECC proposes that global temperature is controlled primarily by this simple mechanism, with minor variations (noise) added by the chaotic intra-atmospheric processes that GWT hopes to model. Conversely, GWT proposes that global temperature is controlled primarily by the chaotic intra-atmospheric processes, with noise added by the solar processes. So GWT must model the chaotic processes even to achieve a mature and testable status, whereas SECC need only model and test the simpler cumulus formation mechanism.

On balance, SECC seems to be well ahead of GWT. We must note, however, once again, that climate science is still in its early days, and that it is premature to declare winners and losers at this point in time. It is just too soon to tell. However, we may legitimately conclude that SECC has as much claim to scientific respect, attention, and research funding, as does GWT. There is no justification for its ongoing marginalization.

Yes 10: GWT Is the Best Theory Overall. All theories must face some negative evidence, and as you say, all theories must face competing theories. But as you yourself have just admitted, none of the competitors to GWT have anything like its scope and depth of development. There are other theories about this or that climate mechanism, but only GWT is a fully fledged theory of climate. None of the contenders are strong enough to dislodge GWT from its position as the consensus among knowledgeable scientists. So unless and until there is some evidence showing that GWT is clearly mistaken, the wisest choice on scientific grounds alone is to accept it.

No 10: Failure of Troposphere Warming Falsifies GWT. The failure of the troposphere to warm as GWT requires falsifies GWT. If this datum were to be generally understood by the global public, humankind would breathe a collective sigh of relief as it tossed GWT onto the heap of false prophecies that have plagued our history. The IPCC, with the help of massive media attention, has told the global public of the *threat* of global warming and claimed that recent warming trends *prove* that the threat is real. Unfortunately, the evidence provided to the public extends only to *surface* temperature, which are rising in agreement with GWT—assuming that the rise is not caused by the urban heat island effect, or natural variation, and ignoring the fact that temperatures have not risen since 1996. The public has not been informed that *tropospheric* temperature trends are the opposite of what GWT predicts. Temperatures in the troposphere (the lower layer of the atmosphere, from the surface to 50,000 feet or so*) are rising more slowly than surface temperatures, which is *impossible* if GWT is right. According to GWT, temperatures must rise more quickly in the troposphere than on the surface. This is another fingerprint of global warming. So the data (if correct) showing that they are not doing so refute GWT. Perhaps the IPCC is afraid that any trace of good news might undermine the public's resolve to cut back on carbon emissions or else purchase carbon credits to pay the penalty for their misdemeanor.

Sadly, the role of tropospheric temperature in GWT is a bit complex and so is not apt to be swiftly appreciated by the global public. Happily, we can put the basic idea in simpler terms, so that the failure of GWT can be appreciated more broadly. Before we do, however, it is crucial to realize that it is GWT itself that predicts that the troposphere will warm faster than the surface. All of the various models of global warming assume that the middle troposphere warms faster than the surface layer.[†] No one, including the proponents of GWT, disputes this point.[‡] Indeed, the IPCC itself affirms this point since it is nothing other than the *central mechanism* of GWT (AR4, pp. 265–271).[§] In its simplest terms, GWT claims that GHGs in the

---

* The layers of our atmosphere are defined in terms of pressure, and thereby ultimately in terms of the mass of the atmosphere in each sphere. Thus, the *troposphere* is defined as the layer of the atmosphere which has a pressure of at least 100 millibar. Since the sea-level surface pressure is usually very close to 1000 millibar, the troposphere is the bottom 90% of the atmosphere by mass. At lower pressures (below 100 millibar) and hence higher altitude lies the *stratosphere*.

[†] GWT models, as well as various calculations, show that the maximum warming should occur at an altitude of about 6 miles (10,000 m).

[‡] For example, Gaffen et al. (2000), NAS (2001, pp. 15–17), Lindzen and Giannitsis (2002), Lanzante et al. (2003), Santer et al. (2003).

[§] Although AR4 never comes right out and states that tropospheric warming must be greater than surface warming, this is presupposed by the discussion in the sections cited below, which are dedicated to arguing that despite numerous scientific data and publications showing that surface temperatures are rising more quickly than tropospheric temperatures, "reanalyses" (pp. 269–270) of these data "*largely* resolves a discrepancy noted in the TAR [Third Assessment Report, (IPCC 2001a)]" (p. 36; my italics). While not mentioning statements by such august scientific bodies as the National Academy of Science and the National Research Council, the IPCC argues that "the satellite tropospheric temperature record is *broadly* consistent with surface temperature trends," as claimed earlier (p. 36). This argument is tendentious in at least two ways: (1) It discounts some data sets, plays up others, and uses the GWT-preferred ranges of its own error estimates; and (2) it downplays the absence of tropospheric warming as a minor matter, assuming that "largely" avoiding outright falsification, and "broadly" attaining consistency with fundamental thermodynamics is good enough for the purposes at hand. Nothing could be further from the truth. Merely dodging sudden death hardly shows GWT to be a good basis for redesigning the human economy. If the IPCC held the welfare of humankind uppermost, it would welcome the possibility of such wonderful news as that GWT may be wrong. Instead, the IPCC (2007) buries the issue in a technical discussion that attacks the data and blurs the problem. This evidences a tendency of the IPCC to take its *advocacy* of GWT more seriously than it takes pursuit of the truth. As any good scientific team will do, the scientific team of the IPCC is busy building and developing a complex scientific theory, and so almost by definition professes and promotes the theory it is building. But usually, this promotion occurs solely within the scientific community, in full expectation of a critical response. However, the advocacy

(Continued)

troposphere will trap heat, and it is this heat trapped above us that will make us warm here at the surface.* The causal sequence of GWT is very clear and the very essence of the theory itself: Tropospheric warming will cause surface warming. Fortunately for the Earth and all of us living on it, whether plants or animals, the very opposite is happening: surface warming is causing tropospheric warming. We can all breathe a sigh of relief—or we could, if only we were allowed to hear the good news.

The role of tropospheric warming in GWT is most easily understood by means of the simple thermodynamic concept at its core. In considering this concept we do not want to presuppose that it is correct.† The point, rather, is to see just what it is that GWT is saying. We can begin with the concept of thermodynamic equilibrium, as presented in Figure 6.24. This figure pictures the concept of thermodynamic equilibrium itself. An object in thermodynamic equilibrium, represented by the center box, maintains a constant temperature because the heat flowing into it is precisely balanced by the heat flowing out. In the simplest case, heat flow in and out is reduced to zero: If the object is perfectly insulated, its temperature will not change. However, if some heat leaks out of the object, its temperature can be kept constant by replacing the lost heat, and if some heat leaks into the object, its temperature can be kept steady by removing the heat gained. The heat may flow by conduction, radiation, convection, or whatever other process will transfer heat, and the principle still applies: If heat input matches heat output, the temperature remains constant. Note that heat only flows from things with higher temperature to things of lower temperature, and that the rate of heat flow

(*Continued*)   approach of the IPCC toward GWT is unbalanced when it comes to the establishment of policy, which is most unfortunate given the *enormous* significance that any scientific error concerning GWT may have for human welfare. It may be that IPCC's *Global Warming Program* is seen as good for the environment, even if it is not good for humankind, and this may soften critical scrutiny of GWT by the IPCC. If that is true, it would mean that the IPCC scientific team is not engaged in value-neutral science but in advocacy environmental science as defined in Chapter 5.

* A concise statement of this aspect of GWT can be found in the definition of the atmospheric greenhouse effect provided by the website for the Central Equatorial Pacific Experiment (CEPEX): "The atmosphere, primarily water vapor and CO_2, absorbs most (70 to 95%) of the surface long-wave radiation and re-emits [it] to space at the much colder temperature of the atmosphere. The effect is to reduce OLR [outgoing longwave radiation, or infrared]" (http://www-c4.ucsd.edu/cepex/index.html, October 10, 2007). This is a description of greenhouse warming in general, the natural phenomenon that we are supposed to be enhancing by adding GHGs to the atmosphere. "Ga [the atmospheric greenhouse effect] then represents the energy trapped by the entire atmospheric column between the surface and the top-of-the-atmosphere. *The atmosphere then heats the surface by emitting absorbed energy back to the surface*" (ibid., my italics). The causal sequence begins with warming of the troposphere, and ends with warming of the surface.

† GWT reduces Earth's climate to heat movement by radiation transfer, and so omits the convection, wind, and weather that actually make up the system. All of the actual mechanisms of climate are reduced by parameterization to their radiation effects in order that they can be handled by computer models. This reduction provides the maneuvering room required to exaggerate the threat of global warming to the point where it has captured global attention. Earth's atmosphere is a horrendously complex system, and without simplification, climate science would be impossible. Good simplifications are precisely what is needed for climate science—although scientists generally recognize that it is very difficult to leave out only what does not much matter and include all that does. Scientists should be, and usually are, the first to warn that simplification is hazardous. This hazard looms large in the current global enchantment with the prophecy of global warming.

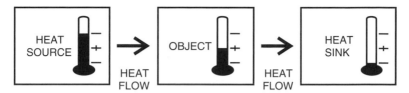

Figure 6.24   Thermal equilibrium. Any object (center) that stays at the same temperature is in thermal equilibrium and must have equal amounts of heat flowing in and flowing out (even if the amount is zero for both). For heat to flow (arrows), the heat source (left) must be at a higher temperature than the object, and the object must be at a higher temperature than the heat sink (right). Other things being equal, the rate at which heat flows from one object to another is proportional to the temperature difference between them.

is proportional to the temperature difference between them. For example, an insulator slows the flow of heat by warming up on the side of the warm object and thereby reducing the temperature difference between itself and the object—something we have all experienced when we wrap ourselves up with a blanket on a cold night.[*] The object itself could be anything, from a simple rock to the planet as a whole. The thermodynamic principles apply no matter how large or small, simple or complex, the system may be. If we draw an imaginary box around anything at all, this principle applies to whatever is inside.

It follows that there are only two ways that something can warm up: Either the heat input increases or the heat output decreases, as shown in Figure 6.25. GWT uses the second of these two possibilities.[†] The important thing for the argument about to follow is that we see that the object in the center of panel B cannot warm up any faster than the object to its right. If it did warm up faster than its heat sink, we would know that it must be getting some extra heat input from somewhere. GWT says that global warming is under way right now[‡] and that on the planetary scale, Earth is like the center box in panel B, while the troposphere is like the right-hand box.

Moving in a little closer, we can see the mechanism in more detail, as pictured in Figure 6.26. In panel A we have taken a vertical section of the atmosphere and

[*] We can think of the surface where one's skin meets the blanket: Heat will flow through the surface only if it is warmer on one side than on the other, and the rate will be proportional to the temperature difference between the two sides. When you first wrap the blanket around you, assuming that it has not been warmed up), it will feel as cold as the surrounding air. You begin to warm up only as the blanket warms up, and it warms up from your own body heat. The blanket slows down the flow of heat from your body; the trapped heat warms the blanket; the blanket then warms you. As we shall see, something similar happens with the atmosphere: Heat flowing from the surface layer is slowed down by the troposphere; the troposphere warms up; the troposphere then warms up the surface layer.

[†] Of course, as soon as the center object warms up, this changes the heat flow some more, a complication that has been left out of this figure for simplicity's sake. As soon as the center object warms up, this will slow its heat input and increase its heat output, thus warming its source and warming its sink as well. These effects have been left out in order to distinguish the two possibilities more clearly.

[‡] According to the IPCC, the globe has warmed by about 1°C since 1750, due to GHGs added by human beings, which are now raising the temperature by about 0.2°C per decade. (The rate of warming will be important in what follows.)

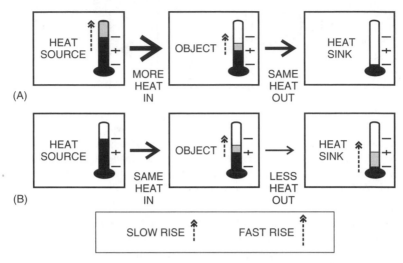

Figure 6.25 Two ways to warm up. To warm up an object, you must either (A) increase its heat input or (B) decrease its heat output—or some combination of the two. Other things being equal, this requires either that (A) the temperature of the source goes up or (B) the temperature of the sink goes up. According to the global warming theory, warming will occur by means of process B: a decrease in heat output. Note that in process B the object cannot warm up any faster than the sink.

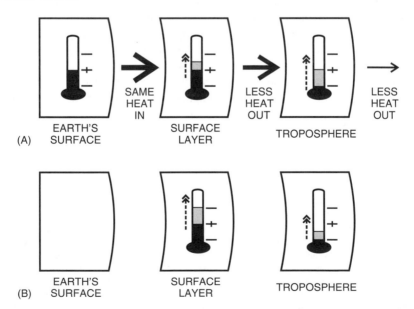

Figure 6.26 GWT is contradicted by the data. (A) What GWT predicts: The addition of GHGs slows down the escape of heat through the troposphere, causing its temperature to rise, and the temperature of the surface layer to rise as a result. Note that surface layer temperature cannot rise any faster than troposphere temperature, as we saw in Figure 6.25B. (B) What is actually happening: Surface layer temperature is rising faster that troposphere temperature, contradicting GWT.

have laid it out sideways to show how the central mechanism of GWT is supposed to work. On the left we have the *surface* of the Earth. This includes every point at which the atmosphere touches the ground, oceans, ice caps, forests, fields, rooftops, or city parking lots, for these are the points at which light from the Sun is converted into infrared and radiated back up into the atmosphere. Most atmospheric heat starts at the surface, as we have noted earlier. In the center we have the *surface layer*, the layer a few meters above the ground. It has no precise depth, but it is at center stage, since it is where we live—and not just us, but all of the birds and bees and grass and trees as well. The warming that is referred to by the phrase *global warming* or *climate change* (in IPCC 2007c the second phrase is preferred) is the warming of the surface layer.

On the right we have the *troposphere*, which is the mechanism of the warming. The troposphere is the glass in the greenhouse metaphor that is supposed to be trapping heat in the surface layer below. The rising levels of GHGs, in particular CO_2, slow down the transport of heat through the troposphere. Because of increasing GHGs, infrared radiation from the surface layer is more likely to be absorbed by a GHG molecule than before. These molecules soon reradiate their heat as infrared once again, and it is reradiated equally in all directions, warming the troposphere. Thus, heat escapes more slowly from the surface layer,[*] warming us up and giving us higher temperature readings in our weather reports.[†]

The problem for GWT, as we see in panel B of Figure 6.26, is that *just the opposite is happening*.[‡] If the warming predicted by GWT had been happening, the troposphere would have been warming faster than the surface. But the data indicate just the opposite: The surface has been warming faster than the troposphere. The National Academy of Science describes the situation as follows: "Although warming at Earth's surface has been quite pronounced during the past few decades, satellite measurements beginning in 1979 indicate relatively little warming of air temperature in the troposphere" (NAS 2001, p. 17). That summarizes the raw data,

---

[*] The IPCC says that since 1750, CO_2 levels have increased from 280 ppm to 379 ppm, which has increased the infrared reradiating down from the troposphere by 1.66 ($\pm$ 0.17)W/m². Thus it is assumed that the troposphere has warmed. This in turn is supposed to have warmed the boundary layer by about 1°C.

[†] Lindzen (1999, 2007) provides a more theoretically elegant model of GWT, which is nevertheless quite intuitive. His model focuses on the reduction of transparency of the atmosphere to infrared radiation as a result of GHGs. The *characteristic emission level* (CEL), the altitude of one optical depth (in the infrared) of the atmosphere from above, is taken as the surface of interest for radiation balance. As far as infrared radiation is concerned, this level is the effective surface of the Earth, and its temperature is Earth's effective emission temperature. Because GHGs make the troposphere less transparent, they raise the CEL (a doubling of CO_2 would raise it by about 50 m; Lindzen 1999, p. 104). Thus, the CEL is at a lower temperature than before, since temperature varies inversely with temperature. Because radiation intensity falls with temperature, outgoing infrared radiation is slowed at the CEL, while incoming solar radiation remains constant, resulting in global warming. Thus, the mechanism of the warming is still a slowing of outgoing infrared radiation by the troposphere (the altitude of the CEL varies with latitude and local conditions, but is usually near 500 hPa, or about 5.5 km). Eventually, the atmosphere will warm up until the temperature at the new higher CEL is virtually the same as it was at its lower altitude, causing tropospheric warming.

[‡] See, e.g., Gaffen et al. (2000); NRC (2000); NAS (2001); Lindzen and Giannitsis (2002); Lanzante et al. (2003, especially Fig. 4); Christy and Norris (2004); Seidel et al. (2004).

but the important issue is the relationship between the rates of change on the surface and in the troposphere: "The committee concurs with the findings of a recent National Research Council report (NRC 2000), which concluded that the observed difference between surface and tropospheric temperature *trends* during the past 20 years is probably real . . .*" (ibid., my emphasis). Thus it is the fact that surface temperatures are rising so much faster than tropospheric temperatures that poses the problem for GWT. Indeed, the NAS summarizes the problem in stark terms: "The finding that surface and troposphere temperature trends have been as different as observed over intervals as long as a decade or two is difficult to reconcile with our current understanding of the processes that control the vertical distribution of temperature in the atmosphere" (ibid.). Once again, we see that climate science is still full of surprises.

The document then goes on to discuss possible sources of surface heat, possible mechanisms of heat storage in the oceans and elsewhere, possible errors of measurement, and possible fundamental errors in modeling. In short, GWT did not explain surface warming. So it is not a simple matter of GWT models having failed to accommodate a phenomenon we are now observing, which would be bad enough. Instead, it is a matter of a failure of the central concept upon which these models are based.

This really should put an end to the prophesized threat of GWT.

Yes 11: Newer Data Have Resolved the Tropospheric Warming Problem. The U.S. Climate Change Science Program (with the blessings of the IPCC) has quite responsibly encouraged a full scientific investigation of the problem you outline, and the results have exonerated GWT (USCCSP 2006). This investigation found errors in the data upon which your objection is based, and by correcting them has shown that measured tropospheric temperatures fall within the ranges of our newest climate

---

* The passage continues as follows: "as well as its cautionary statement to the effect that temperature trends based on such short periods of record, with arbitrary start and end points, are not necessarily indicative of the long-term behavior of the climate system." Since both the NRC and the NAS conclude that the measured trends are probably accurate, indeed that the possibility of their inaccuracy would be "difficult to reconcile with our current understanding," I have removed it to make the conclusion of their analysis of GWT as concerns tropospheric warming perfectly clear, but include it here in the footnotes for its interest to scholars. It is important, however, to recognize that AR4 does indeed attack the data (AR4, pp. 267–271), and that this is its sole response to the problem identified by NAS and NRC. One cannot help but be struck, however, with the fact that despite IPCC's reinterpretation of the data, their graphs of temperature change (Figure 3.17, p. 268 and Figure TS.7, p. 36) still show an increase in surface warming (0.5°C from 1957 to 2005, or 0.1°C/decade) which is *greater* than the increase in tropospheric warming (0.3 °C from 1957 to 2005, or 0.06°C/decade)–leaving the problem as such untouched. So, even after reanalysis, surface temperatures are increasing faster than tropospheric temperatures (indeed, *twice* as fast) according to the IPCC itself (see also Figure 3.18, p. 269)–which contradicts GWT. Just as interesting (especially in light of the discussion in Chapters 5 and 6) is the fact that when it comes to the Technical Summary, where the case for GWT is summarized *for policymakers* with the intent of making them act on its conclusions, the warming trend over the last 50 years is said to be even larger (0.128°C/decade, Figure TS.6, p. 37). As far as IPCC is concerned, the trend of surface warming is larger when it comes to policy than when it comes to its discrepancy with the lower trend of tropospheric warming.

models (ibid.; see Figs. 3 and 4, pp. 12 and 13, for a summary of these results). The USCCSP results have been included in the latest IPCC assessment report (AR4), in which it is concluded that GWT has been proven sound. So the tropospheric warming problem has been resolved, and we must begin the process of eliminating CO_2 emissions, which, although painful, is the least painful path for both the environment and humankind in the long run.

No 11: Tropospheric Data Do Not Resolve the Tropospheric Warming Problem. As the recent history of climate science has shown us, climate data itself can be full of surprises. The USCCSP's "intensive efforts to create new satellite and weather balloon data sets" (ibid., p. vii) to resolve the discrepancy between GWT and the data has borne fruit: The "significant discrepancy no longer exists because errors in the satellite and radiosonde data have been identified and corrected" (ibid., p. 1). The data itself has given way, and thus GWT has been saved from sudden death. However, the USCCSP report does *not* claim to exonerate GWT, only to save it from outright falsification. All it claims is that the estimated error ranges of the reinterpreted data do partially overlap the error ranges of new models of global warming (ibid., see Figs. 3, and 4, pp. 12 and 13), at least when averaged globally.

The mechanics of warming at any particular place on the globe still remains a problem, however. In particular, the mechanics described by GWT fail for the all important tropics. If GWT does not work for the tropics,* it does not work, period. The USCCSP clearly states that the failure of tropospheric warming still applies for the tropics. "Comparing trend differences between the surface and the troposphere exposes potentially important discrepancies between model results and observations in the tropics. In the tropics most observational data sets show more warming at the surface than in the troposphere, while almost all model simulations have larger warming aloft than at the surface" (Ibid., p. 10). In other words, Figure 6.26 still applies in the all-important tropics: GWT models call for warming caused by decreased heat outflow, while the data still indicate an increase in heat inflow at the surface. So GWT does not work for the tropics—so it does not work, period.

Yes 12: Troposphere Warming Has Been Observed, So GWT Is Proven. As the previous argument itself presupposes, the absence of troposphere warming, which had been such a matter of concern for the NAS and NRC, has proven to be a matter of data errors. We should be reassured by the fact that the scientific community has behaved in such a responsible fashion by holding up GWT to the highest standards, and chastened by the fact that their investigations have shown it to be quite real.

No 12: The Relative Rates of Troposphere and Surface Warming Still Falsify GWT. It is the relative *rates* of warming that matter, not the absolute amounts. As the USCCSP report itself states unequivocally in the passage quoted above, GWT

---

* The IPCC defines this as the 40° band over the equator, over which the bulk of solar radiation flux is absorbed by the Earth.

models require that the troposphere warm *faster* than the surface, but the data show that it is warming *slower* than the surface.* So GWT is disconfirmed decisively by these crucial data.

Yes 13: Even If the Rates Are Wrong, GWT Is Still Broadly Consistent with the Data. The latest IPCC report explicitly takes the results of the USCCSP investigation into account,* and it says, quite accurately: "It appears that the satellite tropospheric temperature record is broadly consistent with surface temperature trends" (AR4, p. 237). As long as GWT is consistent with the data, it should be accepted.

No 13: Even If We Ignore the Inadequate Rate of Troposphere Warming, GWT Is Reduced to Insignificance by the Data.[†] An interesting fact lies hidden in the data summary of the USCCSP and adopted by the IPCC.[‡] In the latest GWT models, the surface warms up at about two-thirds as fast as the troposphere.[§] Since we know how fast the troposphere has been warming, we can calculate the amount of surface warming that can be attributed to global warming itself, and it turns out to be only one-third of what GWT calls for.[‖] So even if GWT is right, the actual warming that has occurred *because of rising GHGs* is insignificant (a mere 0.6° by 2100), *less than one-third* of the warming predicted by GWT.[#] If we take the figures of a study better designed as concerns the relationship between surface warming and troposphere warming (Lee et al. 2008), surface warming is between one-third and one-half of troposphere warming, which gives an even lower figure of 0.3 to 0.45° by 2100. If we add in the lack of warming in surface temperatures since 1996, this number will be reduced by another third or so. Sea-level rise and other "catastrophic" predictions must be scaled back accordingly. Life on this planet, including human life, has handled similar climate changes in most of the centuries of its history. So even if GWT is right, the actual warming that has occurred *because of rising GHGs* is insignificant—a nuisance, not a catastrophe.

---

* The crucial "surface minus lower troposphere" trends (USCCSP 2006, p. 13) show that GWT models predict on average that the troposphere will warm faster than the surface by about 0.07°C/decade, whereas observations show that the surface is warming faster than the troposphere by about 0.05°C/decade, just the opposite of what is predicted.

* See especially Section 3.4.1, "Temperature of the Upper Air: Troposphere and Stratosphere" (AR4, pp. 265–271).

[†] I am indebted to Lindzen (2007) for the logic of the following argument, although it uses USCCSP's own data instead of Lee et al.'s (2007) data.

[‡] (USCCSP 2006, p. 13) and (AR4, pp. 265-271):

[§] A little more precisely, the models' 1979–1999 mean tropical warming trend for the troposphere is about 0.22°C/decade, while their mean tropical warming trend for the surface is 0.15°C/decade (USCCSP, p. 13).

[‖] Since the troposphere has been warming at about 0.09°C/decade, the amount of surface warming that can be attributed to global warming itself is about 0.06°C/decade. This figure is *less than one-third* of the warming predicted by GWT (0.20°C/decade).

[#] Here I am taking the very modest GWT prediction of a warming of 2.0°C by 2100. The rate of temperature increase should also trend upward during the coming century, but not by much according to most models or according to most observations.

By the way, the argument above does not in any way imply that GWT has been shown partially correct by the USCCSP report. Indeed, the data reported, even after corrections, is still in stark disagreement with what GWT requires, as shown above. But if we are to get into "even if" arguments, then *even if* some of the warming between 1975 and 1995 was the result of added GHGs, this effect has been insignificant. There is no reason to reduce and eliminate CO_2 emissions. The apocalyptic prophecy is false, and we do not have to sacrifice our economy.*

---

* As this book goes to press (November 2008), news is gradually emerging from the ARGO project. This project is the first to provide information with sufficient range and accuracy to determine whether or not GWT is supported by changes in ocean temperature. By using hundreds of buoys that drift on ocean currents taking temperature (and other) readings at various depths and transmitting them by radio for collection by satellite, ARGO finally provides us a dataset of ocean temperatures that is both uniform in instrumentation and broad in scope both horizontally and vertically (http://www.argo.ucsd.edu/). Data that is accessible to the general public includes the following: "For the period since Argo achieved global coverage, 2004–2008, there is no significant trend in the globally averaged temperature" (http://argo3000.blogspot.com/2008/08/how-much-have-ocean-temperatures.html, September 11, 2008, posted by Argo TC on August 5, 2008). Some unofficial reports say that ocean temperatures are in fact slightly down (e.g., Gunter 2008), a fact hidden by the "no significant trend" rubric. Once again, this datum is both surprising and contrary to what GWT predicts. According to GWT, temperatures should be rising constantly for the oceans, just as they are predicted to do for the atmosphere. In fact, this is not happening. The response of GWT supporters to the fact that *neither the oceans nor the atmosphere are currently warming* has been to retreat to the natural variations plus human forcing model, which as we have seen above, renders GWT untestable (see Yes 2 and No 2, above). Unless natural variations are predicted, there is no way to tell which portion of the current temperature is due to human forcing, and we can only take it on faith that even though temperatures are falling, they *would* have been even lower without human forcing. In fact, GWT modelers claim to be able to predict natural variations through their multiple feedback models. However, these models predict a steady rise in temperature for both oceans and atmosphere. This is contradicted by the current data, which instead is in agreement with the absence of tropospheric warming just discussed.

7 Nature and the Sacred

Praise be to You, my Lord, through our Sister, Mother Earth, who sustains us and governs us.

—Saint Francis of Assisi

One of the first things you miss in the wilderness is the bathroom. My brother John and I were camped on a glacier below the peak of Mount Hector when in the middle of the night the normally simple and civilized business of urinating loomed as a daunting, hydraulic necessity. I had put off leaving the warmth and safety of my sleeping bag as long as I could, but now I had to go outside into the cold and dark. I knew that many a climber had died because putting on boots and crampons seems too difficult in urgent moments such as these. They had ventured outside with only boot-liners on their feet, forgetting just how steep and slippery a glacier is, forgetting that once they begin to slip there is no stopping, forgetting life itself as they smashed into a rock or crevasse down below. As I extricated myself from my sleeping bag inside the cramped little tent, I reminded myself that confronting my mortality was one reason I was there—and then promised myself not to go outside in just my boot-liners. I stuck my head out, searching for my boots outside on the glacier where I had left them, trying not to wake my brother.

Our sort of mountaineering had nothing to do with the strategically planned, testosterone-fueled "assaults" on summits that fill the pages of alpinist novels and extreme sports magazines. We had instead set out from the highway, now many miles away, with no "plan of attack." We had taken our time walking through the broad, increasingly steep forests on the flanks of the mountain, stopping often to admire the terrain or ponder the route ahead. We had free-climbed the first set of cliffs, the ramparts that bar most human beings from the alpine meadows above, a special realm closed to all but a few. Our visit was not an assault on a peak, but our pilgrimage to nature and to the universe that gave birth to her. We were not conquering a mountain, but getting to know it. Rituals of purification had to be followed. Everything we took in had to be taken out again. As far as possible, we would leave no traces in this place. Our mood was one of respect. Joyful, boyish, sometimes gleeful respect, to be sure, but respect just the same. If anyone had asked us why we headed into this wilderness for a few days each year, we would have said that we loved hiking, and that the mountains were beautiful. But just what that beauty consisted in, and how we were able to replenish ourselves with it, is not something easily explained.

Beyond Environmentalism: A Philosophy of Nature, By Jeffrey E. Foss
Copyright © 2009 John Wiley & Sons, Inc.

Weak and sinful creature that I am, I gave up the struggle to stuff my feet into my cold stiff boots, and stumbled out onto the glacier in my boot liners in defiance of the laws of common sense and self-preservation—gripping an ice ax with which, hopefully, to stop myself should I begin to slide down the slope. The wind that had been flapping the tent a few hours before had died. It was dead calm, almost warm. Looking down the slope I saw the top of Mount Andromache, which had towered above our camp the night before. Fighting down my giddiness and fear, I turned round to look upward to the mountain. The glacier shone down on me with a strange light, making it hard to see. My senses were keen, but the silence was deafening. The ice rose up steeply for a few hundred meters, and then crested. Far beyond the crest I could make out the massive stone cairn that was Hector's peak, a black presence towering against the myriad stars. With a dizzying shock, I realized the mountain itself was moving. The peak was plowing through the stars, eclipsing them in bunches as Earth turned on her axis. Suddenly I saw for myself what before I only dimly realized: Earth is a ship hurtling through space. Here I was in a window seat. As I peered out I saw that I and every other living thing was defying the odds just by being here. I was elated yet humbled, terrified yet reassured. I saw that not just my life, but life itself, was vanishingly small—and infinitely precious.

7.1   NATURE AND RELIGION

It is natural to hear the forests, mountains, or seas called sacred, holy, primeval, primordial, worthy of reverence, and so on. These terms are the same as those used by the religious to describe the godhead. Most people intuitively understand experiences of self-enlightenment and oneness with nature that are evoked by wilderness. Indeed, we know that the oldest forms of religion, the pagan religions that predated modern monotheistic religions, were closely tied to these experiences evoked in natural settings. The very first churches and temples were not buildings but natural sites, such as groves, trees, outcroppings of rocks, bits of shoreline, and so on, that most strongly evoked the natural sense of the sacred. People came to such sites and camped in them long before any buildings were erected in them.

The celebrated ruins of Dephi, the holiest site in ancient Greece, are the remains of only the most recent temples built where people had come for thousands of years to experience natural, sacred beauty. As one ascends to Delphi from the plain below, one passes through "... increasingly rugged and remote territory filled with history and myth. Finally, the dramatic scenery of the limestone cliffs, from which gushes the Castalian Spring with its great cleft, comes into view ..." (Ching et al. 2007, p. 121). Unsurprisingly, given the richness of human thought and social life, the "... early history of Delphi is the story of a struggle between different types of religious practices. Initially, the site was dedicated to the great mother goddess in the Minoan tradition. With the arrival of the Dorians, we see the maternalistic element superseded by the paternalistic world concept of the Dorians" (ibid.). But the transition from one religious conceptualization of a naturally sacred site to another does not mean that the former is completely erased by the latter: "Nonetheless, despite the seizure of the shrine by the followers of Apollo the new religion did not obliterate the old, but metamorphosed it into its own mythologies. The mother goddess was transformed into the serpent Pytho who is said to be buried there" (ibid.). Indeed, the sense of the sacred in nature is extremely deep and in this sense primitive, like a root connecting us to our distant beginnings: "Furthermore, the Earth goddess, Gaia, as she came to be called by the Greeks, retained her ancient *temenos* [sacred ground] close to the temple of Apollo, near the rock of the Sybil" (ibid.).

In the Japanese Shinto tradition we see the primordial, natural roots of the sacred still feeding a contemporary religious tradition. The holiest Shinto shrine is set in a beautiful forest on "a narrow, verdant coastal plain" that is "relatively warm even in winter" (op. cit., p. 279). In Shinto, "... every aspect of nature was revered. There were no creeds or images of gods, but rather a host of *kami....Kami* were both deities and the numinous quality perceived in objects of nature, such as trees, rocks, waters, and mountains" (op. cit., p. 278). So, unlike the ancient Europeans, the Japanese did not overlay their sense of the naturally sacred with complex and determinate mythologies. Perhaps because of this, the sites which evoke the sense of the sacred for them still function today as they did in the far distant past: "The *kami* that are still venerated at more than 100,000 Shinto shrines throughout Japan are considered to be creative and harmonizing forces in nature. Humans were seen not as owners of nature, or above and separate from it, but as integral participants in it and indeed derived from it" (ibid.). An additional benefit of the Japanese tradition is the clear realization that human beings are completely natural, a realization that must be an element of any sound philosophy of nature. Environmentalism, by contrast, is based on the tragically misleading—and false—dichotomy between human beings and their surroundings (*environs*); that is, their environment.

Aboriginal peoples from around the globe experience the beautiful and the sacred in nature. The indigenous peoples of North America, for example, confirm the pattern of connections we have seen in the primordial cultures of both the Eastern and the Western parts of the Old World. "In the songs and legends of different Native American cultures it is apparent that the land and her creatures are perceived as truly beautiful things. There is a sense of great wonder and of something which sparks a deep sensation of joyful celebration" (Booth and Jacobs 1990, p. 520). Once again, we see that the sense of the sacred is evoked by the beauty of nature, and once again we see this sense expressed in ways that recognize that we human beings spring from nature and thus are fully natural: "Black Elk, a Lakota, asked, 'Is not the sky a father and the earth a mother and are not all living things with feet and wings or roots their children?'"[*] (op. cit., p. 521). Human naturalness and our connection with all other natural things is experienced as kinship and affection: "Luther Standing Bear ... describes the elders of the Lakota Sioux as growing so fond of the Earth that they preferred to sit or lie directly upon it.[†] In this way, they felt that they approached more closely the great mysteries in life and saw more clearly their kinship with all life" (op. cit., p. 522).

[*] Neihardt 1975, p. 6.
[†] Standing Bear 1933, p. 192.

Unfortunately, some other people's sense of kinship with nature has grown more tenuous. "Standing Bear comments that the reason for the white culture's alienation from their adopted land is that they are not truly of it; they have no roots to anchor them, for their stay has been too short"* (ibid.). Whether or not Standing Bear's explanation is correct, environmentalism aids and abets this sense of alienation by using it as its foundation. Correcting this subversion of our sense of the naturally sacred is the first task of any adequate philosophy of nature.

7.2   CHRISTIANITY AND THE "ENVIRONMENTAL CRISIS"

Standing Bear's comment brings us to a topic about which there has been enormous discussion and publication: the sense that modern industrialized peoples have lost their sense of connection to nature and the sacredness of nature. This loss has been blamed for the so-called "environmental crisis," and countless essays and books have been written lamenting the loss and advising—or exhorting—industrialized peoples to regain what they have lost by learning from their preindustrialized brothers and sisters. In a famous essay, published, in of all places, the prestigious scientific journal *Science* (about which more anon), a historian of science, Lynn White (1967), laid the blame for the environmental crisis squarely upon Christianity. No words more efficiently express his charge than his own: ". . . viewed historically, natural science is an extrapolation of natural theology, and . . . modern technology is at least partly to be explained as an Occidental, voluntarist realization of the Christian dogma of man's transcendence of, and rightful mastery over, nature" (L. White 1967, p. 1206). While there is definitely much truth to White's observation of the Christian doctrine that human's free will (voluntarism)—along with many other things he does not mention, such as their survival of bodily death—is part and parcel of some humans' sense of separation from, and superiority to, the rest of nature. He goes on to elaborate his case against Christianity with a tendentious account of the relationship between science and technology: "But, as we now recognize, somewhat over a century ago science and technology—hitherto quite separate activities†—joined together to give mankind powers which, to judge by many of their ecological effects, are out of control. If so, Christianity bears a huge burden of guilt" (ibid.). How could Christianity be guilty for the ecological harms attributed to science? Christianity, after all, has opposed modern science from the early days when it displaced Earth from the center of the universe right down to the present day when it is replacing our divine origins with evolution and genetics.

---

* Ibid., p. 248.
† The history of modern science clearly shows its close relationship with technology from the outset, as, ironically, much of White's own scholarship shows. It is no accident, for example, that Galileo's great scientific breakthrough was based largely upon technological breakthroughs in lens grinding and telescope design that occurred in The Netherlands. It is no accident that both he and Newton contributed to further advances in telescope design through their scientific investigations into optics. Indeed, this pattern of mutual inspiration and support between technology and science is undeniably one essential element of the historical rise of both of them.

White bases his argument on an observation that is surely correct: There was a struggle between Christianity and paganism that Christianity has more or less won. Of course, Christians do still maintain the pagan tradition of marking the winter solstice by giving gifts and decorating their domiciles with signs of the natural life that has receded in the winter and is expected to return again in the spring: bits of green foliage, red berries. A blazing fire also helps to keep the season bright. But nowadays the pagan gifts have been transformed officially into signs of Christian love. The green tree has become a Christmas tree, has an angel on top, and has been severed from its pagan roots at bottom. In White's words: "To a Christian a tree can be no more than a physical fact" (ibid). This, as it happens, is also the scientific view of the tree. Science has no need for anything beyond the physical, in particular no need for anything *spiritual*, to explain the tree—and in this way happens to be in agreement with Christianity (if we ignore the issue of the tree's ultimate origins). "The whole concept of the sacred grove is alien to Christianity and the ethos of the West. For nearly 2 millennia Christian missionaries have been chopping down sacred groves, which are idolatrous because they assume spirit in nature" (ibid).

Yes, sacred groves were cut down, the old temples were closed, along with the pagan academies of Plato, Aristotle, Epicurus, and the rest. Within a few centuries the early Christians cleared Mediterranean cities of pagan temples and rites, and then over the next 1000 years expunged paganism from the rest of Europe. When vestiges of pagan practices were found still surviving during the Renaissance, witchcraft persecutions finally expurgated them by 1750, the date of the last witchcraft trial in Scotland. But, lest we forget, the Christians were not the first to engage in sacred warfare. The pagans had fought violently over the sacred groves for a millennia before the Christians began cutting them down. White neglects this fact, and denies the fact that for a Christian a tree is more than a physical fact—it is God's handiwork, just like the sacred groves themselves—and to that extent is sacred. Nevertheless, a tree may be chopped down without the slightest stain of sin, since it has no soul of its own. This contrasts starkly with the pagan view that a tree itself could be harmed, and hence that it, or its reigning deity, must be propitiated or recompensed for its demise. For the pagan, trees shared kinship with us as living things, whereas for the Christian, trees fell on the far side of a dichotomy between us and soulless things.

Such niceties aside, however, White is mainly right about the three main claims supporting his charges against Christianity. First, modern science did have one root in natural theology. As White reports, "From the 13th century onward, up to and including Leibniz and Newton, every major scientist in effect explained his motivations in religious terms. . . . Newton seems to have regarded himself more as a theologian than as a scientist" (ibid).

Second, modern technology is partly—but only partly—explained as a result of Christians following God's command in the first pages of the Bible: "And God blessed them, and God said unto them, Be fruitful, and multiply, and replenish the earth, and subdue it: and have dominion over the fish of the sea, and over the fowl of the air, and over every living thing that moveth upon the earth" (Genesis 1:28).

Third, science and technology are powerfully allied, an alliance that really began much earlier than White says, but is nevertheless real and powerful. We have already traced the relationship between pure science and applied science in previous chapters.

THE GAIA HYPOTHESIS

Many environmentalists believe that the Earth as a whole is alive, and they call this supposed planetary organism *Gaia*. The idea originated with a scientist, James Lovelock, who defined and argued for the Gaia hypothesis in a number of works (Lovelock 1991, 1995). He used Gaia, the name of the ancient Greek goddess who was the mother of all the other gods, for Earth itself. So the Earth—or Mother Earth—is reckoned to be a superbeing: not God, presumably, but a being superior to any of the beings to whom she gave life and continues to sustain. Gaia is not supposed to be conscious, but merely alive, her life being the system that integrates life on the planet. Living things maintain a chemical system inside their bodies which extracts energy from the surrounding environment in order to sustain their own structure and grow, and the Earth as a whole does the same thing. Organisms increase order within themselves—or decrease their own *entropy*—by increasing the entropy of the world outside themselves. Again, the Earth as a whole does the same thing, taking in low-entropy white light from the Sun and excreting higher-entropy infrared light into the cold depths of space, and using the energy extracted to create living structures within. Of course, most of Earth is molten rock, iron, and nickel, hence is not alive—but then most of a tree is not alive, either, only the sapwood between the dead bark and the dead wooden skeleton of the tree.

The organized life chemistry, or *metabolism*, of the planet is the chemical system that consists of the metabolism of all of the organisms within the planet interacting with its atmosphere, oceans, soils, and other physical parameters. Just as each cell within a multicellular organism lives within the protected environment inside the surface membrane of that organism, so each organism within Earth's atmosphere, its surface membrane, lives within the protected environment on Earth. Lovelock points out that although the output of the sun has increased by some 30% or so since life first appeared more than a billion years ago, temperatures have remained remarkably constant. Although the salinity of the oceans should have increased steadily as river salts pour into them, it too has remained constant. Similarly, the mix of gases in the atmosphere has remained constant. Although—or perhaps because—we do not adequately understand the processes that maintain these conditions on Earth, Lovelock and other Gaia theorists propose that we treat them *holistically*, the same way that we treat the constants maintained within a living body: Assume that the organism maintains a homeostatic balance, and then study it scientifically to figure out how it achieves this.

Although Gaia has some authoritative supporters, such as the biologist Lynn Margulis (see Margulis 1998), it has been dismissed summarily by many other scientific authorities, such as the biologist Richard Dawkins (1982). Most tellingly, working scientists have found little use for the hypothesis. As the scientist James Kirchner argued (1991), Gaia tends to dissolve into three parts: a restatement of well-known facts (such as that life has influenced atmospheric, ocean, and soil chemistry, and vice versa), falsehoods (such as that the atmospheric temperature or chemistry has remained constant), or untestable metaphors (such as the idea that the planet is a single organism). Christians see Gaia as the recrudescence of paganism: "'Gaia,' the religion of the Earth Mother . . . is now returning, even in 'Christian lands,' in all its demonic power" (Morris and Parker 1982, p. 26).

In any case, the Gaia hypothesis demonstrates that even scientists may experience the sacred in nature. The scientific method is to *reduce* complexity to its parts, whereas Gaia postulates a unifying life for Earth that is more than the sum of its parts, just as each of us is more than the sum of our parts. The idea of *the* ecosystem, the Earth's total and *unified* biological system, is ubiquitous in environmental science. Is the concept of the global ecosystem unstably positioned athwart mechanistic science and our prescientific sense of the sacred?

Insightful though White's account may be, it does not bear up under critical scrutiny. While Christianity helped, despite itself, to make the scientific revolution possible intellectually, it did everything in its power to suppress it. Even today, Christians are being told that Darwin is wrong and that believing in evolution is a threat to their immortal souls. They are told that prayer is physically efficacious, and that human voluntary action is free of the determining chains of physical causation. Nevertheless, despite its efforts to suppress science, Christianity made science possible by conceiving of God as the rational ruler of the entire universe in all of its domains and aspects. The rational way to rule the natural world is to design it so that it follows rational principles, or laws. The universe, since it was created by a perfectly intelligent, rational being, had *logos*. Both our word *law* and our word *logic* come from the Greek word *logos*, which carried the sense of both. In the Roman Christian context in which modern science was born, the notion of Roman law was married to the notion of God's universal domain to yield the idea that everything that happened in the natural world (aside from the voluntary actions of men, angels, devils, or God himself) followed laws laid down for nature by God. If only we had the ingenuity to decipher these laws, we would better understand God's wisdom.

Thus, the scientists of the Christian era were able to look at the universe with the religious conviction that it followed laws which it would be good for them to decipher or divine. White claims that every major scientist "explained his motivations in religious terms." And so they did. But all Christians saw their entire lives in religious terms, so that is hardly an explanation of the rise of science. More relevant is the fact that God was seen as a *methodological* necessity by these same scientists, as they often aver and sometimes explain. One of the deepest, most concise explanations is Newton's own in the very book that was to launch modern science. In that book Newton explains in the "General Scholium" (1687/1947, pp. 544–546) that we can deduce the laws of nature from experimental observations *because* God's mind extends to every point in time and space, and thus guarantees that the same rules apply at every point in time and space.

It is for this reason that Newton famously rejects the method of hypotheses: "Hypotheses non fingo" (op. cit., p. 547). Once we have observed, for example, that the rate of acceleration is proportional to the force applied at one time and place, we know that this same relationship will hold everywhere and everywhen. Under these circumstances of a divine guarantee of the uniformity of nature, we can *deduce* laws from observation. It was for this reason that Newton set out his physics like a geometry text: He believes that geometry itself is an expression of just *some* of the principles of the natural world. He is simply adding some more axioms that will fill out the picture and enlarge our understanding of the universe around us.* Inspired by his own success in exploring the mind of God in his handiwork, the physical universe He created, we can easily understand why Newton then ventured into theological domains, as White notes, thinking by then of himself more as a theologian than as a natural philosopher.

This brings us to another aspect of the history that White does not go into very deeply in his famous essay, although his other work leaves no doubt that he is aware of it: The roots of modern science can be traced back to ancient Greek concepts of rationality much more readily than they can be traced back to the Bible. You will find precious little in the Bible to support the idea that God is a sort of divine geometer. God is no doubt rational, but only because we think reason is a virtue and God must have every virtue. Although God gives many commandments, they are commandments of the heart, not the head. There is no commandment to be rational, although the irrational faith of children is praised and promised a reward. So, yes, White's first premise, that modern science did have a theological root, is true. However, this root was merely the idea that God was a rational lawgiver, an idea which in turn owes much more to pagan reason and logic than it did to the Christian Bible. Christianity itself was and is profoundly suspicious of natural science, which it continues to see as a rival, and even as an expression of paganism because of its faith in the natural order or *logos*. The idea that nature, through a process of evolution, has sufficient creative power to create plants and animals, to say nothing of human beings, still seems idolatrous.

Much the same is true of White's second point: that the environmental crisis (such as it is) is partly the result of the Christian belief that God commanded them to "subdue" the Earth. This is, as White himself says, only a *partial* explanation. In

---

* When Newton comes across problems that his physics does not explain, such as why the universe is not collapsing under the force of universal gravitation, he simply invokes the wise and mysterious ways of God, who placed the stars so far from each other that their gravitational forces were insignificant, at least during the brief lifetimes of such finite creatures as us.

fact, the actual biblical passage (Genesis 1:28) commands us to "... be fruitful, and multiply, and replenish the earth, and subdue it." *Replenishing* the Earth sounds like an ecological command. The very next verse says: "And God said, Behold, I have given you every herb bearing seed, which is upon the face of all the earth, and every tree, in the which is the fruit of a tree yielding seed; to you it shall be for meat" (Genesis 1:29). In other words, we are commanded to be vegetarians—just as so many environmentalists urge us to be (in fact, we are restricted to only certain plants, those bearing seeds). So perhaps when White blames Christianity for the ecological crisis, the blame is partial in another sense, namely that it is not impartial.

On the other hand, White is on to something when he points out that for Christians, living things are not portrayed as divine by their very nature. This removes a barrier to their use which was there for the pagans. On the other hand, unlike the pagan religions, Christian religions explicitly teach that greed, envy, lust, and gluttony are cardinal sins, or vices. Humans are told that the virtuous life is not one of consumption, and that giving priority to satisfying our animal drives is a deadly sin. There is no Christian license to subdue the Earth in order to drive a big automobile or have a cottage at the beach. Instead of a Christian approval of environmental misuse, there is a broad agreement between environmental ethics and Christian ethics in these matters. For Christians, driving a big gas-guzzling car is seen as an expression of *greed*. Rather than drive a big car, you should drive a small car and spend the savings feeding the poor.

So White's accusation of environmental sinfulness against the Christians is unsound. He says we live in a thoroughly Christian mindset, even when we feel we are being scientific, or post-Christian, and he has perhaps unwittingly exemplified this in his own argument, which is all about assigning guilt. Guilt is a matter of values, just as much as it is a matter of facts, and values lie beyond the purview of science. And when it comes to the facts, science tells us a completely different story from White's about why human beings degrade the environment: Human beings are driven by desires and instincts that they have acquired through the ancient process of evolution. From the scientific point of view, religion itself is just another product of these same desires and instincts. Rather than our actions being caused by religion, science sees all of our actions, whether under the rubric of religion or under the rubric of environmentalism, as products of human nature, which in turn is a product of nature via the evolutionary process.

From the scientific point of view, it is because evolution requires us to be fruitful and multiply that we believe God has commanded us to do so. Because evolutionary success requires outcompeting other species, we have come to believe that we have a divinely sanctioned right to have dominion over them. This belief increases our evolutionary fitness. Intelligence, reason, our sense of right and wrong, all of these are themselves in service to the first and second evolutionary commandments: (1) Survive. (2) Reproduce. From this scientific point of view, the sense of the sacred that people—especially pre-scientific people—find in nature is also in service to these evolutionary commandments. Many environmental philosophers profess that wilderness is a place where one gets in touch with the sacred (Abrams 1994, Naess 1995, Snyder 1995, among others). They not only accept that such experiences are real, but believe that their content is accurate as well: The wilderness *is*, in fact, *sacred*. That they have this belief is no surprise from the scientific point of view. Whether the belief has any scientific validity or truth is another matter, however.

7.3   ENVIRONMENTAL SCIENCE AND THE SACRED

It is no accident that environmental scientists in particular have gravitated to the idea that nature is sacred. Biologists especially are likely to discover—or rediscover—this natural attractor in human conceptual space. The eminent biologist, Niles Eldredge, for example, begins his recent book with the words, "If ever a place deserved our attention, even our *reverence*, it is the Okavango Delta" (1998, p. 2; my emphasis). He goes on to call it "the last vestige of Eden" (even though it came into existence a mere 5000 years ago, when geological changes drained a lake that previously covered its locale). Eldredge has a scientific basis for seeing this stunningly and uniquely beautiful part of the world as Eden. He believes that the Okovango Delta is "a remnant of the stage on which our own species, *Homo sapiens*, evolved . . . about 125,000 years ago" (p. 1). This is odd, given that the sacred Okavango came into existence only 5000 years ago, according to Eldredge himself. It didn't exist when our species evolved, when conditions at the Okavango Delta, as elsewhere, must have been very different. Ice gripped the Earth 125,000 years ago, which according to current scientific knowledge rendered the Okavango arid and barren.

If, perhaps, Eldredge leads with his heart rather than his head when it comes to the beauty of the Okavango, we at least have a scientific explanation: It is filled with possibilities for finding the sorts of food, clothing, and shelter that *Homo sapiens* instincts were honed by evolution to find. It is filled with fish, fowl, and game. Although Eldredge would find the idea of sport hunters descending on the Okavango abhorrent, he is driven by the very same sense of beauty as they are. I have never met a sport fisher or hunter who did not see his activity in the light of the sacred. Sport fishers and hunters do not hunt for food but to get in touch with the beauty of nature and—taking a turn that offends many modern sensibilities—to relate to that beauty in a way that they find restorative and a blessing. They *hate* what they call the rat race. That is why they get up before dawn to sit in a boat holding a line and slapping mosquitoes while awaiting the rising sun. They would pay big bucks—they would kill, as the expression goes—to do this in the Okavango.

Eldredge has sublimated this primal desire while still leaving himself open to the sense of beauty it evokes. Like a man admiring a Renoir nude, he is almost beside himself with the chaste but tantalizing joy this evokes. And, to get to the most important point, it is the power of modern science that makes his refined sense of beauty possible. Therefore, Eldredge's sense of the sacred is also rooted in the Christian hypothesis that God (with a capital G) ordered the natural world by natural law. Natural law was enacted by God through His divine command. As Galileo put it in his famous letter to the Grand Duchess Christina (1615), God is the author of both the book of revealed truth and the book of nature. So they cannot contradict. And God, moreover, had shown himself to be the most exquisite and sensitive geometer.

The princess was not convinced, however. Her Bible told her that Joshua commanded the sun to stand still, which was proof enough it had to be moving in the first place, so she resisted Galileo's heretical sophistry.

Putting aside this early enmity between science and religion, we can plainly see here another root of Eldredge's sense of the sacredness of nature. It owes just as much to Pythagoras and Euclid and the natural philosophers they inspired as it does to the Christian Bible. The fathers of the scientific revolution, the great Polish astronomer Copernicus, the great French mathematician Descartes, the great Italian physicist Galileo, and the great English mathematician Newton called themselves *philosophers* of nature. They were—of course!—*Christian* philosophers. But they were philosophers nonetheless, who knew Euclid's *Elements* of geometry every bit as well as they knew their *Genesis*. They believed in God, the rational creator of a world ordered by natural law. It is another two centuries later that they become known as "scientists," during the marriage of science and technology that White brings to our attention.

Newton, Galileo, and the rest are the intellectual ancestors of the American biologists Niles Eldredge and E. O. Wilson, both of whom believe in the sacredness of nature. They rely on reason to understand the way the world is, and in their minds it is reason that convicts humankind of its environmental sinfulness. It is their scientific understanding not only of the way the world is, but their realization that it is startlingly different from the way that it was, that gives them the idea of a new and looming apocalypse, *environmental apocalypse*. In this apocalypse, nature will smite us. In the words of White (whose publication in *Science* may perhaps be the scientific equivalent of the publication of a papal bull), the approaching environmental apocalypse will be in the form of a "disastrous ecological backlash" (L. White 1967, p. 1206). We are hurting nature, and nature will hurt us back, in an ecosystemic version of natural justice.

And lest scientists think that they can avoid the coming apocalypse all on their own, White warns against such arrogance: "I personally doubt that the disastrous ecological backlash can be avoided simply by applying to our problems more science and more technology" (ibid.). It is at this point White is clearly assuming the tone of religious prophecy: He is warning scientists not to play god! He goes on to instruct them that "since the roots of our trouble are so largely religious, the remedy must also be essentially religious, whether we call it that or not. We must rethink and refeel our nature and destiny" (1967, p. 1207). Most important, White warns us we must "reject the Christian axiom that nature has no reason for existence save to serve man" (ibid.).

We should reject this axiom, but not on religious grounds. White's argument that the remedy to our environmental problems must be religious is a glaring *nonsequitur*: It does not follow that if the roots of a problem are of a certain sort that its solution must also be of that sort. The roots of a family's problems may be substance abuse, but it hardly follows that more substance abuse will solve those problems; the roots of an auto accident may be automotive (such as brake failure), but its remedy may be medical (such as emergency surgery). Whether or not religion is at the root of our environmental problems, a solution to our problems must, as always, begin with knowledge and understanding. Of course, nature does not exist to serve us, for *Homo sapiens* is merely part of a greater biological system from which we have arisen and upon which we depend for our existence. We owe everything to nature; it owes nothing to us—the reverse of the axiom White rightly calls upon us to reject.

There is no doubt, however, that environmental science conceives and expresses humankind's relationship to nature in a manner that is—as a matter of observable fact—religious. Environmental science prophesies an environmental *apocalypse*. It tells us that the reason we confront apocalypse is our own environmental *sinfulness*. Our sin is one of *impurity*: We have fouled a pure, "pristine" nature with our dirty household and industrial wastes. The apocalypse will take the form of an environmental *backlash*, a payback for our sins. "Vengeance is mine" says the environment. We are warned not to "play God," not to seek a scientific techno-fix to postpone the day of reckoning. Instead, we must *repent*. We must stop our polluting ways, which are merely the expressions of our materialism and our *greed* (which itself springs from our lust, envy, and pride). We are to reject our greed for modesty, exchange our pride for humility, and respectfully return to our own proper eco-niche within the greater ecosystem. These exhortations are inscribed in textbooks of environmental science (as we have seen in Chapters 4, 5, and 6). And so it has come to pass that environmental scientists have taken on the role of the priests of environmentalism. Just as four centuries ago priests in black robes explained the sacred mysteries of God to lay persons incapable of judging for themselves, so today scientists in white lab coats explain the sacred mysteries of nature to citizens incapable of judging for themselves. And just as priests then told people what they must do to be blameless before God, today environmental scientists tell people what they must do to be blameless before nature (cf. Foss 2006).

7.4 BEYOND ENVIRONMENTAL RELIGIOSITY

On the face of it, the appropriation of the idiom of the sacred by environmental scientists may seem like a benign thing, or at least an innocuous thing. Sure, it is interesting from an intellectual point of view that environmentalists have employed the idiom of religion—but what is the harm? One problem is that the identification of the sacred licenses the broadest range of ethically permissible actions. What is sacred can be, indeed must be, protected at all costs. The concept of the sacred gives emotional substance to the transcendent value at the heart of an ideology. This does not automatically entail any harm, but it does immediately create the risk of harm. The environmental scientist speaks with priestly authority to lay environmentalists who will draw conclusions for themselves and undertake to protect the sacred on their own. We must not forget that environmentalists both scientific and lay see nature as in a state of crisis, as under an ongoing attack by the human species itself. When the sacred is attacked, some response, some counterattack, is required. Attacks on the sacred are sacrilege. Just what form a response to sacrilege takes depends on a great many other things, such as the emotional state of those responding, and, crucially, their other beliefs. The response depends, in large measure, on just what other things are thought sacred: in particular, whether human beings are seen as sacred. If humans are seen as sacred, this rules against their sacrifice for the good of the environment. Sad to say, environmentalists typically do not see human beings as natural, and environmentalists always demand that human interests be sacrificed for those of the environment.

This new form of environmental doctrine is plainly a marriage of convenience between elements that do not really belong together. On the one hand, there is science and reason illuminating the current state of nature, showing us what we have done wrong so far, and forecasting how these wrongs will ramify apocalyptically in the future. On the other hand, the guilt for environmental destruction (or is it environmental sacrilege?) also falls largely upon science in its technological role, with the result that environmental scientists themselves say that the solution is not, in White's words (1967, p. 1206), "more science and more technology." Science is above suspicion as long as it confirms our guilt, but it is not to be trusted with anything to do with expiation of that guilt. We have poked our fingers into the sacred machinery of nature and are being told not to do it again.

As noted in the introduction, environmentalism is a popular movement, so we should not be surprised to find coalitions of convenience on both the practical and the theoretical planes. We lack terms for the current concatenation of elements. It is a sort of neoscientific religion in which the sacredness of nature has replaced the divinity of god, and the reading of environmental omens via scientific measurements has replaced the reading of the Bible. It is equally a sort of neoreligious science, in which the service of nature has replaced the service of truth, and reversing technological intrusions into nature has replaced increasing the power of scientific knowledge. In any case, we need something better if humankind is to see its way to a better relationship with nature. There is absolutely nothing wrong with scientists bringing their professional expertise to bear upon the question of the sacredness of nature. The problem is that their professional expertise does not go beyond the facts as such, whereas the sacredness of nature does. What we take to be sacred is indeed a function of the facts as we see them, but taking something to be sacred is also to adopt an evaluative stance toward it. Scientists do not have any professional expertise when it comes to values—but then, who does? Religious authorities, perhaps? Fortunately, we do not need to answer the question of who the proper moral authorities are in order to move forward philosophically.

Let us move beyond science and religion, directly to the matter of authority itself, which we may approach via the distinction between believing and believing *in*. Someone believes something when they really think it is true, whereas they believe in something when they accept it as true despite reasons to doubt its truth. For example, people believe that Earth is round, or that the toaster requires electricity to work, or that it is raining, because there are no reasons to doubt these things. On the other hand, people believe in flying saucers, or believe in reincarnation, or believe in taking vitamin C to fight colds, which means that they know there are reasons to doubt their truth, but they choose to believe them anyway. Where evidence and logic are insufficient to justify belief, people will nevertheless often exercise what they take to be a legitimate option: believing something because they want to or think they should. The nature of the supposed legitimacy of believing in is partly legal: Democratic countries generally guarantee "freedom of belief." In any case, people will believe things that relieve the painfulness of life's difficulties. A mother whose son or daughter was listed as missing in action in a foreign war may choose to believe that he or she is still alive, despite much evidence to the contrary.

It is true that people in democratic countries generally have the right to believe in whatever they like whenever they like and for any reason they choose. Freedom of belief is a good thing, because its opposite, belief control, leads to despotism and stifles the freedom of thought and openmindedness needed for humankind to gain knowledge and advance understanding. On the other hand, believing in is not always, or even usually, a good thing. In fact, it is generally a bad thing to believe something for any reason other than the high probability of its truth—and even then, it is better merely to believe that the probability of its truth is high. Proportioning belief to evidence and argument is a key tenet of Western philosophy. Sometimes this can be painful; sometimes it takes courage. However, it has also proven itself as the best way to acquire knowledge and understanding. Within science, it has enabled us to solve some of the key mysteries of nature. Within law, it has enabled us to solve crimes and attain a level of justice. Within philosophy, we must avoid believing in altogether, since philosophy seeks truth and understanding as the basis of wisdom. In other words, no claim, whether it concerns facts or values, is to be judged acceptable on the basis that we believe in it. In yet other words, all claims are to be subject to logical criticism.

THESIS 9:   No claim, whether it concerns facts or values, is to be judged acceptable solely because we believe in it. All claims are subject to logical criticism.

CASE STUDY 8: RECYCLING

If you are among the hundreds of millions of city dwellers who have been required by law to recycle tin cans, bottles, and other items, rather than putting them in the trash, you might take a moment to consider what you are doing. Let us consider the cases for and against.

Environmentalist Recycler: Recycling is good both for the environment and for humankind. It is a win–win activity. The total of human economic activity results in the total environmental impact shown in Figure 7.1A. Ultimately the environmental damage due to human economic activity will come back to damage human beings as well, because the current pattern of continuous economic growth means constantly increasing environmental impact, and that increase is *not sustainable* forever. As many environmentalists have famously pointed out, the only thing in nature that follows a policy of continuous growth is cancer. Continuous economic growth implies that

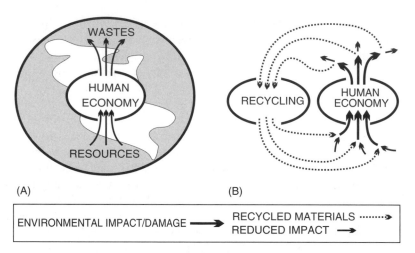

(A)   (B)

| ENVIRONMENTAL IMPACT/DAMAGE ➡ | RECYCLED MATERIALS ┈┈┈▸ |
| | REDUCED IMPACT ➡ |

Figure 7.1   Recycling reduces environmental impact. (A) Human environmental impact. The human economy as a whole takes resources from the environment and puts wastes into the environment. The sum of these inputs and outputs is the impact, or damage, of humankind. (B) Impact reduced by recycling. By taking recyclable materials from the waste stream and returning them to the resource stream, human economic impact is reduced. If 50% is recycled, for example, both the resource and waste streams are reduced by 50%, and human harm to the environment is reduced accordingly.

humans are like a cancer growing on Mother Earth. The ability of the environment to provide resources on the one hand, and to absorb wastes on the other, is obviously finite, so obviously we have no choice but to reduce our environmental impact. As shown in Figure 7.1B, recycling reduces the amount of raw materials and energy that must be extracted from the environment. Recycling also reduces the flow of garbage into landfills. Landfills are bad for the environment in a number of ways: landfills remove land from ecosystems, reducing the available habitat for animals and plants; landfills pollute surface water and groundwater through their formation of leachate, water polluted with bacteria, heavy metals, various chemicals, and so on; landfills pollute the atmosphere through the formation of the greenhouse gas, such as methane.

We know that re-melting aluminum cans uses less energy than making the same amount of aluminum from aluminum ore. As this case shows, recycling reduces resource and energy use, reduces the GHGs released into the environment, and reduces the amount of land used for mining and for waste disposal. This in turn moves us toward sustainability, which is good for *us* as well as the environment. Thus, recycling is a win–win proposition, a no-brainer. We obviously need to recycle for our good and the good of the environment.

Post-environmentalist: The picture you paint begins with a fundamental error: You portray recycling as though it were outside the human economy, but in reality recycling is *part of the economy*. As shown in Figure 7.2, recycling has the same essential characteristics of all economic or organic processes: It takes materials from

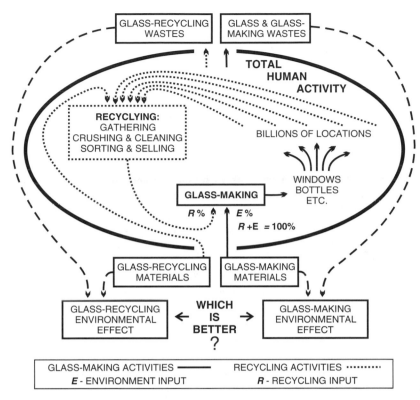

Figure 7.2   Recycling physics. Recycling is a form of human activity intended to improve environmental health. The large oval represents the economy from a physical point of view as the sum of human activity. Some of this activity is dedicated to making and distributing glassware, as represented by the solid box and the solid arrows. Recycling appropriates a portion of human activity as represented in the dotted box and the dotted arrows. Like all human activity, recycling has its own environmental inputs and outputs. The question is whether the total environmental effect of recycling a given percentage of a specific substance—here illustrated by glass—is better than the total environmental effect of producing that percentage without recycling.

the environment and ultimately returns them to the environment as waste products. For concreteness, Figure 7.2 focuses on glass recycling, although it illustrates the physics of recycling in general. One thing it shows about the case of aluminum cans is that the mere fact that re-melting them takes less energy than making aluminum from the ore does *not* demonstrate that recycling aluminum cans has a smaller environmental effect (or impact) than aluminum mining and smelting. We cannot ignore the fact that recycling aluminum cans, just like any other form of human activity, makes *many* demands on the environment, including the demand for energy.

   Household recycling of metal cans, bottles, and other containers begins with rinsing or washing the used container. Flyers, brochures, and television advertising are purchased by government recycling agencies to tell citizens they must recycle various items, including cans, bottles, and plastic containers, and that these items must be washed before they are placed in curbside containers for collection. This use of advertising not only proves that recycling is a business like any other, but places a demand on the environment for the washing of used containers that it legislates. Not only are citizens required to wash thousands of tons of discarded containers, but these waste materials usually have to be washed a second time before they can actually be reused. All of this washing requires water, energy to warm the water, detergent, and so on. Disposal of the resulting dirty water at sewage treatment plants also uses energy and materials, as well as land, air, and water resources.

Once the used containers are washed and put at the curbside, people must drive trucks to collect these recyclables, which burns fuel and emits GHGs into the atmosphere. It also requires all of the resources, energy, GHG emissions, and waste disposal involved in the production and maintenance of the truck fleets used in recycling, their tires, their fuel, and so on. These fleets also are responsible for a portion of the resources used in maintaining the road system. Recycling also requires facilities, such as parking lots for the truck fleets, buildings where the old containers are taken to be sorted, washed, crushed, stored, and so on, which requires land, resources, energy, and waste removal. There are tens of thousands of these facilities around the world.

Does the portion of these facilities dedicated to, for example, glass recycling actually take less land and energy than the pits used to provide the same amount of glassmaking sand? We simply do not know. On the face of it, collecting all of the used bottles and grinding them back into sand again seems like a very high cost way of getting glassmaking materials, and high cost usually indicates greater demands on resources and, hence, on the environment. Moreover, the energy required to melt recycled glass is about the same as that used to make glass from other materials.

We must not forget the people involved—citizens like us—whose labor is required to make recycling happen. All of the people involved in the recycling process, from the citizens who take time to sort, wash, and prepare their wastes for recycling, through to the workers who collect, prepare, and distribute them to the reusers, also make demands on the environment for their food, clothing, and shelter. From the physical point of view, recycling requires human activity, which in turn requires environmental inputs and outputs just like any other biological activity. When recycling workers spend their wages, they make demands on the environment that are an essential part of the environmental cost of recycling. While a person is busy working to recycle some item, they are metabolizing their food, wearing out their clothes, expending their shelter, and so on. Other human activity will be required to provide this necessary input into the recycling process.

Until the total of all these environmental impacts of recycling are shown to be less than the environmental impact of using new material, we have no reason to suppose that recycling of any given material reduces environmental impact. We cannot simply *assume* that recycling will have a smaller environmental impact, no matter what environmentalist sloganeering may have led us to believe. As illustrated in Figure 7.2, recycling reduces the amount of glassmaking materials taken from the environment by a certain percentage ($R\%$) and reduces glass and glassmaking wastes by a similar amount. What is generally overlooked is that recycling also requires other materials to be taken from the environment and creates its own wastes. If recycling glass were good for the environment, the decrease in environmental impact (by roughly $R\%$) would have to be more significant than the increased impact caused by recycling itself.

But this has never been shown. It has not been shown for glass or for the other materials that are recycled. Indeed (as we shall see in more detail below), we have good reason to believe that there is no overall reduction in environmental impact; otherwise, recycling would be done automatically out of a profit motive. Therefore, we are being forced to recycle on the basis of *faith*, not reason. So recycling is an environmentalist form of ritualized penance (cf. Foss 2006).

Environmentalist Recycler: Of course, recycling redirects a certain amount of human economic activity away from extracting materials from the planet or dumping materials back into it. That is the whole point. No one ever said that recycling would be *profitable* from an economic point of view, since that is *not* its objective. Certainly, recycling requires significant numbers of people to expend time and work in pursuit of sustainability, but that is a price well worth paying both for the environment and for future human beings. You are merely being cynical to suppose that the environmental impact of recycling might be greater than the environmental impact of using new materials.

Post-environmentalist: I am not supposing anything. It is the environmentalist who is making the assumptions here. The universal mantra of the environmentalist, "reuse, recycle, reduce," *assumes* that the environmental impact of recycling a material *must be* smaller than the environmental impact of using new materials. There is no mechanism or invisible hand that will guarantee this assumption. For all we know, it may be true in some cases, and false in others. The crucial point is that it has never been shown for any material—indeed, no serious effort has ever been made to even investigate the issue fully. City councils and other levels of government have commanded citizens to recycle without ever *knowing* whether it does in fact reduce our environmental impact, moved solely by the command to recycle that they have heard since the 1980s. Citizens have called for and supported these laws while in a state of ignorance about their actual environmental effects. It is environmentalist faith, therefore, not reason, which is behind recycling.

Environmentalist Recycler: You have completely ignored the question of sustainability. And you have also overlooked the fact that recycling is still at a fairly primitive stage, since at this point only a relatively small percentage of materials are recycled. In Figure 7.2, we must understand that R, the percentage of recycled materials, is relatively small compared to E, the percentage of materials taken from the environment. Until R is closer to 100, and E is closer to zero, the true potential of recycling will not be realized. At this point it is important to keep the recycling initiatives going, so that economies of scale can eventually be achieved. We look forward to the day when recycling technologies, plants, processes, and so on, have become sophisticated and standardized, and will routinely reduce our environmental impact. We can even dare to envision a day when virtually all of the materials we use are recycled and our environmental impact is reduced nearly to zero. Until that day arrives, we must keep pressing forward with recycling initiatives.

Post-environmentalist: Pushing forward without any assurance that your goal is achievable or worthwhile is just the stuff of faith, not reason. When it comes to sustainability, no one has ever shown that we will run out of materials for making containers, newspapers, or packages. Long before we run out of glassmaking sand, we will almost certainly quit using glass to make containers in favor of cheaper plastics or other materials not yet conceived. Nor has anyone ever shown that we will run out of iron with which to make tins, or fiber to make paper, or bauxite to make aluminum. In fact, none of the materials involved in legally enforced household recycling is in any danger of depletion. These materials are being used precisely because they are plentiful. The bigger likelihood is that we will move on to better packaging technology and better information technology long before we run out of the materials used in the current technologies. To say that our use of these materials is unsustainable is merely to recite another unproven article of the environmentalist faith.

Environmentalist recycling is based on two assumptions: first, that recycling will be good for the environment, and second that it will be good for human beings. Note well that these are *value* claims, and hence not open to proof or disproof, although they are open to logical investigation. The scope and difficulty of these assumptions is illustrated in Figure 7.3. Recycling either is profitable in economic terms, or it is not.

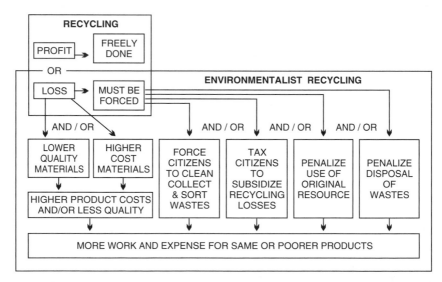

Figure 7.3 Recycling economics. Any form of recycling is either profitable or not. If it is profitable, it will be done freely. Environmentalist recycling comes into play only when recycling is unprofitable. Its unprofitability means either that the recycled material is more expensive or of lower quality than the alternatives. It also means that recycling must be enforced by laws requiring citizens to sort their trash, to pay taxes to cover recycling losses, and to penalize use of resources and disposal of wastes. The ultimate result is that environmentalist recycling is a sacrifice—or penance—for human beings.

Where it is profitable, no other incentives are required to implement it, and it will be done out of the profit motive. There are countless examples of this sort of recycling. For example, people who fabricate sealed window units must cut plate glass to the various sizes required for their products. The pieces of glass that are trimmed away are usually returned to the glassmaker for a profit. The glass trimmings are clean, all of the same sort, and so on, so need no washing and sorting, unlike the case with environmentalist recycling. Window fabricating businesses tend to be located close to glassmakers, and vice versa, in order to benefit from lower transportation costs, which in turn increases the cost-effectiveness of recycling the trimmings. Indeed, window fabricating and glassmaking occur under the same building in most modern plants, so the glass trimmings need merely be taken from the end of the assembly line back to the beginning to be recycled and reused.

Environmentalist recycling comes into play only when recycling is not profitable, but this very fact militates strongly against it being good for human beings or good for the environment. The fact that environmentalist recycling operates at an economic loss implies that either the commodity it produces (recycles) is of lower quality than the original product, or else it costs more, or both. Both of these are negative factors from the human point of view. Glass recycling, for instance, yields a dirty mixture of different sorts of glass. Hundreds of grades and colors of glass are manufactured to meet the technical specifications of manufacturers. Since it is virtually impossible to sort all of the recycled glass into these hundreds of sorts once again, it is generally sorted into two or three gross categories: brown, green, and clear, for instance.

Moreover, since no recycling occurs until the glass is actually reused for some purpose or other, and since recycled glass is of low quality from the glassmaker's point of view, people have to be *paid* in many cases to accept the dirty and ill-sorted glass in what are called materials recovery facilities (MRFs). In the newspeak of environmentalist recycling, these facilities pay a "negative price" for recycled glass to cover their costs to clean, sort, and process it further to make it acceptable to some manufacturer. The concept of negative price is nicely explained in a document produced by The City of New York Department of Sanitation. "Ideally, municipalities sell recyclables, and processors (MRFs) buy recyclables. Yet when the market value of certain commodities falls to zero, MRFs are not willing to buy these materials. Under free market conditions, zero-value commodities would simply be disposed of as refuse" (DSNY 2004, p. 19). To put it bluntly, the materials we recycle are, in many cases, of zero-value; that is, of no value whatsoever to the very industries that are supposed to recycle them. In fact, recyclables may be of less than zero-value; that is, they have a negative value. The rational thing to do with such materials is to dispose of them. "But municipalities with recycling laws or mandates can't just do this— they are required to recycle certain materials, no matter what. In such cases, municipalities 'sell' these materials for a negative price. In other words, they pay processing firms to take them" (ibid.).

As shown in Figure 7.3, taxes must be collected to pay processing firms to accept recycled materials. Taxes are hardly a benefit for the citizen. Other forces are also used by governments to make recycling happen despite its unprofitability. One is to legally require citizens to do the primary sorting, cleaning, and preparation of the materials to be recycled. This work is not a pleasure in and of itself, although some environmentalists may feel a righteous joy in doing it because of its intended—but unproven—effects. Raising punitive taxes, fees, or fines against the usual resources and waste disposal methods also encourages recycling by making the usual methods of production more expensive. Again, this is nothing but a loss from the point of view of the ordinary citizen. As if this was not bad enough, the products made from recycled materials are apt to be of lower quality as well, since the lower quality of recycled materials is one of the main reasons they are recycled at a loss—or if you prefer, at a negative profit.

Nor is the issue of the relative percentage of materials recycled an argument in favor of recycling. Contrary to what you dare envision, we know on basic physical principles that it is impossible that all of the materials needed for life can be recycled. Life physically requires that materials both be taken from the environment and returned to the environment.* However, it is possible that the percentage of a specific material that is recycled may approach 100%. For the sake of concreteness, let us suppose that the percentage of glass that is recycled has been brought from 10% to 100%. Given that there was an economic loss, a loss of money, labor, time, and product quality when only 10% of glass was recycled, this loss will increase at least tenfold when this percentage is increased tenfold. In fact, moving beyond the first 10% of glass that is relatively easily recycled to the last 10% that will be difficult to recycle makes a projected tenfold increase in our collective loss much too optimistic.

If we assume that nothing else changes in the economy except for the redirection of some work and money toward environmentalist-mandated recycling, there is less work and money available for the other things we need or want, so we are collectively worse off all around. So it is just plain false that recycling is good for those of us living today. As we have seen, the claim that it will be better for future generations because it is more sustainable has never been shown. In fact, the only way that recycling will not make us all collectively poorer is if additional economic activity is employed to make up for the labor and money directed into recycling.

So in the end, we are asked to make sacrifices in order to avoid the often-prophesied environmental apocalypse, even though there is no clear understanding of how our sacrifices are supposed to reduce our environmental impact. Environmentalist recycling, therefore, is nothing other than ritualized penance.

---

* The second law of thermodynamics, that entropy must always increase, is behind this fact. Organisms have to take materials from the environment to use in their life processes, and when these materials are returned to the environment their entropy is increased. For example, the food we eat is returned to the environment in a degraded form; that is, its entropy has increased. Organisms locally decrease their own entropy by increasing the entropy of their environment. Universal recycling, the recycling of all of the materials required for human life, would require that we magically decrease the entropy of our waste materials so that they are suitable for re-consumption. Put somewhat differently, universal recycling would require that the human economy become a perpetual motion machine. However it is put, universal recycling is simply impossible from a physical point of view.

Environmentalist Recycler: So your supposed solution is business as usual. Are you denying that the capacity of the environment to provide resources and absorb wastes is finite? Are you not aware that continuous economic growth is therefore unsustainable?

Post-environmentalist: Are you not aware that continuous *economic* growth does not imply continuous increase in the amount of materials we take from or return to the environment? Continuous economic growth refers to growth in *value*, not in physical bulk. We are surrounded by items of small bulk but high value: cell phones, computers, music players, etc. Yes, Earth is finite, but so are we. There is no danger that we will one day need an infinite supply of materials to accommodate our finite human activities. Anyway, the total human population will soon begin to decrease (see Figure I.1 in the Introduction) and that leads us toward decreasing environmental impact. Speaking of trends, should we be afraid that humankind itself is "unsustainable" just because fertility rates are trending downward? Surely not, which shows that extrapolating trends to their furthest possible extent is logically naive.

Environmentalist Recycler: Perhaps your faith in humankind is naive. In any case, you have implicitly granted a point that is on my side of this issue: that we should be looking for greater value in ways that do not make ever greater demands on our environment.

Post-environmentalist: Agreed. In the end, value is what this issue is all about. So the question comes down to this, as illustrated in Figure 7.2: Is recycling better all around, both for us and the environment? Or is it just a penance?

7.5 THE NATURALLY SACRED

The sacred has always been associated with the supernatural. Because the supernatural is, by definition, completely beyond human comprehension, it is necessarily shrouded in darkness and mystery. So the sacred, rather than being essentially a source of light and understanding, has been equally the source of much trouble in human history. Because the supernatural is a matter of believing in, it has been subject to dogmatism and abuse, including such things as despotism and wars of religion. To the extent that environmentalism involves believing in the sacredness of nature, it, too, is cognitively impenetrable and open to dogmatism and abuse. Fortunately, we do not need to indulge in believing in the sacred in order to accept that nature is sacred. First, we can recognize that humankind *finds* nature sacred. Second, we can recognize that it is perfectly rational for us to do so. The conjunction of these two realizations is a concept that we may call the *naturally sacred*.

For untold millennia, human beings have in various ways realized that they are themselves a product of the natural world and that they depend on it for their existence, as do all living things. The source and ground of humankind and life itself is naturally seen as sacred. Science has confirmed this ancient realization by discovering many of the details of the emergence of life and humankind on Earth, as well as the details of the processes of metabolic chemistry whereby life is maintained. Properly understood, this new scientific realization does not in any way reduce life's value, beauty, and awe. We observe that consciousness and life itself come to us directly from nature, regardless of whatever we may speculate lies hidden behind what we can see. The former is a matter of belief, the latter a matter of believing in. Thus, it is a matter of belief that nature is a focus of positive value, beauty, and awe, other essential characteristics of the sacred. Given that we humans value our own lives, it follows that we value their source and their sustenance. Now that science has confirmed in detail how nature has created us, we recognize its value in more detail than before. It is beyond precious, being the source of value of anything within it.

Similarly, science increases the beauty and awe of nature. Whether we observe the countless galaxies wheeling through space or the microcosm of the living cell, we are struck by the beauty and power of the natural world. Scientific understanding of natural processes does not diminish this beauty but enlarges it. And since every mystery resolved is resolved by revealing yet further mysteries, nature remains mysterious, another aspect of the sacred. Indeed, scientific understanding adds texture and depth to the mystery. Respect for nature is a natural consequence of its value, beauty, awe, and mystery. Awe is the mirror image of beauty; beauty and awe are the opposite ends of one dimension of human sensibility. Whereas things of beauty instill desire, things of awe instill fear. Beauty floods at birth and ebbs in death, while awe does the opposite. Beauty is found in flowered valleys, awe is found in glacial mountain peaks. Together they create a tension that we call sacred, and that instills respect.

It is perfectly rational to find nature sacred in the sense just described. Indeed, the aspects of nature outlined in the previous two paragraphs effectively define the sacred: the source and ground of our existence; a higher order of value; a focus of beauty, awe, and mystery. The naturally sacred is not something to believe in but something that can be recognized by humans. The description of the naturally sacred is a description of what human beings *find* sacred. The thesis that we, humankind, find nature sacred is to be understood as a *type truth*.

TYPE TRUTH

The truth that tigers run faster than humans is an example of a *type truth*: the tiger is a *type* of animal that runs faster than the human *type* of animal. Note that it is not true that all tigers run faster than all human beings. The slowest tigers will not run as fast as the fastest humans. But still, at the level of types, tigers run faster than men, giraffes are taller than horses, and dogs chase cats. Type truth is particularly important in the sciences of complex systems, where we do not expect to find exceptionless universalities. In particular, type truth is important in climatology (which is highly relevant to the environmental issue of global warming) and in the social sciences, because both climate and human beings are complex, chaotic systems.* Type truths therefore are crucial within the study of values, ethics, obligation, virtue, vice, beauty, and ugliness. Type truth is the great guide of human life—indeed, it is the guide of life itself.

No doubt some people have no sensibility of the beauty of nature. Certainly, some people find neither beauty nor awe in wilderness. All it takes is a single weekend being "eaten alive" by insects to put some people off wilderness for a lifetime. And almost no one finds value, beauty, or awe in being bitten by insects. Nevertheless, even people who do not like wilderness often love nature just the same. Many people love their pets, a very common human expression of their recognition of value and beauty beyond their own species and its advantage in the struggle for survival. Almost every human home has plants within, under the roof, sheltered from harm as if they were kin, and similarly watered and fed, even doctored. Regardless of whether they find wilderness, nonhuman animals, or plants, virtually all people experience the sacred in the organic realities of life, birth, death, and love. Life flows through us in all of its exquisite complexity, as desire, fear, joy, pain, love, hatred, birth, and death. Despite this, there are some people sadly unable to appreciate the value, beauty, awe, and mystery of life and nature, just as there are some tigers that cannot outrun a human being. Nevertheless, it is a type truth that human beings experience the naturally sacred.

Trying to define the naturally sacred is a little bit like trying to reduce the wonder of existence to a formula. However, it is important for the philosophy of nature that we at least have a working definition, an epigrammatic formula to guide our efforts. I suggest a three-part definition:

1. The cognitive core of the naturally sacred is our recognition of the fact that nature is the source and basis of our existence.
2. The evaluative core of the naturally sacred is our sense that nature is worthy of our respect. Of course, no fact entails any value, but given that we see that nature has produced us and supports us, and assuming we have some self-respect, this urges us to respect nature as well.

---

* Aristotle distinguishes the sorts of truth we can find in ethics and meteorology from the sort we can find in mathematics and basic physics. Whereas the latter may provide us with universal and exceptionless truths, the former yield at best only "truths for the most part." The concept of type truths that I am employing here is, I believe, coextensive with Aristotle's concept of truths-for-the-most-part. In any case, I follow him in thinking that the most important truths we seek within the field of values are not exceptionless, but nevertheless are really true.

3. The emotive expression of the naturally sacred is our sense of the beauty and awe of nature. People often confuse respect with fear, and will say, for example, that bears deserve your respect merely because bears are dreadfully dangerous. To truly respect a bear you must see it as in some way admirable. Both beauty and awe are admirable, and while awe inspires fear, it also inspires respect because it is also admirable. The cognitive, evaluative, and emotive cores of the naturally sacred form a rational whole. It is rational to feel respect, which is itself a blend of admiration and fear, for the beauty and awe of nature, given that it is the source and basis of our being.*

THESIS 10:   We recognize nature as sacred: the basis of our existence, worthy of our respect, and a source of beauty and awe.

7.6   IS ACCEPTING RESPONSIBILITY FOR NATURE SACRILEGIOUS?

That we should accept responsibility for the health and well-being of nature is one major thesis of this book. There are four reasons for this. First, it is undeniable that humankind has had large-scale effects on nature. These effects began long ago, when we lived much closer to nature in a state idealized by many environmentalists. It seems likely that our ancient hunter-gatherer ancestors played a role in many of the extinctions of game species, such as the woolly mammoth and other species of Pleistocene megafauna. We know from the remains of these animals that they were hunted by our species and that they disappeared from their natural ranges shortly after our species arrived in those locales. Our effects accelerated when we turned to agriculture, and significantly large portions of Earth's surface were cleared and plowed for our crops. And, as we have so often been reminded, our impact continues to this day. Second, we are capable of understanding nature well enough to appreciate whether these effects are good or bad not only for us but for living things in general. This is presupposed by environmentalists whenever they identify any form of environmental damage or suggest a remedy. If we accept, for instance, that dumping toxins in rivers has a bad effect on living things, we accept that we have the capacity to appreciate what is good or bad for living things. Third, we therefore have the power to effect large-scale good or harm for nature as a whole. Fourth, we are the only species with this understanding and power. It follows, therefore, that we have a large-scale responsibility for the wellbeing of nature as a whole. Developing this argument will be one task for the rest of the book.

---

* The rationality of the emotions (or the lack thereof) is a topic that we cannot delve into here, but many excellent philosophical studies are available. A brief sense that the concept at least makes sense can be gotten from the observation that it would be irrational to love someone for the harm they have done you, or to be angry at someone for being good. Rationality is a concept that embraces not only logic, but values and feeling as well.

The toughest obstacle to accepting responsibility for nature will be the objection that we should not "play God." This is just one form of the idea that the sacred is to be left alone, that not leaving it alone is *sacrilege*. But since it is the strongest form, if we can lay it to rest, the other forms will be rebutted as well. The presupposition behind the warning not to play God is that God is responsible for nature, and taking this responsibility from God is arrogance, or the sin of pride. I sympathize with the humility and self-doubt that this attitude represents. Nature is, after all, filled with awe. It can extract an awe-full toll should it so decide, as floods, pestilence, plague, and famine have testified since the dawn of history—not to mention the meteorites and comets that have collided with Earth in prehistoric times. Still, the awe-fullness of nature does not justify rejecting our responsibility for our effects on nature, so it follows that there must be a way to accept that responsibility that is not arrogance, not pride, and not sacrilege.

The "Do not play God!" principle is never used to make people do something, only to *stop* doing something. The principle commands us to let nature take its course. "Do not play God!" might be said, for example, to a mother considering surgery for her unborn fetus in order to correct a heart defect, remove a tumor, or prevent hydrocephaly. The mother is being told to let nature take its course. It may be argued that the course of nature represents the will of God, and its acts are said to be the acts of God (the phrase "acts of God" still serving as legalese for natural disasters). In short, doing something about the fetal defect is sinful arrogance, while doing nothing is virtuous humility. The objection to this is logical: the distinction between doing and not doing something is purely verbal. Therefore, even when we choose the course of what we take to be inaction, we are playing God just the same. The mother who refuses fetal surgery thereby chooses to bring an unhealthy baby into the world. Either way, she plays God. There is no getting away from it. From a mainstream theological perspective, every voluntary action or inaction (or "forbearance," as inaction is sometimes termed) is an exercise of one's God-given gift of free will, which is precisely the power to determine, if in only a small way, the course that creation will follow. To have free will is to be given responsibility by God for the course of events insofar as they are affected by one's own choices. God has assigned us a role to play in the unfolding of the universe. We are, to put it poetically, assigned the role of small-g gods. *The question is not whether to play God, but what sort of god one plays.*

This being said, we must hasten to recognize that the warning not to play God is generally given when it seems the responsibility in question is too large to handle. This might be sound advice, regardless of any logical or theological flaws in its expression. Someone who is not a surgeon should not take responsibility for taking out someone's appendix, for example. This, we should agree, is correct: We should not take on responsibilities that are beyond our capacities. On the other hand, we must realize that what is or is not too large is a complex matter that calls for a judgment, and that the judgment cannot always be avoided or averted. We are responsible in part for the state of some of Earth's rivers, forests, and fields. What we do or do not do affects them, their natural inhabitants, and their (geographical or eco-functional) neighbors, too. What should we do? We cannot avoid our responsibility at this point.

No matter what we do or do not do, we will produce effects that are under our control. It would be wisest, then, to accept out responsibility.

THESIS 11:   It is consistent with the sacredness of nature for us to assume responsibility for nature to the extent that we are able to affect it.

Fortunately, the proponents of global warming theory support this thesis, since it is presupposed by what they argue. They argue that we *are* responsible for a catastrophic warming of the climate and then conclude that we must stop doing whatever we are doing that is causing the catastrophe. As we saw in Chapter 6, this argument is not above criticism. Nevertheless, it presupposes that we can affect nature for the good as well as the bad. It urges us to take control and *assume* responsibility for the climate. So whether the argument is sound or unsound, its supporters grant this thesis.

8 Nature and Romance

One impulse from a vernal wood/ May teach you more of man,/ Of moral evil and of good,/ Than all the sages can.// Sweet is the lore which Nature brings;/ Our meddling intellect/ Misshapes the beauteous forms of things—/ We murder to dissect// Enough of Science and of Art,/ Close up those barren leaves;/ Come forth, and bring with you a heart/ That watches and receives.

—William Wordsworth (1798a)

Many years ago, my 9-year-old brother Bob showed me something he held hidden in his hand. It was a tiny robin. It had a badly broken leg and seemed on the verge of losing consciousness. I can still see the dull look in its eyes and the tears in Bob's. He had saved it from some kids who were about to kill it, and now he desperately wanted it to survive. He had an exaggerated opinion of my powers, since I had attained the august age of 12. We dug up some worms in the garden, but the robin would not even look at them. It had decided to shut out the world. I had a sinking feeling, but could not give up. We finally managed to put a bit of bread soaked in milk into its beak. It just sat there, motionless, while bubbles formed around its nostrils, and a drop of milk ran down its throat. Then it swallowed—and everything took a turn for the better. It ate quite a lot of bread soaked in milk, and then switched without difficulty to worms the next day. Already it seemed happy as could be, wiggling his wings and opening his beak anytime a hand approached him. Or was it her? We never knew.

I put his tiny leg in a tiny matchstick splint the first night, but could not fix his foot. His thumb claw stayed limp and folded under his toe claws. But this was a peppy little animal, and soon he was hopping around and perching without any trouble. We have videotape copies of dad's home movies of the robin, begging for food as baby birds do, eating worms in the garden as we dug them up, and taking a bath in a bowl of water that our mom had thoughtfully provided. He lived in an old birdcage in the house, which we would put on the back porch every day so he could be outside with us. We would leave the cage door open so he could hop out, or when he was older, fly out. He would leave on his own and return when he felt like it. Once we were eating lunch outside in the back yard while the robin begged for bits of food. Suddenly a pair of robins made an angry scolding attack on us in a determined effort to rescue this little bird. We were amazed. The idea that these birds were his parents seemed

Beyond Environmentalism: A Philosophy of Nature, By Jeffrey E. Foss
Copyright © 2009 John Wiley & Sons, Inc.

farfetched. He had been rescued a half-mile away, after all. But, parents or not, this pair of shrieking robins made a much more fearsome attack than you might think, and we retreated to see what would happen. The adult birds seemed agitated, while our little robin was perfectly calm. Clearly, there was some sort of communication going on between them. But in the end they left quietly, and he stayed with us.

Eventually, he began to sing the strangely evocative song that robins sing at dawn, answering the robins outside from his cage in the house. He began to sound and look like a proper robin. He turned from a hungry, mottled pudge-ball into a sleek, swift, sassy, sharp-beaked bird. He instinctively knew how to hunt worms—he seemed to listen for them to surface, and when one did he just hopped over and fiercely tugged it out of the ground. Given his uncanny skill, there was no need to dig worms any more. So he got his own meals while I guarded against cats. His flying became stronger, and his forays from the cage became longer. The last time I saw him, he was sitting high in a poplar tree with another robin that he had, apparently, befriended. That night he did not return to his cage on the back porch. We never saw him again. For a few years I checked every robin I saw for a crook in its right leg, but was always disappointed. We never did give him a name.

My family's rescue of this bird was, from a rational point of view, very strange behavior. What advantage could we possibly find in rescuing a bird? No other species of animal would ever have saved that bird. Why did we? It makes evolutionary sense for animals to act altruistically toward members of their own species, since that increases their species' fitness. But being Good Samaritans toward members of other species must then reduce fitness. The Good Samaritan himself might have scolded us for wasting our time helping a bird instead of a person. What evolutionary quirk might lead *Homo sapiens* to undermine its fitness by spending its time and energy on members of other species? Obviously, a 9-year-old boy does not have a mothering instinct that is triggered by small birds. So what made him shelter the bird in his own hands?

If you feel that questions like this one can only be answered by the heart, not the head, then you have a romantic side. If you feel that the heart's answer to the question is more or less obvious, you are inclined toward the romantic side. If you feel that anyone that has a heart would never ask a question like this in the first place (especially in the presence of children), you may well be a full-blown romantic. What is a romantic? There is no explicit definition, no set of words that captures all, and only, romantics. Romantics must surely see this as poetic justice, since they believe that real life escapes the analytic cuts and synthetic fences that intellectuals use to divide and conquer it. The romantic soul is free, and the proof of this is that it will not submit to analysis or be hedged in. Nevertheless, despite the lack of a generally accepted definition, there is general agreement that there was a romantic movement in Europe and North America in the first half of the nineteenth century. It is also generally accepted that there is a romantic way of looking at things, a romantic philosophy, a romantic "take" on the world as a whole and human life as a whole. It is even generally accepted that a person of romantic temperament (i.e., whose personal philosophy is romantic) will understand why a couple of boys would rescue a robin, bind its wounds, nurse it back to health, and set it free.

8.1   ENVIRONMENTALISM AND ROMANCE

You cannot understand environmentalism (nor yourself if you are an environmentalist) unless you understand romance and the romantic, and you cannot fully sympathize with environmentalism unless you have a romantic side. Most people do have a romantic side, of course, and so most people can fully sympathize with environmentalists. Environmentalism is a modern echo of a strand of romanticism that has run through human cultures since pagan times. In particular, the roots of the environmentalist idea of humankind's proper relationship with nature can be traced back to that of the romantics. At the very least, there is a large and significant overlap between environmentation and romance, as I hope to show briefly in this chapter.*

The environmentalist ideal is not a nice apartment in the city, but a simple cabin in the woods. Environmentalists want to *reduce their impact on nature*. They want to make themselves smaller, not larger, when it comes to the natural world. They wish they lived in simpler times, with simpler problems, simpler joys, simpler values—before their biological kindred abused and subdued the Earth. There was a time, they believe, when *Homo sapiens* lived in harmony with nature, in small kinship bands. Environmentalists believe in the noble savage, at least in principle. The "noble savage," the untamed human being, is capable of every sort of virtue that the civilized human being is capable of: bravery, loyalty, love, creativity, intelligence. Virtue does not depend on analysis, but something deeper: human sympathy, the ability to imagine yourself in the other person's place, the need to protect the weak . . . in a word, the human heart. Civilization puts pavement between the human heart and the natural world that gave birth to it, and to which it answers. Human nature itself has been cut off from its roots and has gone bad in the cities, with their violence, drugs, and various forms of perverse behavior. So-called "primitive" people may not have been able to make nuclear weapons, but their inner nature hummed in tune with nature outside. They were free.

I cannot say whether every environmentalist will accept this characterization. They too, are free. No doubt there are some who prefer a more scientific idiom. Science is certainly more authoritative than poetry in any case, and to that extent useful for environmentalism. Even scientists, however, will agree that in the nineteenth century there was something called "the" romantic movement, in which nature and the concept of the noble savage captured the popular imagination. Scholars tell us that there were various romantic movements in various places as romanticism waxed and waned through history, but I think that a good case can be made for thinking that "the" romantic movement of the nineteenth century was special: It rose up in reaction to the most intense wave of industrialization that the world had seen up to that time. The steam engine appeared in 1712 and then in 1763 was launched by James Watt on a long course of improvements that continue to the present day. It

---

* I will not state a formal thesis to the effect that our feelings about environmental issues in general and our modern environmental ideal of the human–nature relationship are largely derived from romanticism, partly because the only thing that can be sketched here is the overlap between these sets of ideas, and partly because the history of ideas is as much an art as a science.

replaced muscle power, so it replaced people and animals. Unemployed people went to the cities looking for work, along with the animals' unemployed breeders, keepers, and handlers, launching a wave of urbanization that continues to the present day. In 1764, James Hargreaves invented the spinning jenny, a machine that would replace spinners of yarns and threads. In 1768, a bunch of spinners wrecked a bunch of his machines in a doomed attempt to avert their impending unemployment. We might, with considerable justice, date the start of the romantic revolution to that historical event.

Romanticism was the poetic revolution that inspired and traveled along with the industrial counterrevolution.* By the time the Luddites were squaring off against the British Army in 1811, the major English romantic poets—Byron, Coleridge, Shelley, and Wordsworth—were sweeping away the poetic competition with their love of nature, their love of common folk, and their rejection not just of what Blake called "the dark Satanic mills" of industry, but of the science—and the scientific intellect—that gave birth to it. The Luddites' industrial counterrevolution was doomed; its leaders were to meet their end on the gallows, but romanticism conquered the hearts of Europeans even as the factories won their minds and pocketbooks. Romantic poetry, novels, plays, opera, music, dance, and painting dominated the media of the times. Architecture returned to the time before the industrial revolution and even before the scientific revolution that made it possible: the age of King Arthur, Lancelot, Robin Hood, and Friar Tuck. Medieval castles and churches were restored by Violet le Duc, part of a massive medieval revival in architecture. Gothic cathedrals regained their aesthetic currency and were once again being built. The amazing gothic cathedral of Cologne was the tallest structure in the world from 1880 to 1884. Gardens and parks surged in popularity during the romantic period, as nature made its way back into the cities.

Hiking and mountaineering are romantic inventions from this same period and place, being favorite pastimes of the romantic poets. You can see several romantic elements together at once—the garden, the park, the hiking, the mountains, and the medieval architecture—in the hotels that sprung up in the Alps and other mountain-vacation destinations to accommodate the surge of romance in the hearts of Europeans.† These hotels, and the expansion of the tourist trade that they represented, owed their profits in large part to the poets who created the utopian romantic vision that they embodied. Among the most famous of romantic poet hikers and mountaineers were Wordsworth, Coleridge, Byron, Shelley, and Mary Wollstonecraft Godwin (Mary Shelley). Their dedication to hiking and mountaineering confirmed a growing sense among the industrializing peoples of their time that nature was beautiful, a source of value and inspiration *in its own right*. Although we assume without question

---

* Environmentalism still contains an element of industrial counterrevolution, or may even be a species of industrial counterrevolution.

† The Banff Springs Hotel in Banff is a North American example. The building is Scottish baronial style, one of the main forms of the medieval revival. Resting at the foot of towering mountains, it overlooks manicured gardens and lawns that lead down to the dramatic junction of two rivers. Hiking trails head off in all directions along these rivers and up into the surrounding mountains.

nowadays that natural beauty is an obvious reality, this was not a common opinion prior to the romantic movement. For thousands of years mountains had been seen as impediments to human habitation and travel, not as things of beauty. The romantic age changed all of that. Suddenly, landscapes and seascapes were being painted and purchased, and scenes of simple country life were being painted for city dwellers who had a newfound, indeed newly invented, nostalgia for the rural.

Obviously, the surge of romanticism in the nineteenth century represented a surge of recognition of the naturally sacred. In this way, romanticism is a reconnecting of Christian Europeans and Americans with pagan sensibilities. The revival of interest in classical Greece and Rome that had begun as early as the Middle Ages, and had grown steadily since then, flourished in the romantic era. Along with a new interest in Arthur, Lancelot, and Robin Hood, we find a strangely anachronistic interest in Zeus, Hera, Mars, and Venus. However, to think of romanticism as mainly, or even solely, as a return to paganism would be a gross simplification, for Christian themes (such as medieval themes) are equally present in romantic literature and thought. Romanticism is, instead, primarily a counterrevolution to the industrial revolution, inspired in large part by a reawakening sense of the naturally sacred. From a pagan point of view, nature is sacred in its own right, whereas from a Christian point of view, nature is sacred as a work of God—but either way, in romanticism nature came into prominence as good, beautiful, and inspiring.

This change in attitude toward nature continues to the present day. The romantic movement officially came to an end somewhere in the late nineteenth century, when it was replaced by "realism." But unofficially, romanticism was not replaced, but rather integrated into the sensibilities of industrialized city dwellers. Perhaps because they are in closer contact with the natural world, rural people are more realistic about nature than are city dwellers. Rural people see the weeds, ticks, fleas, death, and disease that are the darker side of nature. Contemporary urbanites, however, accept the beauty of nature unquestioningly. In so doing, they embody the romantic values that sprang into life around the year 1800. This romantic fiber running through the contemporary concept of nature gives environmentalism strength and character. As we shall see in more detail, the romantic movement created a fairly complex image of the proper relationship between humankind and nature, and of humankind's guilt for renouncing that proper relationship and cutting itself off from nature. This guilt is still prominent within environmentalism and goes a long way toward explaining the environmentalist's rapid and uncritical acceptance of the need to rescue nature from human science and technology.

8.2 ROMANCE AND SCIENCE

The only female mentioned among the list of famous hikers and mountaineers of the romantic era was Mary Godwin, who as Mary Shelley would create a classic romantic figure, Frankenstein. Her "monster," who in the popular imagination mistakenly bears the name of its scientific creator, is usually thought of as epitomizing the horror genre, which in turn is usually thought of as not only distinct from romance, but its polar opposite. In fact, horror as a literary or artistic genre was also a creation of the romantic age. Byron, who of all the romantic poets had the strongest interest in the gothic and the grave, launched the vampire as an image of horror, while Mary Shelley did the same for the monster of Frankenstein. Her monster crystallizes in a single image the romantic vision of what the scientist makes of humankind. Her monster still states with brilliant clarity what her now-forgotten male counterparts struggled to express in their many volumes of published poetry. It is by no means a mere coincidence that environmentalists today instinctively call genetically modified crops "Frankenfoods." You can almost picture the genes of these crops stitched together from bits, just like Frankenstein's "monster."

It was Mary who understood that for the romantic, scientists are sorcerers. She understood that the romantic line of descent of the post-Newtonian scientists of her times was from Goëthe's *Faust*, and before him the medieval alchemists characterized by the legendary Merlin,* and before them the ancient scientists characterized by the legendary Prometheus, who stole fire from the gods and gave it to humankind. From the outside, the scientist and the sorcerer are indistinguishable to those who are neither. The scientist, like the sorcerer, must study arduously for years under the masters of his art to gain control over the powers of nature. The scientist, like the sorcerer, can leash or unleash the powers of nature by means of acts that seem to have no resemblance or relevance to their results. To the uninitiated, a scientist halting a plague by sticking needles in people's arms is just as mysterious as a sorcerer halting a thunderstorm by blowing magic dust into the wind.

Modern science is accepted in large part because of the magic it does. It is natural, not supernatural, magic, but magic nevertheless. Fire spurts from a match, electric fire is released from dammed-up water, and a fiery rocket puts a camera in orbit so that we can see the Earth below. Scientists' control of the elements, particularly fire, is proof of their power. The modern scientists' release of nuclear fire in an atomic weapon is proof that the magic in their books and formulas is more powerful than the magic in the books and incantations of the sorcerers of common legend. The fact that scientists use formulas, never incantations, is precisely what distinguishes them from sorcerers. Incantations are supernatural instruments. God, the most super of all supernatural beings, created heaven and Earth simply by commanding them to exist. An incantation is a magic command to the powers of nature to do one's bidding; incantations are intrinsically supernatural. But while the book of the sorcerer shows how the elements can be controlled by means of incantations, the book of the scientist shows how to control the elements through the runic formulas of natural laws.

Thus, modern scientists have intensified and consolidated the powers of the sorcerers of old. They are no longer hostage to the will of supernatural agents who must be propitiated. Instead of making imprecations and incantations, they pull the switches and levers that control the machinery of nature, using the natural power of their own hands and heads. The scientist, Dr. Frankenstein, did not need to call upon

---

* Morris Berman's *Reenchantment of the World* (1981) is a wonderfully readable, detailed account of this lineage. As Berman shows, Newton was just as much the ultimate alchemist as he was the first physicist.

higher powers to create his "monster." In modern movie representations scientists do not call down the thunder of the gods but capture lightning with lightning rods. Scientists *take* what they need from nature. It is the scientists who tell the story: They set out to perfect human nature, to create a perfect human being, but instead, created the monster out of bits of human beings. The monster grew out of Dr. Frankenstein's intellect, the intellect that "murders to dissect" but cannot put the pieces back together again. Dr. Frankenstein's attempt is not only doomed to failure, but makes a creature that is pathetic.

An anatomy text divides the body into its scientific components: bones, muscles, tendons, cartilage, veins, nerves, heart, and brain. The intellect is directing this dissection of the body, just as it directs the dissection of nature in general. The intellect carves up nature precisely where the knife finds the going easiest: at the joints. Scientists *say* that a good theory is one that cuts nature up at the joints. Knife and intellect are natural allies, hence it is inevitable that "our meddling intellect misshapes the beauteous forms of things." But where Wordsworth invites us to "close up the barren leaves" of the anatomy text, Mary Shelley boldly shows us the monsters that science will create. Her monster, the first Frankenperson, is a harbinger of Frankenfood (genetically modified crops) and Frankensheep (the first genetically engineered animal)—and all the Frankenpeople that environmentalism warns us we will become in the brave new world of science and technology.

8.3 ROMANCE AND ALIENATION

Romanticism gives expression to the emotional tensions that arise from modern man's alienation from nature. The romantic sees that industrialized life separates us from nature and infers that in doing so, civilization cuts us off from our own nature. This is a plausible hypothesis today, even though our concept of the genesis of human nature is different from that of the romantics of the early nineteenth century. Darwin ironically proved the romantic's idea that human nature is rooted in nature at large, and by implication that *Homo sapiens* has moved beyond its roots. Human nature was shaped by evolution to survive in a world very different from the world we have made for ourselves and now inhabit. We have the instincts that evolution provided us, instincts that were adapted to the life of the hunter-gatherer. Agriculture and city life (the two always go together) have provided us with more food than hunting and gathering ever could, but it has also pent up our hunter-gatherer instincts. We are free neither to fight our urban adversaries nor to flee from them. Our lusts may not go where they list.

City life suppresses our drives for fighting, fleeing, and sex in exchange for satisfying our drives for feeding and families. Food and shelter have always been the main promises of city life, and it has now delivered both in sufficiency and even excess. Whereas starvation was the specter raised by Malthus, it is actually obesity that threatens us now. Scientists describe a *global epidemic* of obesity. Evolution has taught us to eat whenever we get the chance. Our ancestors necessarily include those who were plump enough to survive the periodic famines of preindustrial life, and their hungry genes are inside us now, telling us to eat. The genes inside us sing out of tune with the world outside: This is alienation of a sort many know intimately.

The Garden of Eden is a complex and powerful image that also says that our natural place is in a natural setting, not in the city. However, in this image, shared by a number of religions, both nature and human nature are romanticized. In the Garden we, like Earth's other animals, had no need to reap or to sew. We neither hunted nor were hunted. There was no disease. In short, we had nothing to fear. We did not even need to fear stepping on thorns or bashing our foot on a rock. We were barefoot like the other animals, indeed totally naked just like them, immune to biting insects and inclement weather. Sex is not mentioned, though presumably Adam and Eve obeyed God's command to multiply. Assuming they enjoyed sexual union, Adam and Eve felt not even the slightest trace of shame—quite unlike alienated city dwellers, who are ashamed merely to be outdoors naked.

But in the Garden we had nothing to fear and nothing to hide. This defines freedom in its most romantic form: the harmony of one's inner nature with one's environment. Romantic freedom has nothing to do with freedom of the will, the metaphysically mysterious ability to act without being caused to act. Romantic freedom is freedom from compulsion, freedom from ever feeling forced to do anything, and especially freedom from ever *forcing oneself* to do anything. Freedom is never having to exercise self-control, of always doing what one is inclined to do, like the wind that bloweth where it listeth. We have lost our freedom, according to story of the Garden, because of our sin of disobedience to God, which in turn grew out of our arrogance and pride. God punished us by making nature our opponent, our antagonist. From now on, nature would not willingly give us the necessities of life, and so we must plow and plant and herd and harvest to *force* the Earth to yield us a living. Eating, a perfectly natural process in itself, has been turned from a pleasure into a chore. Giving birth—also a natural process if ever there was one—becomes painful. The romantic concept of our alienation from nature corresponds quite closely to the curse placed upon us as we were expelled from the Garden of Eden: From now on we would find nature to be our opponent rather than our friend. The first casualty of this alienation is our freedom from death, suffering, and work.

To some extent the science confirms the romantic image. We *have to* get up in the morning and *have to* work. Work, by definition, is activity that we do not like to do. The birds of the field, by contrast, never have to do anything that they do not like to do. They never have to force themselves to do anything. They don't wake themselves up with an annoying alarm clock and drag themselves off to look for food. Instead, they wake up when they feel like it, and when they feel hungry, off they go and eat. The birds of the field do not experience compulsion from without or within, so they have a sort of freedom that we can scarcely imagine. But if we consider animals more like us, science disconfirms the romantic image. Chimpanzees, our nearest genetic neighbors, are highly social animals who must therefore confront a wide variety of social forces that constrain and mold their instinctive behavior. From an early age, their eating, playing, fighting, wandering, sexual behavior, sleeping, and waking are controlled by their elders and shaped to meet the needs of the group. So, long before agriculture or civilization, our ancestors already felt the pressure to conform. Freedom was compromised long ago in our evolutionary ancestry. Alienation from one's own instinctive nature runs deep in our branch of the evolutionary tree.

8.4   THE CLASSICAL ROOTS OF ROMANTICISM

Shelley based Frankenstein on the classical Prometheus. She learned of the Prometheus myth from Byron, who was fond of Aeschylus's *Prometheus Bound*. In Aeschylus's treatment of the myth, there is no action, only Prometheus chained to a rock where he is being tortured for giving us the fire he stole from the gods. He stole the fire so that we would have the power to resist the gods themselves, and the gods are punishing him by having his liver eaten each day by an eagle (his liver is magically restored, so Prometheus can be tortured again the next day, in a foreshadowing of the Christian image of eternal damnation). From his eternal deathbed Prometheus speaks his eternally truthful confession to us, his human audience. Prometheus apologizes to us eloquently for angering the gods, reminding us that he brought us out of the caves we had formerly inhabited—assuming an image of human origins that we now take to be scientifically correct. Aeschylus knew that his audience knew the myth of Prometheus before they ever sat down to watch his play. They knew that the reason that Prometheus stole fire in the first place was to prevent Zeus from exterminating the human race. They also knew that this resulted in both good and evil. The good was that we escaped God's plan for our extinction, and the evil was that the gods stopped looking after us.

The romantic sense that there was a golden age in the past when human life and humankind was far happier and nobler was expressed in the Promethean legend. In that legend, the gods used to look after us like parents look after their children, tending our every need and nursing our every wound. We had been thrown out of the parental nest because Prometheus had stolen fire and given it to us. In Aeschylus's retelling of this story, Prometheus seizes the chance to explain to us that he is not our enemy. He reminds us that caves are dank, dark, and dirty. He reminds us that the gifts he gave us are agriculture, architecture, writing, metallurgy, mathematics, astronomy, and medicine—a list that recapitulates the historical trajectory of the human species from its hunter-gatherer past into the cities his audience now inhabits. He reminds us that we are warmer, dryer, better fed, and healthier than when we were in the caves. He reminds us that we have come out of the darkness and into the light spiritually as well as physically, that we can now understand how the heavens go and can interpret our dreams.

Whether or not Aeschylus's human audience accepts this apology, it knows in its bones that it is not related to nature in the same way that other animals are. According to some strands of the Prometheus myth, Prometheus himself was the creator of the human species, which he had constructed from clay and water, and that his fatherhood of humankind was the reason he bore it such affection. Humans have a different creator than does the rest of natural creation. Aeschylus's audience knows that it is out of joint with the natural world, that its genesis involved some sort of crime against God, with the result that humankind is an uneasy hybrid of the natural and the divine—exactly what Mary Shelley's audience also knows. So it is no accident that Shelley sees the scientist as a new Prometheus—the bringer of fire, science, and civilization. It is also no mere coincidence that Mary Shelley's father, the philosopher William Godwin (1756–1836), taught that human nature is perfectible, since the perfect man is precisely what Frankenstein was trying to create when he, instead, stitches together his monster. Godwin inspired not only his daughter's creation of Frankenstein, but Malthus's theory of unending human misery as well. The very idea of the perfection of humankind—or even its improvement—met with instinctive rejection on both romantic and religious grounds. The myths of Prometheus, Frankenstein, and human condemnation express a sense of human guilt and alienation that outweighs any message of hope.

Godwin's rebuttal, that human population was steady in Western Europe despite increasing food supplies, was mere data that disconfirmed Malthus's theory but was no match for its mythical resonance. Religion reaffirmed humanity's sense of unworthiness. Lucifer, one of the names of Satan, means bringer of fire.* Both Satan and Prometheus lacked respect for their god. One way or another, humankind's alienation from both the Earth and the heavens is always explained as caused by pride or arrogance or both. The Prometheus legend, and the story of Satan, provide humankind some explanation—and justification—of their need to work, suffer, and die. The only *justified* suffering is just punishment. The explanation that both Prometheus and Satan provide is that fire was stolen from the gods, and that our continuing use of it is arrogance. They have given us a power that we lack the virtue to possess. Our science and technology, symbolized by fire, alienates us from both nature and God (or the gods). This makes a lot of sense, inasmuch as the things that distinguish us from the rest of nature are also the things that separate us from it. In any case, the sense that humankind is out of joint with the natural world is pervasive and ancient. The romantics knew this, and they knew that the industrial revolution was merely the latest wave in a tide that had been carrying their species progressively farther and farther away from the rest of nature for millennia. They knew that our feeling of alienation was as old as civilization and agriculture, and that every step forward created its own echoes of self-doubt.

---

* Mary Shelley's husband, Percy Bysshe Shelley, infamously argued that Milton's Satan in his *Paradise Lost* was a more heroic character than Milton's God (P. B. Shelley 1821): "Milton's Devil as a moral being is as far superior to his God, as one who perseveres in some purpose which he has conceived to be excellent in spite of adversity and torture is to one who in the cold security of undoubted triumph inflicts the most horrible revenge upon his enemy, not from any mistaken notion of inducing him to repent of a perseverance in enmity, but with the alleged design of exasperating him to deserve new torments." Percy Shelley also argues that Prometheus is, in turn, better than Satan (P. B. Shelley 1820, Preface): "The only imaginary being resembling in any degree Prometheus is Satan; and Prometheus is, in my judgement, a more poetical character than Satan, because, in addition to courage, and majesty, and firm and patient opposition to omnipotent force, he is susceptible of being described as exempt from the taints of ambition, envy, revenge, and a desire for personal aggrandizement, which, in [Satan,] the hero of *Paradise Lost*, interfere with the interest. The character of Satan engenders in the mind a pernicious casuistry.... But Prometheus is, as it were, the type of the highest perfection of moral and intellectual nature, impelled by the purest and the truest motives to the best and noblest ends."

CASE STUDY 9: IS ENVIRONMENTAL APOCALYPSE A REAL DANGER?*

Yes: Environmental degradation has been unchecked over the last few decades and is inevitably going to come to a head, with devastating results. Human beings have proven historically that they are psychologically incapable of dealing with crises until too late. Warnings about pollution were ignored for centuries until human beings began literally dropping dead in London streets in the pollution crisis of the 1950s. That crisis was staved off with techno-fixes, new "environmentally friendly" pesticides that do not immediately cause massive death of innocent plants and animals, and "pollution control technology," which captures only the grossest pollutants from our industrial wastes. Meanwhile, technology has charged ahead like the headstrong adolescent that it is, giving us nuclear power, Frankencrops, Frankenfoods, and a spike in population that is clearly unsustainable, especially if humans in the developing world persist in their doomed attempt to reap the levels of "prosperity" already being squandered by the greed-driven developed world. Now there is a new threat, bigger than all the rest: global warming. True to form, humans have ignored repeatedly the warnings to cut back on their manic use of fossil fuels or have made promises to do so that have proven hollow. Global warming is the keystone crisis, the one that will assemble into a formula for disaster the individual environmental crises that we touched on so briefly above.

No: Dire warnings for humankind have been the norm throughout the ages. If people are psychologically incapable of heeding apocalyptic warnings, as you say, that may be because they have been producing apocalyptic visions which in the end were proven false, one after another, for centuries. Something in the human psyche, perhaps the recollection of the lost security and innocence of childhood coupled with the knowledge of the inevitability of death, makes us project our fears upon the larger world around us. Apocalypse has been the stuff of myth and religious zealots for millennia, but in this scientific age we would be wise to let it go.

Yes: You misunderstand the current state of play completely: It is science, not myth or religion, that is the basis of our current "apocalyptic" warnings.

No: Some scientists have, admittedly, gotten into the doom-saying business just like their religious predecessors, but that alone does not make their dark prophecies realistic. As we saw in Chapter 6, scientists are not perfect, and the use of scientific methods does not guarantee that human fears, desires, and other pre-judgments (prejudices) will not also play a role. We need only look at the case of Thomas Malthus (1766–1834), who was both a priest and a scientist, to see that this was so. His scientific model was seen as irrefutable at the time, like the greenhouse model today. His Principle of Population (Malthus 1798) is elegant and easily understood, as illustrated in Figure 8.1: The food supply grows arithmetically (1, 2, 3, 4, 5, . . .)

---

* Note that this question is distinct from the question whether "the environmental crisis" is real (cf. Chapter 2).

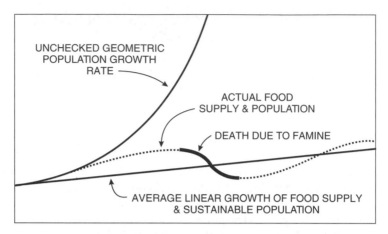

UNCHECKED GEOMETRIC
POPULATION GROWTH
RATE

ACTUAL FOOD
SUPPLY & POPULATION

DEATH DUE TO FAMINE

AVERAGE LINEAR GROWTH OF FOOD SUPPLY
& SUSTAINABLE POPULATION

Figure 8.1 Malthus's apocalyptic vision. Malthus argued that if unchecked, human population growth is geometric but that food supply prevents geometric growth and thereby keeps it in check. Since the average growth in food supply is linear, the sustainable growth in human population is also linear. When temporary conditions (good weather or inexpensive imports, for example) permit a short-term above-average growth in food supply, population grows as well. Human misery is guaranteed because eventually the food supply must fall again, causing death by famine and eliminating the unsustainable human population.

and so obviously cannot keep up with unchecked population, which grows geometrically (1, 2, 4, 8, 16, . . .). However, population is ultimately kept in check by starvation—assuming that disease, pestilence, and poverty fail to do the job on their own. So population is kept at a sustainable level by, in a word, misery (which, argued Malthus, in turn breeds vice) vindicating the biblical vision of human life as he understood it.

Malthus thought of his scientific theory as an explanation of misery and vice and in effect a proof of Christ's dictum that we shall always have the poor with us. Like you, chastising the human race was an essential element of his apocalyptic vision. Malthus argued that his model proved that it is pointless to try to increase the food supply to relieve human starvation, because starvation, along with other forms of misery, is inevitable. This was an early argument against what environmentalists call the "techno-fix." As you can see in Figure 8.1, if for whatever reason the food supply happens to grow faster than the average linear rate, this will cause a spurt in population growth which cannot be sustained in the long run (for, by definition, in the long run the food supply grows only at the average rate). This will, in turn, cause an increase in the death rate, and the good news of the temporary spurt in food supply and population will be followed by deaths due to starvation.

Malthus warned that people must rein themselves in instead and accept their inevitable suffering and limits. His call for self-control and self-sacrifice is a feature common to apocalyptic warnings down through the ages, including the contemporary apocalyptic vision portrayed by the global warming theory. Malthus professed that population should be controlled by "positive checks" such as starvation and disease, and encouraged "preventive checks" such as sexual abstinence. In the same vein,

Malthus argued against those who wanted to reduce the misery of the British poor by reducing import barriers to foreign grain (i.e., by repealing the English corn laws), saying that this artificially created spurt in the food supply would only lead to greater misery in the end, just like a natural spurt in the food supply. While Malthus's view prevailed, people stole bread, with the result that misery bred vice just as he predicted, and when they paid the penalty for it through imprisonment or hanging, this bred yet more misery, again just as he predicted.

Malthus's model is generally thought to have been disproven: Misery has decreased, despite a constant growth in population, due to such scientific and technological advances as the "green revolution" and birth control. People such as Jeremy Bentham, who argued for repeal of the corn laws, were vindicated. The corn laws were repealed, food did become cheaper in Britain, and starvation was alleviated—without an apocalyptic backlash. Globally, food supply has also accelerated to keep up with the geometric growth of population. Malthus's mistake, it is generally agreed, was to overlook the inventiveness of human beings.

Yes: We do not yet know that Malthus was wrong. All we know is that the day of reckoning has been postponed—so far. The "inventiveness"—or more properly speaking the craftiness of human beings—and their techno-fixes cannot save them forever from the day of reckoning. That simply does not make any sense, and we can see the gathering storm on the horizon. Obviously, the human population is going to reach a crisis level when it gets to 10 or 15 billion people in this century. It is not possible for Earth to sustain that many people at the levels of material prosperity we have come to expect as our right in the developed world. Obviously, the human economy cannot persist in its policy of endless growth. Only cancers grow without constraint, and they kill their hosts in the process. We, too, are in danger of killing our host, Mother Nature, as global warming proves—unless we change our ways.

No: World population will rise only to 9 billion, and global warming is merely the latest expression of the human need for an apocalyptic vision. Those of us who have been alive long enough remember that back in the 1970s, when lower-than-normal temperatures caused record low winter temperatures, warnings that the next ice age was upon us were very common (Gribbin 1973, Ponte 1976). These warnings were also based on the most sophisticated climate science of the day concerning ice ages, namely the *Milankovich theory.** That theory, which was accepted through the usual

---

* Among scientists, the most popular theory of ice ages was first persuasively proposed by Milankovich (1941): Changes in the earth's axial tilt and its relationship to its orbital eccentricity cause the sequence of ice ages via alterations in solar heating (distance and angle) and its relationship to the seasons, or *orbital forcing*. Still, other theories, such as solar variability and changes in ocean currents, have their firm supporters. The main problem with orbital forcing is that it should have been at work for billions of years, although the cycle of ice ages has appeared only during the last few millions of years, far less than 1% of the total. So some other mechanism is needed to support the orbital forcing hypothesis, and about this there is no general agreement. In fact, the sun is supposed to have *increased* in brightness by some 30% since Earth was formed. The rapidity of the temperature changes involved in ice ages is another problem, since orbital changes are extremely gradual. Still, these and other problems are generally discounted given the generally good temporal agreement between orbital forcing and the ice ages, and the absence of a strong alternative to the Milankovich theory.

scientific consensus, said that an ice age is overdue (see also Ruddiman 2003). Yet instead of an ice age, we got 20 years of rising temperatures. Global warming is also said to be the consensus view of the most sophisticated climate science of the day, but that does not prove the theory to be true. The fact that the global warming theory is tied intrinsically to thundering rebukes of human greed (in this case because they use fossil fuels) actually puts the theory in a worse position than that of the overdue-ice-age theory of the 1970s. The casting of blame is a trademark of the same old doom-saying tendency that has kept prophets in business as far back as our historical records can take us.

Yes: Modern prophets do not rely on reading sheep's entrails and fear-mongering. To take a prime example, Al Gore has won a Nobel prize, the most prestigious scientific award in the world. As he shows in his movie *An Inconvenient Truth* (Guggenheim 2006), global warming is *already happening*, and is based as firmly in scientific fact as tomorrow's sunrise. He shows us that the seas will rise 30 feet or more, devastating the great coastal cities and other low-lying areas. The Netherlands will disappear. The lowlands of Bangladesh will disappear. Where are the hundreds of millions who live close to sea level going to go? And Gore is not alone in his warning. To cite just a couple of examples, the prestigious journal *Science* recently published an article (Kerr 2006) which predicted a sea-level rise of 6 m. The prominent science journal *Scientific American* featured an article with the telling title "Defusing the Global Warming Time Bomb" (Hansen 2004), which warns that global warming will cause "explosively rapid" (p. 74) disintegration of glaciers and sea-level rises of "several meters or more" (p. 68). These scientific warnings are everywhere—all we have to do is open our eyes and face up to what we see.

No: There is no denying that some scientists do make these warnings, but we should not jump to dire conclusions just yet. A closer look at Kerr's article shows that he merely *speculates* that there *might* just conceivably be a 6-m sea-level rise, while actually forecasting a rise of a mere 0.2 m over the next century on the basis of the evidence he presents. That comes to only 2 mm (about one-twelfth of an inch) per year, which is hardly apocalyptic. As for Hansen's article, we might remember that *Scientific American* is a popular science magazine that has to sell copies to nonscientists in order to stay in business. Anyway, a close reading of his article also shows that he does not actually predict these events, but instead presents them as within the range of possible effects of global warming. As for Al Gore, he does not predict catastrophic floods either. Instead, in his movie he draws a "what if" portrayal of the worst-case scenario. And by the way, Gore is not a scientist, and he did not win a Nobel prize for science, but for promoting peace.

Yes: The IPCC *is* a scientific body, a Nobel prize–winning body at that, and it also predicts a catastrophic rise in sea level. This body of internationally acclaimed scientists represents the work of yet thousands of other scientists as well. It is their work that supports people like Gore. He is just their messenger.

No: The actual change in sea level cannot hope to match the rate of change in the IPCC's own predictions. The IPCC predictions of the coming apocalyptic "flood" started out small—contrary to their own media releases and their exaggerations in the resulting media frenzy—and then shrank again with each new IPCC assessment report. Its first report in 1990 forecast a sea-level rise of only 25 inches (0.64 m) by 2100. Obviously not the Biblical flood, although interesting enough from a scientific point of view. Five years later, in 1995, the forecast fell to 20 inches (0.50 m), then it fell again in 2001 to 19 inches (0.48 m). In the latest IPCC report in 2007, in the face of evidence that it simply could no longer ignore, the forecast plunged to a mere 8 inches (0.2 m). Since the background rate of sea-level rise over the last few centuries is estimated to be 7 inches per century, the effect of global warming itself is tiny: 1 inch by 2100. Given that we have been easily accommodating a rise that is seven times that for centuries without anyone even noticing any problems, the additional inch will not be noticed either. This is not the apocalypse! And by the way, the Nobel prize awarded to the IPCC was also for peace, not for science (in fact, it is half of the prize given to Gore, which they shared).

Scientists are not immune to hysteria, but the scientific community cannot be blamed for the current hysteria about rising seas. The media is in the business of selling newspapers, magazines, television programming, and advertising for all of the above, so they have capitalized on the most newsworthy: that is, the *worst*, scenarios science can give them. But the scientific facts do not support their apocalyptic vision. When it comes to melting ice caps, the bulk of the arctic ice cap floats on the sea, and when floating ice melts, it decreases in volume to occupy precisely the same space as before, thus causing no rise in sea level.* As for glaciers on land, basic science tells us that even in the worst-case scenario of global warming, the bodies of ice on Greenland and Antarctica would require many centuries to melt (Bentley 1997), if they melted at all. Anyway, many of the IPCC's own climate models predict a buildup of glaciers, especially in Antarctica, as evaporation, and hence precipitation, increases. The small sea-level rise predicted by the IPCC is not predicated on melting glaciers in the first place, but on the thermal expansion of ocean waters as they warm up—assuming they do (and apparently they are not).† Some scientists calculate that the loss of ocean water to the growth of glaciers will counterbalance the thermal expansion and yield no appreciable change in sea level (e.g., Bugnion 1999). Finally,

---

* You can see this for yourself by letting ice cubes melt in a glass of water that is full to the top; the water level does not rise as the ice melts; ice floats only because water increases in volume as it freezes.

† Early releases of the results of the international ARGO (http://www.argo.ucsd.edu/) project indicate that the oceans have either remained the same over the last five years (http://argo3000.blogspot.com/2008/08/how-much-have-ocean-temperatures.html, September 11, 2008, posted by Argo TC on August 5, 2008) or have actually *cooled* slightly (Gunter 2008), not warmed, thus confounding the predictions of global warming theory. ARGO is the latest, most thorough, and first truly scientific measurement of sea temperatures. It involves a network of hundreds of buoys that take readings at various depths at various points in the world's oceans and then beam the results up to satellites, and thence to scientific teams for analysis (replacing the old patchwork system of thermometer measurements of the water intake of the engine-cooling systems of commercial ships). Given that the ARGO measurements are by far the best available, this negative result on the crucial issue of ocean warming must be taken very seriously.

as far as actual measurements are concerned, the rate of sea-level rise has remained unchanged since the early twentieth century (Nerem (NASA) 1997, Luick 2000, Luick and Henry 2000).*

Yes: Rising seas are just one threat among dozens or hundreds that now confront us. Even if some of these threats do not come to pass, there are still plenty of others left to cause us grief. For example, there is no doubt that there have been hurricanes of unprecedented power and fury in recent years, and this fulfills the predictions of global warming. The IPCC predicts an increase in hurricanes and hurricane intensity (IPCC 2001b, Sec. 9.5; 2007a, p. 74). If you trust the IPCC when they predict a moderate rise in sea level, then to be consistent you must trust them when they predict catastrophic storms.

No: Just because someone is right about one thing does not mean that they are right about everything. Anyway, the point being argued is not that the IPCC is right, but merely that it does not predict the apocalyptic flood that Al Gore pictures in his movie. In any case, we have to look to the evidence when it comes to prophecy of any sort, and simple observation shows that both the number and intensity of hurricanes has decreased slightly over the last century (Landsea et al. 1996, E. Smith, 1999).[†] And we might do well to remember that in 2005, Landsea resigned from the IPCC over what he called its continual misrepresentations of this issue in the global media (see Section 2.1).

Yes: When it comes to the evidence, there is no denying the warming in the Arctic, and no denying it is the harbinger of global warming. Arctic sea ice is thinning at an alarming rate. Polar bears are dying because they have no ice floes on which to hunt.

---

* An easily accessible account of the research Luick reports is given by BBC News online, November 22, 2000; see http://news.bbc.co.uk/hi/english/sci/tech/newsid_1035000/1035489.stm. According to Wolfgang Sherer, director of the National Tidal Facility (NTF) of Flinders University, South Australia, which undertook the research: "There is no acceleration in sea level rise—none that we can discern at all." Nerem came to the same conclusion after having earlier published mistaken results of NASA's Topex/Poseidon satellite showing an anomalous rise in sea level (1995a) of about 1 cm per year (100 cm per century). Then, after learning of a software error in the earlier computations, this was corrected to 0 mm per year. Finally, yet another source of error was discovered (Hancock and Hayne 1996), yielding a new result of "roughly 2 millimeters per year," well within the long-term background average of about 18 cm per century. It is of more than passing interest, given the huge sociopolitical and environmental significance of the global warming hypothesis, that whereas NASA's earlier results showing accelerated rise were given prominent publication (Minster et al. 1995, Nerem 1995b, White and Tai 1995, Hendricks et al. 1996), the corrected results were confined to a letter to the editor of *Science*.

[†] In somewhat more detail, Smith says (p. 2720): "Past hurricane intensity and frequency research indicates that hurricane frequency is on the decline. However, data from 1900–98 for East Coast landfalling hurricanes support the notion that the frequency of hurricanes is beginning to increase once again after bottoming out in the 1970's, while average hurricane intensity has remained essentially steady the last 3 decades." Landsea et al. instead consider the last 50 years, finding the same falling frequency, and that maximum intensity of hurricanes has remained constant, while average intensity has decreased. Both essays comment that these data disconfirm the greenhouse effect picture according to consequences drawn by its own proponents.

Arctic sea ice has broken up during recent summers, and the Northwest Passage has opened up for the first time in recorded history.

No: Arctic ice often breaks up in the summer: It is, after all, floating on a large, windy, and moving sea that is warmed by the Sun 24 hours per day during the summer.* In addition, the prophecies of the collapse of the Arctic ice sheet have been withdrawn. The latest IPCC report (2007a) has abandoned the claim of catastrophic Arctic warming, and now says merely that "Arctic temperatures are highly variable." In fact, it even admits that a ". . . slightly longer warm period, almost twice as long as the present, was observed from 1925 to 1945 . . ." (AR4, p. 37). (We might suspect, by the way, that describing a warm period that is almost twice as long as only "slightly longer" indicates that there may be some rhetoric involved in the IPCC assessment reports.)

Yes: The breakup of the Ross ice shelf and other enormous bodies of ice in Antarctica shows that something unprecedented is going on with the climate: It is warming up, just as gobal warming predicts. We seem to have reached some sort of tipping point for global climate.

No: The Ross ice shelf, like all ice shelves, is formed by glacial ice flowing out over the open ocean. Since these shelves can extend for 100 miles or so and are only hundreds of yards thick, their width is thousands of times greater than their thickness. They are called shelves, but if so, they are extraordinarily thin, like a shelf made of paper. Ice shelves must, and do, break up periodically, as they get too large to be structurally stable. Anyway, the ice shelves are just a tiny bit of the Antarctic ice. Satellites show that the total amount of Antarctic sea ice, including the Ross Sea, continues to grow year by year (Parkinson 2002).†

Yes: The very credible World Health Organization (WHO 2003) warns that heat waves will kill thousands. We have already seen this begin to happen in Europe in recent years. This is completely analogous with the deadly London fogs of the 1960s. Surely you cannot ignore the deaths of your fellow human beings.

No: Heat waves used to kill even larger numbers before the introduction of air conditioning, and death rates fall below normal immediately after heat waves, revealing that they only hasten the deaths of those already mortally ill and too poor to afford air conditioning (White and Hertz-Picciotto 1985, pp. 184–189). In addition, we know that cold weather causes more deaths than warm weather, and that warming would therefore decrease the total number of fatalities (White and Hertz-Picciotto

---

* Greg Holloway, a scientist with the Institute of Ocean Science in Victoria, Canada, discovered that reports indicating a thinning of Arctic ice may be based on measurement errors. See the Cambridge Conference Network report: http://abob.libs.uga.edu/bobk/ccc/cc042501.html.

† An easily accessible report of this research is given on the Goddard Space Flight Center (NASA), as its Top Story for August 22, 2002. See http://www.gsfc.nasa.gov/topstory/20020820southseaice.html.

1985, p. 184; McMichael and Beers 1994). So even if global warming happens (and in this context we should perhaps consider the various sorts of evidence for and against marshaled in Case Study 7), it may do more good than harm as far as death due to extremes of temperature is concerned. According to the proponents of global warming themselves, maximum temperatures will not rise significantly anyway. The bulk of the predicted rise in temperature is supposed to occur in minimum nighttime lows and winter averages. In short, global warming theory predicts that extremes of temperature will decrease, not increase. Since health problems occur in temperature extremes and rapid temperature changes, the overall effect on human health would be positive.

Yes: But global warming will bring a modern-day plague of tropical diseases into Europe and North America (IPCC 2001b, Sec. 9.7.1). The species of mosquito that carry malaria are now restricted to the tropics, but once the climate changes they will be able to move north. Surely a malarial plague counts as apocalyptic even for skeptics.

No: There is no evidence of this happening. In fact, there is evidence indicating that there is no link between malaria and such small temperature changes (Hay et al. 2002). Disease is linked far more tightly to poverty than to weather (Taubes 1997, Hay et al. 2002), so if we want to prevent a modern apocalyptic plague, the best thing we can do is to make sure the economy is not harmed by programs such as the Kyoto initiatives, which are supposed to (merely) slow down global warming by stopping our use of fossil fuels. Sometimes we might do better to remember history rather than to make forecasts, whether scientific or otherwise. It is a common, garden-variety historical fact that malaria was widespread in the temperate zones of Europe, America, and Asia all through recorded history up to the mid-twentieth century. Temperate-zone malaria gradually disappeared in the late nineteenth century, although we do not know why. We do know, however, that the most dramatic increase in malaria in recent history was due to the interruption of mosquito control programs over environmental concerns about DDT some 25 years ago (as we saw in Case Study 4).

Yes: When you isolate the individual signs of the coming apocalypse, doubt can be cast on each. Doubt is easy to come by. However, when they are taken together, they comprise a very credible threat. Even if each threat has a small probability, the probability that at least one of these threats will be realized is very high. In any case, the question we are considering is not whether the environmental apocalypse has been proven, but whether or not the *danger* of environmental apocalypse is real. Danger does not have to be proven to be real. And danger demands that we do not become complacent.

No: Agreed.

8.5   LOGICAL ANALYSIS OF ROMANTICISM

Romanticism is too complex to be captured fully in a simple analysis, but a rough outline of its main features is both possible and desirable. The heart of romanticism is a sense of lost innocence—a longing for the peace, security, and loveliness of an idealized childhood. Guilt is implied by lost innocence, so we conceive of various sins that we have committed. We have taken, or at least have accepted, something stolen from the gods; or we have disobeyed God. There were two paths, and we took the wrong one. One path, the path of innocence, is one of humility, faith, patience, and submission. The other path, the path of knowledge, is just the opposite, a path of confidence, knowledge, impatience, and defiance. Since we have taken the wrong path, the path of knowledge, repentance is required: We must go back, return to our state of original innocence.

In our hypothesized state of primordial innocence, we were magically free of care, want, fear, strife, suffering, or work. We lost this ideal state through using our knowledge to assert our own desires and to relieve our own fears. In other words, we used know-how, craft, and technology defiantly to take what we were impatient to get: food, clothing, shelter, art, science, entertainment, and so on. So we can regain our innocence only by rejecting technology, rejecting both the physical fires that enable our technology and the psychological fires of impatience and defiance that motivated us to employ technology in the first place. Romanticism asks us to go back, to reverse the entire evolutionary trend of *Homo sapiens*, to repeal not just the industrial revolution but also, civilization, agriculture, science, and technology. The first mistake, from the romantic point of view, was the use of fire by our distant forebears. We must get off the path of knowledge and get back on the path of innocence.

ROMANCE AND GLOBAL WARMING

From the romantic point of view, humankind now finds itself at a fork in the path, a fateful point in the unfolding of history where the choice it makes will define the age to follow. We have employed the craft of our science and technology to lift our curse of work, suffering, and death, if only partially and temporarily. We have used fire, the fire that the mythical Prometheus stole from the gods, to provide us with *energy*. It is energy that does our work for us, and that transforms the world for us so that we can live more easily and at greater length. In this process of turning fire into our servant, the process of industrialization, we have committed crimes both against nature and even against ourselves. In addition to many other crimes, such as leveling too many forests and gobbling too many fish, we have fouled the air with our smoke. Although fire is our servant, fire makes smoke, and smoke is an injury and an insult to all of nature, including us. The proof of our crime is that we are even injuring and insulting ourselves as natural beings, killing off our brothers and sisters through urban smog. Our crime is not only an injury we have perpetrated upon our own members, but proof that we have placed ourselves above nature, proof that we have even forgotten that we have a biological nature that can be poisoned. This is pride, and pride goes before a fall.

It is from this charged womb that the threat of global warming emerges. The global warming threat is that unless we remove not just the poisons (acids, heavy metals, particulates) from our smoke, but the CO_2 as well, a catastrophe will befall us and all other life on the planet. Although this catastrophe is perhaps not quite the full environmental apocalypse that some of the more fervent supporters of global warming theory (GWT) portend, it is nevertheless forecast to be a catastrophe in the full sense of the word, a catastrophe that we will bring on ourselves as a just punishment for continuing to offend nature by fouling the air with our chimneys and exhaust pipes. From the point of view of our reason, GWT is all about the greenhouse made by our smoke and just how much it will throw the atmosphere out of kilter. But in our hearts (i.e., from the romantic mindset) we know that we are at a moral, not (just) an intellectual, fork in the path.

The first choice is whether or not to accept GWT. It is open to the romantic, just as to anyone else, to reject the theory and to breath a sigh of relief that we are innocent of the latest, and perhaps greatest, crime against nature that humankind is alleged to have committed. However, romantics have a motive to accept the charge, since it confirms that their sense of alienation is justified. The fact that people who doubt the truth of GWT are called *deniers* is perhaps an expression of this romantic mindset: Denial is an inability to admit a misdeed or a flaw. The romantic believes in the crime of humanity against nature, so the thought that GWT may be false is experienced as a threat to the entire conceptual system of the romantic, a threat to the romantic faith, so to speak. Whereas from the point of view of reason, evidence that the criminal charge against humankind laid by GWT is false should bring relief and optimism to human beings in general, the romantic rejects such evidence as sacrilegious. If GWT is false, the romantic should breath a sigh of relief just like everyone else rather than cling to a mythical sense of guilt. Although it may be validating to the romantic soul to be told by environmentalists that we have harmed nature, the romantics who truly love nature would rather hear that it had not come to any harm in the first place.

If we suppose that the romantic accepts GWT, then the romantic accepts that we must, swiftly and massively, reduce and eventually eliminate the quantities of CO_2 that we release into the air. This, note, is precisely the conclusion that the IPCC proposes. The IPCC has both a scientific position, GWT, and a political position, the Kyoto accords. The core idea of the Kyoto accords is the value judgment that we ought to stop releasing CO_2 into the air. In the accords,

CO_2 is identified as a *pollutant*. So from the point of view of the romantic, or the IPCC, the next fork in the road looks like this: To the left we stop using fossil fuels, to the right we remove the CO_2 from our smoke. Let us consider these in turn.

The Path of Innocence. We stop using fossil fuels. Although I have heard of no one who advocates an immediate halt to the use of fossil fuels, even those who advocate a gradual halt nevertheless believe that ideally, in the best of all possible worlds, we would stop immediately. Since our release of CO_2 is causing harm, we should, ideally, stop doing that harm right now. According to standard GWT, the sooner we quit releasing fossil fuels, the more the approaching catastrophe may be abated. It is impossible to avoid catastrophe completely, according to GWT. All we can do at this point is limit how long the globe will warm and by how much. Most of the warming that is expected by 2100 (the IPCC's best guess is 2°C) will happen no matter what we do given the CO_2 we have already put into the air. Nevertheless, the sooner we quit dumping CO_2 into the air, the better. If we can swiftly begin to slow CO_2 use and then brake it to a halt by 2075 or so, this will limit the harm we do, and in the long run eliminate it altogether, whereupon we can once again live in harmony with nature. This is the approach recommended by the IPCC.

Why has there been no loud call for an immediate halt to the use of fossil fuels? Both the romantic and the IPCC accept that our continuing fossil-fuel use harms the environment continuously. So why do they not call for an immediate ban? The reason they do not is that we cannot stop using fossil fuels immediately without causing massive economic disruption and human suffering. Environmentalism itself would be among the casualties. If, tomorrow, all the tractors that work the fields and trucks that carry our food and clothing and shelter to market were to stop running forever, billions of human beings would perish. We have no idea how to secure enough food, clothing, and shelter for all of us without using fossil fuels. Quitting fossil fuels would lead to famine, disease, death, and war on a scale that would make the conjunction of the black death, religious warfare, and witch burning in Europe during the little ice age look mild. It definitely would make the industrial revolution look mild.

The Path of Knowledge. We remove the CO_2 from our smoke. If our scientists and technologists can remove the CO_2 from our smoke, they will not only avert the catastrophe, they will transform human nature as well. Removing the CO_2 from smoke leaves behind harmless water vapor, thereby *eliminating* smoke as a human waste. It is perhaps not common knowledge, but from its very inception, smoke has been a waste product of humankind. The earliest *Homo sapiens* not only used fire for campfires and cooking, but also to flush out game and modify the environment. We might envision a romantic biological classification, *Bestia incendiae*, Latin for "animal of fire," which would designate only those species that use fire. As far as we know, the use of fire for light, heat, cooking, sharpening points, and so on, was developed by *Homo neanderthalensis*. But it was *Homo sapiens* who really capitalized on fire, using it first to yield metals from ores, and then to yield energy from engines, and then electricity, and finally, the entire postindustrial, postmodern civilization that we now enjoy (assuming that "enjoy" is the right word). If we can transform our smoke into water, we will have transformed fire, and thereby ourselves.

From the romantic point of view we will have transformed the fire that Prometheus stole for us from the gods, the fire of Lucifer, the fire that has characterized *Homo sapiens* from its inception and confirms the romantic vision of the alienation of humankind from nature. But if we transform fire, the question remains: Would that absolve us of our guilt? Transforming the thing that we have stolen does not normally imply a return to innocence. Removing the CO_2 from fire may be nothing other than a bit of craft, a bit of cunning that has allowed us to escape justice, in which case we will have only elaborated our crime and delayed the day of reckoning. Removing CO_2 may, in other words, be nothing other than a techno-fix. On the other hand, taking a more scientific view, the crime against nature is the CO_2 itself, and removing it removes the crime, which is always the first thing to do in order to absolve guilt and return to a state of innocence.

The news is that it takes no magic to achieve this transformation of fire, humankind, and the fate of the natural world—other than the sorcery of physics and chemistry. Even as I keyboard these words, CO_2 is being taken from fire and locked up deep in the depleted oil wells from which it came in the first place. With yet another one of the ironic twists that characterizes human history, it is the petroleum industry that is spearheading this technology and is preparing to scale it up to "meet the challenge of global warming." It turns out the industry looks upon this technology as another way to make money. Who could have guessed? Perhaps this will permit us to stop releasing CO_2 into the air, or perhaps some reason will emerge for not using this technology (perhaps acidification of former oil-bearing formations?). Another answer might involve an even more profound transformation of fire: namely, the direct conversion of sunlight into electricity via solar cells. Those who loathe the techno-fix will prophesy the day when overuse of solar-electrical power brings its yet dimly envisioned environmental crisis. In any case, the technologies of CO_2 sequestration, solar-electric conversion, wind power, tidal power, and so on, prove that we can generate smokeless energy. The only question is: How fast, how efficiently, to what extent, and at what expense to both humans and nature.

Note well that even the main proponents of GWT and the Kyoto Accords do not call for the immediate halt of the use of fossil fuels. This implies that they accept the relevance of human welfare even when it is directly opposed to the health of the environment. From their point of view, if human beings were to vanish magically from the face of the Earth, this would halt our emissions of greenhouse gases, eliminate our warming effect on climate, and give the environment a chance to heal. This would represent the ideal outcome for the environment. So the only reason that they do not call for the immediate elimination of fossil-fuel use must be concern for human beings. Just how heavily they weigh the fate of humankind against the fate of nature is an open question in which we all have a natural interest. Assuming that humankind and nature are opposed in this way, as GWT implies, we should all be keen to investigate and debate just how heavily the fate of each should be weighed.

8.6   THE BALANCE OF NATURE IS A ROMANTIC FANTASY

The most common conceptualization of our environmental guilt is that we have upset the "balance of nature." Indeed, it was quite common to hear environmental scientists use the concept a generation or two ago, even though it was already known by then that a "balance" did not exist. From a scientific point of view, there is no such thing as the balance of nature. Nevertheless, the concept has such mythical resonance that many of you reading this will be jumping to its defense, (some scientists included, if only in a qualified sense). Certainly, the notion is so useful to environmentalists rhetorically that they are loath to give it up. The idea carries with it the idea of justice, which is also symbolized by a balance. Children reduced to tears by seeing a coyote kill and devour a rabbit can be soothed by the notion that the victim's suffering and death are all part of the balance of nature: In the larger scheme of nature, its suffering is all for the good, and so is just. By the same logic, the charge of upsetting nature's balance instantly carries with it the charge of injustice, of upsetting nature's scheme.

The *balance of nature thesis* (BNT) is that if left alone—that is, if not affected by human beings—natural systems (ecosystems, etc.) find the ideal balance between predator and prey populations, between wetlands and dry, between warmth and cold, and so on. One implication of BNT is that, among all of the elements of the natural world, the human beings, and only human beings, upset this balance. This is a dubious aspect of BNT, inasmuch as it is an expression of antihuman bias. Since the balance of nature is good, it follows that human beings will be bad, since every effect of human beings upon natural systems will by definition upset the supposed balance. Aspects of global warming theory (GWT), for example, show this antihuman bias, since human effects on climate are immediately characterized as "perturbations"—or disturbances—of natural climate. Thus GWT implies that whatever the climate does or would do independent of human influence is what it *ought* to do.

There simply is no reason to accept this value judgment. It is merely a romantic expression of antihuman bias on the one hand and of the innocence of nature on the other. We know that the ice ages drastically reduced the quantity, range, diversity, and security of life on Earth. Thus, in terms of environmentalist values, they were *bad*, not good. The ice ages were also bad in terms of romanticism and its values. Indeed, over the span of millions of years, the history of life on Earth has totally falsified any notion of a time of primordial innocence. From the beginning, life itself has involved external competition between living things with the elements themselves, and internal competition between living things and other living things. We know that the natural world is not like the balance arm of a set of scales, which tends to return to equilibrium if disturbed. It is, instead, like a ball rolling down an endless bumpy hillside (a multidimensional energy surface, to be a bit more precise), bouncing headlong into adventures, and never returning to quite the same state twice. The Earth and its living cargo comprise what physicists call a *disequilibrium system*, a dissipative system "pumped" by radiation from its local star, our sun. The resulting "system," like any very complex dynamical system, is most unsystematic. In fact, it is chaotic (i.e., deterministic, but unpredictable) and will periodically generate massive and abrupt (global and discontinuous) change. We know this, or at least the scientific facts are established beyond any reasonable doubt, but we have not come to terms with this reality.

The knowledge that nature is chaotic is like our knowledge of our own death. We all know that we are going to die—and we do not *realize* the fact. Wisely, perhaps, we put it out of our minds. So when the time comes for us to die and the doctor tells us that we have only a small time left, we are shocked and saddened. Tears come to our eyes and to the eyes of those who love us. Now we are beginning not merely to know that we are going to die, but to *realize* it, to truly comprehend it, both factually and emotively. In a parallel way, and for parallel reasons, we have been slow to realize that nature is chaotic. For one thing, that would entail admitting that the forms of life we now see are not eternal. Just as each living thing must die, so, too, one day, must all species die, and all the "ecosystems" that we have come to know and love must vanish along with them. Just as the species of old, like the multifarious dinosaurs, are no more, one day there will be no mammals, no coyotes, no rabbits, no *Homo sapiens*. Wisely, perhaps, we do not dwell on this fact. But whether we dwell on it or not, it is surely unwise to cling to, to believe in, its opposite, the notion of the balance of nature. If we need not dwell on unpleasant truths, we need not deny them either. The return to innocence cannot be purchased so cheaply as by a simple denial of knowledge.

In the biology texts of a few generations ago, the balance of nature was standard fare. Even though it was well known that there had been a recent series of ice ages, and that before this, eras in which none of the current life forms existed, it was nevertheless believed that somehow or other nature struck a balance. For example, it was thought that prey and predator populations were kept in balance. Inasmuch as virtually all animal species play one or both roles (herbivores prey on plants and are in turn preyed upon by carnivores, etc.), this amounted to a massive balancing act. But it could be seen as consistent with casual observation, and the natural tendency to romanticize nature did the rest. Thus, the balance of nature gained the status of an unstated presupposition that was effectively unfalsifiable. Any apparent imbalance could be explained as a temporary disturbance, or perturbation. A series of warm, wet summers might lead to a lot of grass, which then leads to a lot of antelope, and thus a spike in the population of lions—followed by a massive die-off of all three when conditions return to normal. So biological thinking evolved to include the idea that the balance of nature is sometimes perturbed but then returns to the norm, like a balance arm of a scale moved temporarily away from its point of equilibrium.

However, this craftier version of the balance of nature hypothesis conflicted with more careful observation, which revealed that predator and prey populations continued to oscillate even in the absence of known perturbations. So the notion of a natural balance was abandoned in favor of the idea of natural cycles in animal populations. Unchecked impala populations, for instance, would tend to grow, which in turn would cause a growth in the population of lions. The increase would then cause impala numbers to decrease, which in turn would cause a decline in the number of lions, bringing us back to where we started and setting the stage for another cycle. Hudson Bay records of fur trapping were broadly consistent with this cycling model in the case of snowshoe hares and lynxes.

On the basis of this data set and some others like it, the model of natural cycles of population was generally adopted. The model can be elaborated to include potentially all species, whether plants or animals. The lion–impala model, for instance, can be enriched by the addition of a species of grass upon which the antelope feeds, providing a sort of Malthusian check on both the lion and antelope populations. This gives a three-layered predator-prey model, with lions at the apex, impala in the middle, and the grass species at the bottom. Assuming that a model could be devised which actually tracked these three populations, other species could be added, such as cheetahs and leopards at the top, wildebeest and giraffes in the middle, and other herbaceous species at the bottom, without limit, in principle. In principle this model can also be enriched to include biological relationships other than that of predator and prey, such as populations of insects that feed on animal dung, flowering plants pollinated by those insects, and so on. Now the model is not one of a balance arm, but of a series of interlocked cycles linking the populations of all the species, like an enormous set of gears meshing with each other in a massively complex clock.

This model, the standard of professional biologists a generation or so ago and still presented in introductory textbooks, was also discovered to be simplistic and disconfirmed by the data. What the data showed was that cycles in animal populations are inherently unstable. A simple two-layer predator-prey experimental model of microbes in a test tube will sustain only a few cycles before it collapses, often with the death of both species as the predators become so numerous as to exterminate their prey, thereby sealing their own doom. Sometimes larger systems are more stable, sometimes not, at least in the sense that they do not collapse with the death of the organisms involved. Presumably the global totality of life on Earth is stable in this sense. In any case, all of these systems, whether large or small, are chaotic.

So there is no such thing as the balance of nature, but that means that the environmentalist conception of environmental health is based on a nonentity. Environmentalists and their organizations consider the ideal to be the state of nature when not influenced by human beings. Thus, to take a telling example, they believe that human beings should reduce—or eliminate—their ecological footprint. This is sheer romanticizing, an obvious call for us to return to the path of innocence and leave behind the path of knowledge. It is based on belief in human guilt for having upset the balance of nature. However, no argument has ever shown that untouched nature is ideal. Indeed, as the ice ages prove, no such argument can be given. It is merely *presumed* that human influence always disturbs the natural world. However, to influence nature is simply to change its course in some way, large or small. But change is not bad in and of itself. Yes, some of the changes we have brought about are bad, such as the extinction of some species of plants and animals. But it is simply false that any change that humans bring about is necessarily bad (intrinsically bad, bad as such, bad per se), even if we bring it about through agriculture or industry.

The history of our misty blue–green planet is one of constant unpredictable change. There is no guarantee that had we not devised agriculture, cut down the forests, hunted animals for food, driven cars, and so on, through the usual catalogue of our environmental sins, the natural world would now be in a better state—or even be more to the liking of environmentalists.* For one thing, given that we have followed the path of knowledge, we have achieved a level of understanding that permits us to consider and care for the health of the living things on this planet, something that otherwise would have been impossible.

THESIS 12:   The "balance of nature" is a romantic fantasy. The course of nature untouched by human beings is not necessarily good, and often leaves room for improvement.

8.7   THE ECOSYSTEM IS A ROMANTIC FANTASY

The concept of an ecosystem, or *the* ecosystem, is a logical cousin of the balance of nature and is used by environmentalists, including applied environmental scientists, for precisely the same purposes: to chastise human beings for their effects on nature, which then will count as "disturbances" or "perturbations" by definition. At a park not far from my house there is a sign saying that the park contains a fragile ecosystem, and exhorting those in the park not to disturb it, to stay on the paths, not to pick the flowers, and so on. The words *fragile, delicate*, or *threatened* almost always go just before the word *ecosystem*. This portrays the natural life of the park as a complex piece of machinery, and at the same time warns us not to be poking our fingers into it. Just as often, we are told that the ecosystem is precious or invaluable. Could it be that *every* ecosystem is fragile? Life itself, as we saw earlier, is anything but fragile. It is powerfully tenacious. Yet we always find ecosystems described as though they were playing-card towers, ready to collapse at the slightest touch. Human beings are

---

* The ancient and sacred forests that environmentalists are now trying to save from logging might never have been there at all if it had not been for us. Perhaps, as Ruddiman argues (2003), there would, instead, have been an ice age that would have destroyed the forest.

by definition a disturbing influence on these systems, and so by definition are clumsy, insensitive, greedy, and arrogant.

Admittedly, ecosystems are sensitive—that is precisely what is meant by saying that they are chaotic: Everything that happens to them or within them will have effects, some of which may be quite unexpected. But sensitivity is a completely different thing from fragility. Like all living things, a rhinoceros is sensitive, but it is certainly not fragile. The rhino is so sensitive that you cannot sneak up on one without being detected. Similarly, the life in the park of which I speak is sensitive: It responds to everything that happens to it. The main thing that has happened to it in recent years is that it has been identified and protected by humans as being a "fragile ecosystem." It shows many effects as a result, thereby proving its sensitivity. For one thing, there are many ducks, geese, red-winged blackbirds, and other birds nesting comfortably near the shoreline, raising their families. Were it not for the presence of humans there would be far fewer of these birds, because predators such as coyotes, lynxes, weasels, and raccoons would have gobbled some of them up and driven many of the others away. Clearly, this ecosystem is sensitive and so has responded to the presence of humans—albeit in a positive way as far as nesting birds are concerned.

Actually, it is very misleading here to speak of the ecosystem as sensitive. It is really the birds that are sensitive: They can tell that this bit of shoreline is predator-free. It is predator-free precisely because this bit of Earth's surface is frequented by humans, which scares away the predators. From the point of view of those who tend to think of human beings as unnatural, the ecosystem of these birds has been upset, affected, harmed by the presence of the humans that wander about on the pathways every day. But it hardly follows from this that the ecosystem is worse for the experience. What is wrong with ducks, geese, and blackbirds finding one more safe place to reproduce? Naturalists break into rhapsodies when natural conditions provide protection from predators, as sometimes happens on islands, allowing prey species to flourish. How, then, can this be wrong when human beings provide—however unwittingly—the same protection? It is wrong only if it is assumed that every effect of human beings on nature is automatically bad. Of course, since this stretch of shoreline is an ecosystem, any human-induced change will be readily characterized as a disturbance.

The idea that a piece of Earth's surface defined by municipal property boundaries and the organisms on it automatically forms an "ecosystem" is surely more poetry than science. Environmental scientists who employ the concept of the ecosystem can draw an analogy with the physicists' practice of treating any isolated set of elements as a system. A good example is the solar system. For all practical purposes, the solar system is isolated from the effects of the bodies around it, which permits it to be treated *in isolation* from the rest of the universe. The Earth and Moon can also be treated as a system, and so can a pendulum or a pot of water boiling on a stove. But arbitrarily chosen bits of Earth's surface will almost never be isolated in this way. In the case of the park in question, most of its elements are not only affected by things from outside the park but are only passing through it in the first place. The air, the ocean waters, the rain, and the birds themselves come and go in their migratory wanderings, as do the seals, herring, deer, insects, and so on. So it is not a system in the sense used in physics.

Anyway, environmentalists would not want us to think of an ecosystem as just any arbitrary isolated physical system. They want us to think of an ecosystem as a system in the biological sense, not the physical sense. In biology, the paradigm cases of systems are the digestive system, the circulatory system, the nervous system, and so on. These are systems in a very strong sense: They have *functions*. A function is a process that achieves a goal. The function of the digestive system, for example, is to transform foods into a form that can be put into the bloodstream and then to put them into the bloodstream in that form. Those very words also describe the goal of the digestive system. It can only achieve its goal by virtue of its organized structure, in which hundreds of components cooperate. But an arbitrarily chosen bit of the natural world will not have a function in this sense. A bit of shoreline plus its inhabitants do not make up an organized set of components that achieve a specific goal. All of the inhabitants do have goals, for they are organisms. But although each organisms has goals, it does not follow that the group of them share some higher goal, purpose, or function. A postman's function is to deliver the mail and the function of the key in his pocket is to open his car door, but he and his key do not form a system with the function of delivering the mail and opening his car door.

To think that an arbitrary subsection of the natural world forms a system is just like thinking that an arbitrarily chosen subsection of a body or a machine would be a system. But an arbitrarily chosen chunk of a body will contain bits of many systems, a bit of the circulatory system, a bit of the nervous system, a bit of the limbic system, and so on. But these bits of systems are not themselves a system. Real systems are composed of a set of parts that cooperate to achieve a goal. But an ecosystem contains a set of contestants that compete for the same goals: food, shelter, survival, and reproduction. Certainly, there are striking instances of symbiosis among organisms, and that is a form of cooperation. The paradigm of symbiosis is that plants breathe in carbon dioxide and breathe out oxygen, whereas animals do the reverse, thus completing a cycle. Given this cycle, it makes sense to speak of the corresponding functions and purposes. But for every example of cooperation there is just as good an example of competition. The animals also eat the plants, and predators eat their prey. Bacteria prey on both animals and plants, and viruses prey on all three. You will not find one set of cells within the nervous system preying on another set.

When it comes to *the* ecosystem, the total of all of the local ecosystems (however demarcated) taken together, the situation is, logically speaking, even worse. To put it very bluntly, to suppose that there is a function for the entire planet is to accept the Gaia hypothesis. What is the function of the entire planet? Surely it cannot be to maintain conditions so that they are optimal for life. Are we to think that all of the enormous numbers of different conditions that Earth has sustained over the eons were all, every one of them, optimal for life? That would be a lucky coincidence, to say the least. Are we to suppose that ice ages are optimal for life when the total biological activity on the planet is drastically reduced? And if we do suppose that prior to our arrival on the scene the ecosystem was optimal, this has the consequence that the specific set of species that existed at any given instant was optimal. In addition to being improbable simply on logical grounds, alone, this seems incredible on biological grounds. It seems quite likely that the removal of some species would increase the overall life of the planet. Removal of anthrax, for instance, might increase the health of life on the planet.

If in situations like this we are inclined immediately to suppose that even a known pathogen with no obvious purpose *must* serve some purpose, even if we do not yet know what it is, then we are obviously under the false impression that the planet as a whole is optimal in some way, that it serves some purpose with perfect fidelity. How could we know this when the function of the planet is unknown and the functions of its parts are unknown? Anyway, if we are to suppose that each organism and species performs a function that in mutual cooperation performs an overall function of life itself, how did it happen that we alone among the species are not optimal? If, in the face of these problems, we are inclined to defend the idea that everything has a natural function and that nature itself has one, too, and that we are guilty of upsetting this natural system, we have left the scientific attitude behind and are engaged in romantic fantasy. If, on the other hand, we respect the scientific demand for clarity and observation, we will not assume that the planet is a system in the biological sense. The very idea of the ecosystem is confused. There simply is no such thing.

THESIS 13:   The ecosystem is a romantic fantasy.

8.8   OUR SENSE OF ALIENATION FROM NATURE

As we have seen, romanticism begins with our sense of alienation from the natural world, our sense of lost innocence. This amounts to a form of guilt. Do we have anything to be guilty about? Is our sense of alienation justified, or is it also a romantic fantasy? Most of us grew up with pavement under our feet, industrially produced food in our stomachs, and vaccines in our blood. It is a fact that we are *separated* from the rest of nature, whether or not we are alienated from it. We place layers of technology between us and nature: clothing, houses, umbrellas, automobiles, and so on. We ingest medicines and vaccines to keep nature at bay even inside our bodies. Does this separation entail alienation and guilt?

We are not alone in separating ourselves from the rest of nature. All organisms do it, since life requires distinguishing self from other. In fact, we are not nearly as separated from the outside world as some other species are. Many species of termites, for example, seal themselves up inside the wood in which they live, thus combining food, clothing, and shelter in a triumph of efficiency and simplicity. Some termites build towers, which are termite cities, complete with nurseries and gardens and transportation routes and ventilation channels and termites specializing in the various trades required to tend to them. Termite cities are built, as are ours, to keep the natural world at bay, to deflect the elements and defeat the predators. So, separating ourselves from nature-at-large is not unique to *Homo sapiens*. But despite this similarity between us and other animals, our separation from nature is more profound than theirs. Like other animals, we have eyes and ears and mouths and noses. We are born, draw our first breath, and eventually our last, as they do. But the works of humankind stick out from the rest of the natural world like red sticks out on green. An automobile and a hamburger are not very similar as physical objects go, but they are identical in one way: They are completely unlike anything else in the natural world. We are far more *adaptable* than other organisms. Our constructions and our behaviors change continually, and continually become more effective.

But this still does not explain our sense of *alienation* from nature. Alienation is much, much more than the simple fact of separation. Alienation involves a sense of guilt, a sense of having willfully destroyed a friendship, perhaps for personal gain. Our sense of guilt is easily explained. Our sewers flushing our bodily wastes into our rivers, our factories belching smoke, our fields dewy under our pesticides, all these things that the romantic sensibility instantly sees as the very image of our alienation, involve *injuring* nature, generally for our own gain. Reason does not jump to our defense on this score. The charge that we have damaged nature is not easily dismissed as a romanticized fancy. Reason tells us, furthermore, that damaging nature is the height of folly, for in damaging nature we damage ourselves. Reason tells us that as far as our self-interest is concerned, the waste-removal technology of civilization has obviously lagged behind the food, clothing, and shelter technology. As late as the nineteenth century, cities such as Paris, London, and Rome were dumping untreated human wastes into the Seine, the Thames, and the Tiber, respectively. Not only did this injure other species, but injured us as well. Our sense of injury to nature is justified, for it is proven by our injury to ourselves as part of nature.

The fact that our waste technology still had not advanced past that of ancient Rome well into the middle of the twentieth century demonstrates our native stupidity raised to the level of criminal negligence. If we had been prepared to overlook the fact that the wastes of the tanner, the painter, and the chemist had been added to household wastes in the nineteenth century, we were apparently prepared to overlook not only the poisoning of fish, fowl, and game mammals, but the poisoning of our own kind as well. That would surely have been criminal negligence in purely human terms, even if not a crime against nature. It was a crime against nature as well, since we are part of nature. We feel horror and guilt over the fact that the naked chimney was still being used as the means to vent gases from factories and power plants in the 1950s, thereby killing uncounted thousands, although the chimney first came into common use during the Renaissance. Surely we can hope that the still-industrializing peoples of the twenty-first century do not have to repeat those tragedies—although it is dismaying, to say the least, to see their apparent rush to do so. At this point the romantic heart cannot help but feel some pain, and reason can at most reassure us that it is not too late.

So the romantic concept of our alienation from nature does point to something real and horrible, just as environmentalists and romantics aver. Our sense of alienation grows out of our sense of guilt over the harm we realized we were doing (even to ourselves) and a sense of horror over the worse harms we suddenly realized we were capable of doing (and not only to ourselves). Happily, beginning in the early nineteenth century citizens of cities such as Paris, London, and Rome undertook a series of improvements in waste technology that continues to this day. In general, our sewage, garbage, and chimney technology has been responding to the challenge set by the environmental problems of the nineteenth and twentieth centuries. We can breathe a sigh of relief that the levels of pesticides in wildlife body tissues has fallen to a fraction of what they were in the 1970s (see Figure 2.4). Note well that it was *technology* that solved, and continues to solve, this problem.

Our restlessness, like our intelligence and our technological expertise, is a product of evolution, just as the traits of other animals are products of evolution. It is perfectly natural for us to seek and devise technological solutions to the problems we face, just as it is perfectly natural for bees to make hives and for rabbits to dig an exit tunnel from their burrows. Our guilt is not ingrained in our nature, but derives from specific things we have done, in particular the pollution problems we have created. As we have already shown, we are able to recognize what we have done wrong and to take steps to repair the harm that resulted. Our intelligence and technological facility are not wrong in themselves, and rejecting them is not the right way to address our guilt and alienation.

The romantic notion that we must repent, reject the path of knowledge, and return to the path of innocence is out of touch with reality. To reject our intelligence, technology, and science would be to reject the natural processes that gave birth to human nature, an act of self-mutilation that would be every bit as much an insult to nature as to human nature. We did not steal fire from the gods. That is just a romantic myth. In reality, nature gave us the spiritual fire of intelligence that enabled us to employ physical fire. While this capacity creates hazards both for us and for the rest of nature, it also creates opportunities for both. Every power implies a corresponding responsibility. We should face up to responsibility, not run from it. Our *sense* of alienation is real and understandable. It does not follow that we should undo ourselves.

PASTORALISM

Like romantics everywhere and in every age, environmentalists ask us to go back to a simpler lifestyle, to the technology of the past. We are to repent traveling down the path of knowledge and return to the path of innocence. Generally speaking, romanticism prefers the comforts of childhood to the challenges of adulthood: We are to go back to the safety of obedience and not take responsibility for ourselves and our fate. Many environmentalists imagine that there was once a golden age, prior to technology, when human beings lived in harmony with nature. Just how far back one must go varies with the environmentalist in question. Moderates ask that we return to preindustrial times, to a time before genetically modified foods, pesticides, automobiles, fossil fuels, megacities, artificial fertilizers, petrochemical plants, and factories—to a time before the "dark, Satanic mills." Radicals ask that we return to the ways of small hunter-gatherer groups, before our numbers were large enough to hunt any species to extinction.

And now, many environmentalists (but not all) preach the abandonment of our technologically advanced way of life for a more primitive lifestyle; some "green" farmers, for example, have abandoned tractors for horses, and chemically fixed nitrogen fertilizers for horse manure. The irony is that horse-and-plow agriculture is also a human technological innovation, with roots going back to the first few millennia after the end of the last ice age, which caused far more environmental disruption than do modern agricultural methods. Most of the farmland now in use was put under the plow long before industrialization, using the ax and the ox. In many cases the land quickly became unusable due to preindustrial agricultural methods, and remains so to this day. In both the Old World and the New, remnants of these old, depleted, eroded fields can now be seen as depressions in land now given over to forests or brush. There is no question about it: There is nothing intrinsically better about preindustrial technology over our own, no matter how romantic horses and thatched roofs may seem. If the romantic environmentalists opt for a time prior to all technology whatever, they run up against the fact that *Homo sapiens* never had a nontechnological form of life at all. From the start we had fire, tools, and weapons.

It is becoming clearer and clearer that the bulk of humankind's effects on the environment occurred before the industrial age—often long, long before. Many species were hunted to extinction by prehistoric hunter-gatherers, such as the mammoth and the mastodon, and all of the large game species in Australia, as well as those of Madagascar and smaller islands. Some environmentalists themselves are aware of this. Eldredge says, "modern humans . . . reached Australia about 40,000 years ago, triggering a die-off of the larger native species of Australian mammals and lizards" (Eldredge 1998, p. 35). This is not a unique event, but part of a larger pattern: "Everywhere you look . . . the same pattern appears over and over again. As soon as modern humans arrive, there is a quick die-off, especially of the larger mammals and birds. . . . We modern humans were clearly like bulls in a china shop, disrupting ecosystems wherever we went" (ibid.). Or was the trouble that these "modern" humans of so many tens of thousands of years ago were not nearly modern enough?

The way forward for the environment is not to be found in a return to the past: neither the near nor the distant past. Our proper relationship with the environment is to be found in the future, not the past. If the fossil-fueled automobile is to be replaced, it is not by the horse, but by a better automobile or something better than an automobile. Rather than returning to horse and plow, agriculture must instead gain every bit of advantage from technological innovation that it can, in order to feed the billions of human beings already on the way with the fewest possible negative effects on nature or humankind.

This is not to say that every technological innovation must be embraced. There is wisdom in wariness of things that have not been seen before, but it is unwise to reject the new automatically. Cautious openness is the best attitude. Romantic efforts to stop scientific progress, to return to the past, are based on a romanticized vision of the past and a romantic rejection of technology. The solution does not lie in the past, and returning to the past is simply not going to happen. The issue is not whether to accept technology, but what the best technology would be.

9 Nature and Values

Nature has placed mankind under the governance of two sovereign masters, pain and pleasure. It is for them to point out what we ought to do, as well as to determine what we shall do. On the one hand the standard of right and wrong, on the other the chain of causes and effects, are fastened to their throne.

—Jeremy Bentham (1789)

Death eventually intrudes, even in this postindustrial, postmodern age devoted to the ideals of safety, comfort, health, and happiness. We do try to keep death hidden behind closed doors as much as possible, where it can be dealt with by professionals: doctors, priests, veterinarians, and killers on the slaughterhouse floor. But people who are dying need their friends and loved ones more than ever, and in answering this need we eventually confront death. I looked death in the face during my final watch over Chico, my brother-in-law and friend, a lovely man who was dying, too young it seemed, of cancer. In the middle of the night Chico suddenly jerked up into a sitting position in his bed, gasping, his eyes bulging. I reached out to him and he grabbed my wrist with cold, hard desperation, and began to breathe as hard and deep and fast as humanly possible, as though he were trying to inhale every molecule of air in the room. I asked him what was wrong, shouting his name, but he did not respond. Then his eyes looked into mine, and I saw that he was struggling to live with every ounce of his will and strength. He was breathing like a horse at full gallop, and yet he was suffocating. But terrified as he was, he was not giving up. He struggled to live with every ounce of his being. I knew Chico; I knew immediately what he was feeling, as if his face were my own. I felt his horror of impending death. I stared into the darkness that he stared into, and felt that I was falling into it. I confess that I wanted to run away from the sheer terror of the moment. I could not have left if I had tried, since he gripped my wrist so hard I thought he might break it. I began to feel sick as we both heaved, locked together in his mighty struggle to draw air.

Somehow I managed to keep my stomach down, and Chico managed to get enough air to survive until a medical team arrived to connect him to the hospital's oxygen system. He had suffered an embolism that blocked blood flow to one of his lungs, they told me, though why a liver cancer patient would get an embolism was never explained to me. He survived the night, and the next one too, and so got to see his best friend, who was traveling from afar, before dying. I got to ask him the question I was

Beyond Environmentalism: A Philosophy of Nature, By Jeffrey E. Foss
Copyright © 2009 John Wiley & Sons, Inc.

dying to ask: What did he see when he was staring into the darkness? He, a former nuclear submariner, answered, "It was unfathomable. Unfathomable." I understood.

The process of natural selection favored animals with a lust for life and a dread of death. Lust and dread are famously insensitive to the dictates of reason. Reason cannot quite comprehend death, much less control our feelings about it. Try holding your breath and you know firsthand the dread of death and lust for life in the form of a bodily system that commands you to breathe and forces you to obey. Every cell in our body is programmed to survive at all costs, and the body, their creation, is designed to obey at all costs. Being made of these cells, it is difficult for us to accept death. We are designed to fight or flee as required, never to surrender, never to give up. We are not designed to know and understand, but to fly in the face of the odds and always, always bet on our own survival. We have been designed by natural selection to deny the reality of death. Romanticism capitalizes on this denial in its portrayal of nature.

But if we do face the reality of death, it is plain that nature takes our life, just as it gives us life. If we take the life of another human being, that is murder, a truly horrible crime, yet nature will take the life of each and every one of us. Nature is our mother, but it is also our executioner. It is our friend, yes, but it is also our foe. It is the sum of all our blessings and all our curses. For us, as for all living beings, nature is good *and* bad, beautiful *and* ugly.

9.1 NATURAL VALUE

The most fundamental evaluative error in environmentalism is the belief in the goodness of nature—or even the *perfection* of nature. It is because nature is romantically believed to be good that it must be saved, even at the expense of human suffering or death. But it is fairly easy to see that nature is neither good nor bad in itself. To see this, we need to reflect on the origins of goodness and badness themselves, or in other words, the origins of *value*. We can begin by imagining the world without any pain and pleasure, and we will discover that it cannot have any good or evil either. The easiest way to conceive the world without pain or pleasure is to imagine it at the time before life and consciousness emerged. According to our best current conceptions of the history of the universe, it existed for billions of years before conditions right for life-capable planets such as Earth emerged, and planets such as the Earth itself existed for hundreds of millions of years before life appeared on them. For the purposes of this exercise, it does not matter whether this scientific history is accurate, but rather, that it is conceivable. In such a world, before life itself exists, and hence before consciousness, pain, and pleasure exist, nothing is either good or bad. Why? Because it simply does not matter what happens, because absolutely nobody cares. A beautiful, albeit lifeless planet might come into existence, but its beauty exists entirely unappreciated and entirely unable to be appreciated. This planet might then crash into another and be destroyed, but nothing and nobody is hurt. The only way any of this could possibly matter is if some being might at some time and place possibly care.

NATURE AND INTRINSIC VALUE

It seems plain that a universe completely devoid of consciousness of any intensity or sort would also be devoid of value. Many environmentalists have argued in favor of values that do not require consciousness at all, in large part, I suspect, because they feel that this is one way to guarantee, at least in principle, that plants and landscape will be accorded some value by their fellow humans even if both are unconscious (as seems almost certain).* Thus, it is proposed that nature has *intrinsic* value, as opposed to instrumental (or extrinsic) value. From a logical point of view, it seems that value is at a minimum a two-place predicate: X is valuable to Y. From a logical point of view, value is just like the concepts of food or visibility. All food can nourish some organism or other, and everything that is visible can be seen by some organism or other. Just as it makes no sense logically to speak of something being intrinsically food or intrinsically visible, it seems that it makes no sense logically to speak of something being intrinsically valuable. A one-place value predicate seems logically confused and so therefore does the concept of intrinsic value. As we saw in Chapter 3, the idea that nature has intrinsic value is part of the environmentalist tendency to see the good of the environment as a transcendental objective, which in turn dangerously subverts all other human values. We thus are well advised to reject the idea that nature is intrinsically valuable both on logical grounds and on the basis of human values in general.

But the minute there exists a single conscious being with the capacity to feel pleasure or pain, the world contains both good and bad things from the point of view of that being. If the entity experiences pain, that is a naturally bad thing. The natural world is worse than it otherwise would have been because of that pain. Similarly, if the entity experiences pleasure, that is a good thing, and it makes the universe a better place. When we speak of the universe becoming better or worse, this is to be understood as an evaluation from the point of view of the sole conscious being that it contains. By hypothesis, our imaginary universe contains only this single conscious being. Without this being, there would be no pain and pleasure, and nothing that is either good or bad. Clearly, then, the goodness or badness that results upon the arrival of this being is *relative* to this being. Since it is the *only* being, there is a hint here of absolute goodness or badness, inasmuch as there is no other being whose evaluation

---

* On the other hand, I do not think that there is any sharp boundary within nature between things that are alive and things that are conscious. As I see it, all life requires some sensitivity to surroundings, and sensitivity is a form of consciousness. Nor do I think that there is a sharp boundary between the living and the nonliving. This is not to say that these distinctions cannot be drawn, only that they are distinctions of degree rather than distinctions of kind. A mountain is distinct from a molehill, even in the absence of a sharp boundary between the two. A rock is not conscious, and a person is. Between them lie many grades and gradations of consciousness, which again shows that the distinction between the conscious and the unconscious is valid.

matters. Nevertheless, the logic of the situation is clear: Natural good and natural evil is always relative to the point of view of some conscious being that is capable of pleasure and pain.

THESIS 14:   Pain and pleasure are the basis of natural values.

Bodily pain is the paradigm of natural evil, and bodily pleasure is the paradigm of natural good. However, this is not to say that only *bodily* pains and pleasures, the ones that have a location in the body, count as good or bad. There are all sorts of nonbodily pains and pleasures which create their own sorts of good and evil, but all of them depend on bodily pains and pleasures for their existence. For example, witnessing your child suffering may cause you enormous pain, although there is no particular place in the body where the pain is felt. We call this emotional pain or psychological pain, to distinguish it from bodily or physical pain. Of course, emotional pain does normally cause us some physical discomfort as well: We may feel tightness in our chest; our heart may pound: our stomach may be in knots. However, these pains are not the pain of knowing that your child is suffering. Jogging on a full stomach might cause the same cluster of physical discomfort, but that would be a very different pain, indeed a much less intense pain, than that of knowing your child is suffering. On the other hand, most emotional pain is a function of physical pain. If someone threatens us with a weapon, the threat of the pain it would cause would in turn cause us emotional pain. Since it is physically pleasurable to eat when we are hungry, anticipating eating may give us psychological pleasure. But whether or not all pains and pleasures are ultimately a function of physical pain and pleasure, they still provide a basis of natural values.

Those things that an organism seeks as part of its nature will have a positive natural value for the organism, while those things that it must avoid will have a negative natural value. Natural values can be observed empirically from both the first-person point of view (we can experience food, sex, warmth, and other sources of bodily pleasures and pains) and from the second- and third-person views (I can see that you value freedom from pain; we can see that other animals value freedom from pain as well). Like other natural properties, natural values are tied to natural kinds and are captured in type truths. Just as it is a natural type truth that rocks are hard and dandelion fluff is soft, it is naturally type true that photosynthesizing plants need sunlight and human beings need food. That human beings value food is also a natural type truth. Natural values are matters of degree, such as warmth, light, nutritional content, toxicity, or other natural properties. Too little food is bad, but so is too much food. Just as plants need a certain level of sunshine, so there is a certain level of food that is good for human beings. Natural values are also typically in tension with each other. Animals require both food and shelter, and typically cannot pursue both by the same course of action. Therefore, a harmonious resolution of the tension is required.

At the core of natural values of any species are its natural necessities, the needs it must satisfy to survive and reproduce. Again, there is no sharp boundary separating needs from other values. The coyote clearly needs food, so it kills and eats a rabbit, but whether it needs every mouthful of that rabbit is not clear. Natural necessities gradually shade off into natural desires. We should not make the mistake of thinking that any vagueness in the distinction between needs and desires means that the distinction itself is false. There really is a difference between mountains and hills, even though there is no nonarbitrary height to distinguish the two. Similarly, there really is a distinction between the human need for food and the human desire for caviar: You can get by without caviar but not without food. Many animals are capable of pursuing their desires even after their needs are satisfied. Cats may hunt even when they are full, and beavers continue to dam streams even when they have more than enough pond habitat already. However, the gap between needs and desires tends to be small except for *Homo sapiens*. The fact that human values have a vast range and even go beyond the natural to the supernatural and the transcendent is part of what makes humankind appear to be somewhat unnatural or even beyond the natural world altogether.

Nevertheless, the fact that people desire such supernatural things as pleasing God, and aim for such transcendent objectives as the health of the environment, does not eliminate the fact that they, too, have natural needs, and that the satisfaction of these needs is a natural good. While illuminating the goals that human beings *should* have, whether natural or not, is an ongoing philosophical project that cannot be addressed profitably in this book, recognizing the natural good of satisfying the natural necessities they *do* have, and that nonhumans have as well, is central to the philosophy of nature. The natural goods of the human species are food, clothing, shelter, and natural freedom (freedom from excessive work, suffering, and death). Other things being equal, it is good for human beings to have food, clothing, shelter, and natural freedom. So, given merely that we ought to act, as far as we can, to bring about the good, we ought to act to realize natural human goods. This assumes that the achievement of natural human goods does not cause any outweighing natural harm, for the range of natural goods is not restricted to human beings alone but extends as far as consciousness extends within nature. Pain, the fundamental natural harm, is bad whether it is the pain of a human being or the pain of a mouse or whale. Given the same quantity of pain (roughly the product of its intensity and its duration), it is just as bad no matter whose it is.

Natural values, then, are plainly not anthropomorphic, or human-centered, or biased toward the interest of human beings. For human beings to work for the increase of natural goods and decrease of natural evils is not for them to be selfish in some way. To repeat, natural good and evil extend as far as consciousness of pleasure and pain extend within the natural world. Just how far they extend is not entirely plain, although we do have clear cases. Clearly, human beings are capable of conscious pleasures and pains, but so too are many other animals. Cats, dogs, horses, and pigs do feel conscious pains and pleasures. The evidence for this is manifold: Their bodies, in particular their nervous systems, are similar to ours. Pain serves the same purpose for them as it does for us from an evolutionary point of view: avoidance of damage and death, nursing of wounds, prevention of further damage through normal activity, and so on. Pain is also indicated by the same sorts of behavior: involuntary interruption of activity; involuntary spasms, contortions, and cries; determined efforts to escape the cause of pain; and so on.*

Whether or not something is capable of feeling pain or pleasure is a vitally important question because natural good and bad extends only as far as pain and pleasure do. A cat cannot be kicked around like a football, because doing so would cause it pain, which is naturally bad. And while we do not get the luxury of proof concerning any matter of natural fact, including whether or not a given organism can feel pain, we nevertheless can be sufficiently confident about the extent of consciousness to act without uncertainty or remorse in most important cases. We have no reason to suppose that we have caused massive pain to untold thousands of microorganisms when we take an antibiotic, and little reason to suppose that cutting down a tree causes it any pain. We have good reason to suppose that it does hurt warm-blooded animals (and some other animals as well) to be physically injured. Assuming that we are correct about the potential for pain and pleasure in these cases, it follows that microorganisms and trees cannot be harmed, but many types of animals can be harmed. In general, only those organisms that can feel pain can be harmed, and only those that can feel pleasure can be helped.

THESIS 15:   Only those beings that have the capacity for pain can be harmed, and only those with the capacity for pleasure can be helped.

Therefore there is nothing wrong in itself with damaging or destroying an organism that is incapable of feeling pain. The only way that such damage or destruction could be wrong would be via its impacts on other organisms that are capable of pain. Most human beings are aware of this, so they treat inanimate or insensitive things very differently from the way they treat those beings that are capable of pain or pleasure. Roughly speaking, our care and concern for other natural entities is proportional to their capacity for pain or pleasure. Human beings have by far the largest range of pains and pleasures, since we invest most things in life with our basic biological interests for survival, flourishing, and reproduction. Such biologically abstract things as a rise in the price of gasoline might hurt one person because she owns a trucking firm, and give joy to another because he has invested in fuel companies.

---

* It is sometimes argued that we can never know that nonhuman animals feel pain, or are even conscious, since they lack language with which to communicate to us their pains or their conscious states. This argument unreasonably privileges linguistic communication over other sorts. Indeed, we have the power to describe our pains to others in words. However, verbal avowals that we are in pain are not convincing in the absence of the more universal behavioral and physiological expressions of pain. If someone calmly tells us that he feels excruciating agony, we will not believe him. People in excruciating agony should be agitated, covered in sweat, unable to pay attention, and struggling to escape their pain. Simple verbal avowals mean nothing in the absence of these more important forms of expression when it comes to pain and pleasure. Although we have words such as *ouch* to express pain, they do so mainly by their emotive rather than their lexicographical content. A calm "that was extremely pleasurable" is not half as convincing as an emotionally charged gasp or sigh.

However, here we can concern ourselves only with natural good and evil at the type level. This, in turn, requires that we deal with the most basic and universal types of human desires, or human needs. This is not to deny the importance of the particular desires of each of us or our own particular conceptions of the good in a larger sense than being considered here. The philosophy of nature is continuous with ethics in the full sense, and the border area between them need not be formally, and officiously, defined. On the other hand, its main concern is the ethical or evaluative place of nature within philosophy as a whole. The answer, I suggest, is that there is a type of good and bad that is completely natural. It presupposes nothing supernatural, and can be observed like other natural phenomena and even studied with the aid of the natural sciences. Natural values extend as far as conscious pains and pleasures do, and we should continue to explore just how far that is. Indeed, we have seen that there are various sorts of bodily pains and pleasures, each with its own range among organisms. Assuming that this summary is correct, then natural good and natural evil does exist. If we merely assume what is generally taken to be platitudinously true, that we ought to promote good and decrease evil, it follows that we should support natural goods and oppose natural evils.

The values entwined in our very nature are our values. They are not only our values, since we share these values with every conscious organism capable of feeling pain or pleasure. And they are not our only values, since we also value justice and courage and the willingness to forgo pleasure on behalf of higher causes and even to suffer and die for them. Nevertheless, natural values are our values. They are in touch with fundamental realities. Bodily pain has the function of indicating bodily damage, and bodily pleasure has the function of making us satisfy our biological necessities. Natural values are an expression of our fundamental human nature and our human love of life. They grow out of our union with life itself and entwine our interests with those of living things everywhere.

CASE STUDY 10: IS NATURE GOOD?

Yes: Nature is not only good, but good in itself. No doubt the case can be made that nature is good for human beings as well, but to conceive of the goodness of nature in those terms is to make the same human-centered mistake that has led to so much environmental damage in the past. Yes, nature is good but not just from the anthropocentric point of view, not just in terms of human interests. It is good on its own terms, without reference to us. In other words, nature is intrinsically good.

No: Intrinsic values do not really make any sense. Suppose for a minute that you are right, that nature is intrinsically good, which means that its goodness is not a matter of being good for us human beings. Well, assuming that we are not being singled out here as the only organisms whose interests are not served by the intrinsic goodness of nature, it follows that none of the other life-forms on the planet are being served by nature's intrinsic goodness. Indeed, that is precisely what is meant by saying something is intrinsic: namely, that it is not relational or relative to anything else. In other words, being *intrinsically* good is distinct from being *good for* something else. But that means that the intrinsic goodness of nature is not good for any living thing—or indeed any thing at all. So even if nature is intrinsically good, this is hardly of any interest as far as the question at hand is concerned. To put it baldly, goodness that is not good *for* anyone or anything can only be of interest to metaphysicians and no one else (see also the box "Nature and Intrinsic Value" above).

Yes: Perhaps the concept of intrinsic goodness is too abstract to capture the goodness of nature accurately. But nature is nevertheless good, and the only reason that the concept of intrinsic goodness was used here was to make it obvious that the goodness of nature is not merely a matter of its serving human interests. Maybe the best way to conceive the goodness of nature is that it is good for *all* natural beings rather than none. Indeed, it is good not only for living things but for the purely physical things upon which living things depend: the oceans, the land, the atmosphere, and so on.

As a matter of fact, this way of looking at the goodness of nature helps us to get at the heart of the environmental crisis. Nature has been good to us human beings, but we alone, among all of the life on the planet, have not appreciated nature's goodness and have repaid it with neglect and even contempt. We call nature "Mother Nature" for a very good reason: It has brought us into the world and has nurtured us. But we have repaid her with pollution, destruction of habitat, and unrestricted growth of our population and its environmental impact. Nature *is* good, but we are not. We have returned nature's goodness with evil, and it will surely extract its revenge unless we change our ways, and soon.

No: You are now speaking of nature as though nature were a person, and that raises a different sort of problem. Values can be either personal or impersonal. The use of the good–evil contrast usually indicates personal values. Good and evil flow from people's intentions, while impersonal good and bad are not intended by anyone. The kindness of one person to another is an example of personal goodness, while the healthy effects of vitamins are examples of impersonal goodness. Cruelty is an example of evil, while bad weather is merely bad, not evil. But nature is not a person and does not intend to do us either good or harm. In terms of good versus evil, the question of whether nature is good can only have a negative answer. Nature is not good, but not because it is evil. Nature is not good because it is neither good nor evil. Nature does not want anything for us, either beneficial or harmful.

Yes: Don't get carried away with my use of the poetic image of Mother Nature! Of course nature is not literally our mother, and does not literally love and care for us—at least it would be difficult to argue that it does without sounding farfetched. So let us just grant that nature has no intentions to either help or harm us or any other thing. Even so, it is good, for it has given us life, and life is good. As Wordsworth once said, "tis my faith that every flower enjoys the air it breathes" (Wordsworth, 1798b).

No: It seems to have gone generally unnoticed that there are far more bodily pains than bodily pleasures in life. Part of the reason it is not noticed, perhaps, is that we

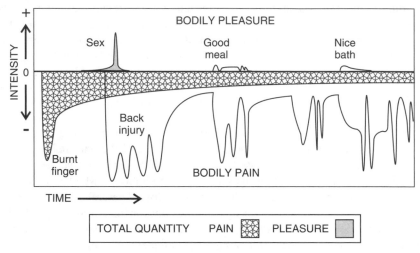

Figure 9.1   Bodily pain and pleasure. Bodily pleasures are brief and mild compared to bodily pains, which tend to be more intense and last longer. Consequently, total bodily pleasure is massively outweighed by the total bodily pain for virtually everybody. The total quantity of pain or pleasure corresponds to the area under the curve for a given sensation. The total quantity of pleasure for sex and the total quantity of pain for a burnt finger have been graphed to make their relative sizes plain.

are in the habit of lumping bodily pleasures and pains in with pleasures and pains generally, rather than considering them in themselves. A bodily pain is one that is felt in, or at, some location in the body. If you burn your finger, the pain is in your finger. Similarly, bodily pleasures occur at some location in the body. Now, as you can confirm for yourself, the most intense bodily pleasure, orgasm, is not nearly as intense as the most intense bodily pain, nor does it last nearly as long as equally intense bodily pains, as shown in Figure 9.1. In fact, a fairly mild bodily pain, such as that produced by being stabbed by a pin, will normally overpower an orgasm. Given that the most intense bodily pleasure is outweighed by a modest bodily pain, we should not be surprised to find that generally speaking, our pleasures are small relative to our pains.

It is common for people to suffer from prolonged (even intractable) pains. A car accident may cause a back injury that is intensely painful day after day and can only be relieved by strong medication. Moreover, just from hefting a bag of groceries, you can experience nearly paralyzing back pain, a continuous dull throbbing that lasts for days, punctuated by agonizing stabs when you try to move. It is perfectly routine for people to call their doctors or their pharmacists to get pain medication to get relief from pain. But you will never hear someone calling the doctor in search of relief from paralyzing pleasure: "I bumped my knee and the pleasure it is causing me is driving me crazy. I can't concentrate, can't get my work done, can't drive safely. Please, Doc, you've got to give me something for it!" Of course, if we did feel constant intense pleasure, we would think twice about making it stop, but the point is that the issue just never comes up.

Each of us knows what it is like to suffer protracted and intense pain. Some of us experience pain of truly epic, life-spoiling dimensions. Yet none of us has experienced pleasure of anything like the same degree or duration.* None of us has any trouble imagining the torments of hell, and descriptions of hell abound, yet the pleasures of heaven are vague and insubstantial—because we have plenty of experience of prolonged and intense pain but no experience of prolonged and intense pleasure.

Yes: There is a perfectly good explanation for the imbalance: Nature is protecting us against the hazards that surround us. Pain is certainly awful but it serves a natural and necessary function, as is proven by the fact that those rare individuals who are born unable to feel pain usually do not live very long, because they tend to injure themselves and not look after their injuries when they do. Pain protects us from all of the things that might harm us, although we naturally do not find it very nice. On the other hand, pleasure is nice, but it also just serves a function: It draws us toward the things we need, such as food and sex. You are being self-centered, and frankly, rather childish to complain about the amount of pain in life, when both pain and pleasure serve our best interests in the end.

No: There is no denying that pain and pleasure serve functions, but that is not relevant. The question is whether nature is good, and that brings us back to the fact that the natural world in which we live gives us much more pain than pleasure. Food and sexual partners, as it turns out, form only a very small proportion of the objects with which nature confronts us. Aside from these two sources of significant pleasure, there are innumerable sources of significant pain. Snow chills us, fire burns us, thorns stab our flesh, insects pierce us and suck our blood, attach themselves to us, or bore right under our skin. Plants poison us; larger animals kill us to eat, and sharp stones gash our unguarded feet as we walk. Clothing is not an option for us. We need clothing just to avoid physical harm and discomfort. We need footgear just to walk about.

There are sharp pains, dull pains, stabbing pains, shooting pains, burning pains, throbbing pains, aches, stings, prickles, itches, and nausea: a wide variety of negative stimuli designed to help us distinguish one noxious source from another. There is no matching range of bodily pleasures of similar variety, force, and vivacity. The intense pleasures that we do feel are fleeting and can be enjoyed only under ideal circumstances. Sexual intercourse leaves us very vulnerable to predators or competitors; hence sexual pleasure is fragile and easily interrupted. Eating is a pleasure, but not intensely so, for that would be dangerous and distracting. Our fear of predators and of other dangers must generally be stronger that our pleasures, or else we would be eaten long before we had a chance to perpetuate our species. Nature has designed us so that our pains generally outweigh our pleasures. That is hardly a good thing.

* Drug users may make an exception to this claim. Since like the ones surrounding it, this claim is meant as a type truth, as in Section 7.5 and as explained below, its truth is consistent with occasional exceptions. For more on how drug use fits within the range of pleasure and pain, see Foss (1996a).

Yes: If you are right, you are implying that life is not worth it, which reduces your argument to absurdity. You have been given the invaluable gift of life, and yet you complain that there is too much suffering. Life is good, and therefore the struggle to live is a worthy struggle. Your argument undermines our morale in this struggle and therefore should be rejected as not only misguided but as dangerous. If you were right that there is more pain than pleasure in life, people would not want to continue living; but people do want to continue living, so obviously there is more pleasure than pain.

No: People do in fact cling to life even when it obviously gives them much more pain than pleasure. Natural selection has provided us with a powerful drive to survive, a drive that persists in the teeth of even quite severe pain. That is why people want to continue living. The simple fact of the matter is that we will strive to keep living as long as we are free of prolonged and intolerable pain. Because we have been outfitted by evolution with a burning desire to stay alive at all costs, we are vulnerable to what amounts to self-torture: We are inclined to keep ourselves alive even when we are in a state of unrelenting pain, and even when there is no hope in sight for its relief, rather than die. To keep someone else in a state of pain without any hope of relief would be a crime, but nature routinely does this to us as well as other animals. Nature and human nature cooperate to make us not only suffer but to make us unable to escape our suffering.

Yes: You are exaggerating the pain and ignoring all of the joy. What you say is not strictly speaking false, but it is not balanced. Nobody has claimed that nature is entirely good. Nobody is trying to deny that nature can cause living things pain, even intense and prolonged pain. But that is not to say that nature is not good.

Common ground. Nature is not wholly and entirely good.

THESIS 16: Nature is not wholly good. It is both good and bad.

9.2 THE VALUE OF LIFE ITSELF

What about the value of life itself, the joy of living? Is that the reason we continue to struggle to survive even when our bodies no longer provide us with an excess of pleasure over pain? There is something to this idea, but it is not easy to define just what the value of life itself might be. Certainly, death is of extreme negative value to us. Death is the object of fear and loathing. Usually when death is near we are racked

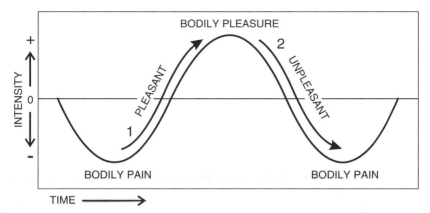

Figure 9.2   The pleasant and the unpleasant. We find the relief of pain pleasant, and the loss of our pleasures unpleasant. 1, even when in a state of pain, we will find it pleasant when the intensity of the pain decreases; 2, even when we are experiencing pleasure, we will find it unpleasant when the intensity of the pleasure decreases. This shows that our sense of satisfaction and dissatisfaction can extend beyond mere pleasure and pain. This possibility permits us to find satisfaction and happiness in life even though our pains outweigh our pleasures.

with bodily pain and painful emotions.* Since the removal of pain is itself pleasant, the avoidance of death will be pleasant. However, pain is still pain, and its relief can only lead to its absence. The absence of pain is not itself pleasure, and there is no bodily sensation of pleasure implied by the mere absence of pain. How, then, can we speak of the value of life in such a situation? Part of the answer seems to lie in the fact that the movement of our bodily state away from pain and toward pleasure is plea*sant*, whereas movement toward pain from pleasure is unplea*sant*, as illustrated in Figure 9.2.

Pleasure and pain are *states* of our body; pleasantness and unpleasantness are *trends* toward pleasure or pain. So we feel a sense of disappointment and loss when our bodily state trends downward toward pain, even if we do not actually feel pain, and we feel a sense of satisfaction and gain when our body trends upward toward pleasure. As Mark Twain observed: "Do not undervalue the headache. While it is at its sharpest it seems a bad investment; but when relief begins, the unexpired remainder is worth four dollars a minute" (Twain 1897).

This also shows, moreover, that there is more to our values than bodily pain and pleasure. Natural value does not exhaust the domain of human value, although it is the basis of that domain.† Complex creatures that we *Homo sapiens* are, we have

---

* Our experience of death seems to be in part a function of the *amygdala*, a small subunit of the brain that is essential for the experience of fear and loathing. Amygdala activity results in a state of the nervous system and body similar to that in the experience of bodily pain. By means of the amygdala, natural selection has enabled many animals to react to the perceived danger of death with much the same response that would be generated by actual damage to the body.

† In logical terms, natural value is *presupposed*, or is a *necessary condition* of the other values.

evolved a structure of values, ethical rules, systems of obligation, concepts of virtues and vices, and notions of beauty and ugliness that defy any brief analysis. When it comes to life and death, we have buttressed our biological imperative to stay alive at all costs with ethical principles against suicide and murder. We have also developed concepts of heroism in which risking or sacrificing life in order to save the lives of others is deemed a great good. These complex values in which life is sacrificed or pain is tolerated still attest to the positive value we place on life itself. Life is required in order that pain and pleasure can exist, and in this way life is the foundation of natural value. Even though life may not be worth it for various individuals at various times, even though it may be hellish, life itself is often of positive value.

In addition to the value we find in life collectively, each of us can find value in our own individual lives. Humans have developed countless forms of life, activities, aspirations, amusements, arts, athletics, and so on. Our interests extend beyond ourselves to our kin, our countrymen, our fellow human beings at large, and even living things in general. Of course, there is also the mirror image of all of these possibilities: We get bored when we are denied the chance to be active, our amusement may become insipid, and our aspirations can be crushed. Just as we take pleasure in what pleases others, we may also be pained by what pains them. We may then speak in more general terms of our capacity for satisfaction and dissatisfaction. Although climbing a mountain or getting a university degree may be extremely difficult, and exhausting both physically and emotionally, we may nevertheless get great satisfaction out of these activities, or even in just attempting them. In these ways, we may find our own life either satisfying or not, and thus feel happy or not. This is yet another way in which life itself may have a value for us. Even when we live with almost continuous physical pain it is possible for us to be happy. Given that most of us achieve a level of satisfaction and happiness in life, life in general has a positive value in this sense.

Thus, humankind not only places a high positive value on life, but is allied with it. We have common cause with everything that is living and feel a sense of solidarity with living things. Our solidarity with life is not just with currently living things here on Earth, but with living things in general. We belong to the domain of living things and depend on that domain for our own life. We are united with life in general, and through it to the natural world at large.

9.3   ARE WE ALIENATED FROM NATURE?

Environmentalism emerged in the industrialized nations in the postwar 1950s, taking root in the romanticism still latent and manifest in that era. We think that the romantic age died out in the late nineteenth century and was replaced by realism. To some extent, that is true, and to some extent romanticism only went underground. The new modern ideals of human rights, along with freedom from despotism, want, and ignorance certainly have a romantic ring to them. In any case, despite whatever realism had injected itself into people's thinking by the 1950s (and heaven knows there were sufficiently many horrors to kill romantic fancies), environmentalists believe in the beauty and goodness of nature. Belief in the goodness of nature is a romantic trait, and quite different from both pagan and biblical views of nature. The pagan view was that nature was animated by good and evil personalities (even apparently inanimate things such as rocks and streams were thought to be alive and conscious), hence that nature was both good and bad. The pagan gods explained both the good and the evil in the world. The biblical view is that nature was originally good, the Garden of Eden, and then became bad because of human sinfulness, which, in turn, was due partly to Satan's even more original sin. Because of our offense to God, the natural world we see, which has been changed to include work, pain, suffering, and death, is no longer only good. And Satan continues to dwell among us on this Earth, in this natural world, ensuring its corruption.

But in the romantic view, nature is our mother, for nature gives us life. In the romantic picture of nature, we are *alienated* from nature, like a child cut off from its mother, and this alienation is a major source of our unhappiness, our moral failures, even our warfare. The romantic typically believes that we are the only species that kills its own kind (or is willfully cruel, or sexually adventurous, or whatever else she or he disapproves of), and is inclined to explain this aberration by reference to urban crowding or some other unnatural aspect of city life. Any aspect of city life that the romantic imagines was not part of the life of pre-urban humans will count as unnatural. Although our alienation from nature may not explain all the evil we suffer or do, it is certainly thought to explain much of it. This is also the view implicit in environmentalism. Prior to agriculture, humankind lived in harmony with nature in small hunter-gatherer bands, in harmony with local ecosystems. In those days we played a proper role in nature, keeping some species in check and encouraging others. We had respect for the natural world, and recognized its sacred value.* Nature, in turn, kept us in check. Since that time we have used technology to escape the natural bounds that were set for us and have treated nature with contempt. We have created enmity between ourselves and nature. We are thus alienated from nature. Natural justice will come due (or is upon us) in the form of an environmental crisis.

There is no reason to believe that the romantic view is true. Nature is magnificently indifferent. It is beautifully and fearsomely unmoved by our concerns. Nature neither loves us nor hates us: It does not care. Nature is the basis of all that is good and all

---

* An environmental organization in my community (Garry Oak Restoration Project, www.gorpsaanich. com) is dedicated to restoring stands of local trees to their pre-European state. These "restoration ecologists," as they denote themselves, have discovered that the local native populations used to routinely set fires among the trees before the arrival of European and other settlers. This native behavior they describe as tending the trees, and they would like to resume the practice, if and where feasible, to return the stands of trees to their native state (GORP 2007). This presumes that the native populations were in harmony with nature, that they were not significantly agricultural and technological to have alienated themselves from nature. It indeed seems likely that the native practice of setting fires did prevent the occurrence of the very large forest fires that would have otherwise occurred if the underbrush had been allowed to accumulate. But even if another form of fire prevention could be assured, these restoration ecologists would still like to resume the regular burnings to encourage the larger trees and specific mix of native plants and native animals they believe were present prior to colonization. Again, the presumption is that this is the proper or natural state of the trees, which in turn assumes that the native populations were in harmony with nature, and that the modern population is not.

that is bad. It gives us life and joy, and but it also gives us death and sorrow. It is presupposed by all that is either good or bad, but it is neither in itself.

Whatever our problems may be, they will not be solved by reuniting ourselves with nature and reversing our technological achievements. We never left nature—that is impossible. We are natural creatures just like everything else. Although nature is, from the point of view of our interests, both good and bad, nature never intends to be good or intends to be bad. It waters your fields or floods them with complete indifference. Ice ages or meteorite impacts are served up by nature without any cruelty on its part. Nature is not a person.* We cannot have a personal relationship with nature, so there is no relationship with nature that could have ended in alienation. Nature was not our friend, and never has been. It has never threatened vengeance, and never will.

THESIS 17:   We are not alienated from nature. Nature has never been our friend or our enemy. We cannot give offense to nature, and it cannot seek or get vengeance.

9.4   OUR SEPARATION FROM—AND UNION WITH—NATURE

Consider the fertilized egg inside the womb. It has a most amazing power: It can turn into an animal. This power is not unconditional, however, for if the egg is removed from the womb, this power is lost. The egg can take the raw materials of life and assemble them into a complete animal, but only if these raw materials are provided to it, within a certain range of temperature, pressure, and so on. So the power of life does not reside solely in the fertilized egg itself, but also in the womb. The womb, of course, does not have its power to nourish the egg unless it is inside a functioning animal, and this animal does not have the power to support the womb and the egg unless it is in the sort of natural order that provides it with food, shelter, and other animals of its own kind. All living things on Earth are structured to function within the parameters provided by the nonliving physical system of this planet: specific ranges of temperature, oxygen, sunshine, rainfall, salinity, and so on. Finally, the physical parameters that Earth provides for the life it harbors is dependent on the conditions outside itself in the solar system and beyond. Thus, every living thing on Earth is tied to the Earth by metabolic or biological dependency.

Nevertheless, the biological union of living things with the larger environment around them is not unconditional. The fertilized egg is necessarily separated from the rest of the contents of the womb by a membrane. A membrane, or skin, separates each living thing from the rest of the universe, and in this way defines it as a separate entity. The womb, too, is separated from the body by a membrane, the body is separated from

---

* If it were, its intelligence and power would place it beyond our comprehension and place us beneath its concern: it could no more have a personal relationship with us than we could have a personal relationship with a bacterium.

its environment by a membrane, and so on. The membranes have to be permeable in a functional way: they have to let certain things in while keeping everything else out and let certain things out while keeping everything else in. So the fertilized egg, which has the conditional power to create a living being, is separated from the rest of the universe outside by a series of membranes, each of which separates some living entity from the rest of the universe. Thus, living things are not only united with the rest of nature, but *separated* from it as well. This separation is just as essential as the unity. Separation of the individual organism from the rest of nature is the logically necessary (realist) complement to its (romantic) union with the rest of nature. Life is a tension, logically speaking, between union and separation. The resolution of the tension lies in balance and selectivity: Traffic through the dividing membrane must be controlled in both directions.*

Feathers, shells, burrows, hives, nests, all provide another layer of separation, another membrane between the living entity and the environment outside. Our clothing, houses, and cities are just an extension of this salutary trend. Cities are common in the organic world. Organisms of a given species make characteristic structures. Corals make coral reefs, cement cities in which they dwell. Even plants make structures. Prairie aspen trees form small forests of interwoven clones, cities of wood on which their living sapwood and leaves lodge, to gain an advantage in the struggle for existence. Animals make nests and hives, structures distinct from their own bodies, with specialized substructures for specific purposes. For instance, the wax combs of a beehive are used as nurseries and for honey storage, while the outer paper skin is used for protection from the elements and predators. Protection, of course, implies the presence of threats. The nonliving elements of nature present threats. The weather can kill you. Living threats, such as predators, not only can kill you but have been naturally selected for that very purpose. Ants and bees make nests and hives *because* that is a way to gain protection *from their environment*. The knowledge of how to make the nest or hive is hardwired into their nervous systems.

Humans are in the same overarching natural scheme as the trees and social insects: Life is maintained just as long as the struggle for survival is won. Like them, we struggle with the elements: the wind, rain, lightning, and frosts that can and do kill us. Like them, we struggle with predators large and small that feed on us or even move inside us to take up residence while dining. Our cities are nothing other than a new wrinkle on the nest and hive building of other species. Rather than hardwiring the structures we make right into our nervous systems, natural selection has hardwired the ability to learn right into our nervous systems. Unlike the social insects, our behavioral repertoire is not fixed at birth. Instead, we are provided with the ability to acquire behavioral skills from our parent generation, and to modify those skills as well. We are programmed at birth to be program*able*. If we are born among first-century Romans we become a first-century Roman, and if we are born among thirteenth-century Eskimo, we become a thirteenth-century Eskimo. Our brains have been the same for about 100,000 years, but our cultures evolve much more quickly.

---

* Global warming theory claims that the flow of heat through the membrane of the Earth's atmosphere has become unbalanced.

Because we are organisms just like any other, nature is our foe, just as it is our friend. Because we have evolved to evolve, our species can acquire new behavioral possibilities continuously without having to wait for the eons-long process of genetic evolution. We can and do collectively acquire new ways of living not only from one generation to the next, but within a single lifetime. My grandfather traveled by horse and buggy when he was a boy, and lived to see a man walk on the moon. So all of the things that separate us from nature—our clothing, shelter, and cities—keep changing. The pace of change was much slower long ago. For hundreds of thousands of years our ancestral hominid species made identical stone axes one generation after another. *Homo sapiens* made only small changes too, at first, and even a few thousand years ago most cultures aimed for stability from one generation to the next, rather than change. Constant change has become accepted since then, and now many people not only expect change but demand it. We want a cure for cancer, better television coverage of international news, a manned mission to Mars, and so many other things that will permanently change the world we live in.

So our adaptable, programmable brains have speeded up our own evolution relative to that of other species, and this has given us an enormous competitive advantage when it comes to defending ourselves from our environment. Nature is our foe as well as our friend, and in our struggle with nature we have gotten the upper hand by virtue of our programmable brain. We have discovered and developed agriculture and civilization: country and city life. Our cities protect us so well from the elements and predators that we have had the leisure to reflect that the elements (in particular, the climate) and the predators might need protection from us as well.

THESIS 18:   Every organism, including us, must achieve separation from the rest of nature even while remaining biologically united to it. Separation is not alienation.

9.5   THE VALUE OF FREEDOM

Freedom has been the main political ideal of those people who have acquired democracy. It is noteworthy that democracy emerged gradually alongside industrialization in Western Europe and North America, a pattern that was repeated—and is still being repeated—in the rest of the world. This is no accident, because industrialization provides freedom from constant work, the foundation on which democratic freedoms are built. This foundational freedom from constant preoccupation with the struggle to survive, or *natural freedom*, is the form of freedom on which all of the other freedoms listed among human rights are based. The facts concerning freedom are almost exactly the opposite of what the romantics tell us. Humankind gained freedom because of industrialization, not because of its absence. As long as muscle was the source of power, slavery was inevitable. The steam engine harnessed fire and provided a new and abundant source of energy, thereby sounding the death knell of universal slavery.

The democratic freedoms we take for granted would have no value if it were not for natural freedom, freedom from constant work. Freedom of movement, for example, is of no worth if you have to work all day long just to obtain food, clothing, and shelter. You will not feel much like working even more hours to raise the fare for the bus or the plane if you are struggling just to get by, and even if you did, you would be too tired to really enjoy the trip. Much the same sort of limitations applies to the freedom of speech, freedom of conscience, freedom to run for office, and so on. The industrial revolution made it possible for people to gradually reduce the amount of work they needed to do just in order to survive, and this in turn sharply raised the value of political freedoms. Freedom to travel means a lot to someone who has the time, energy, and money to use it, whether to visit her parents across town or the national park across the continent. The freedom from excessively burdensome labor is the main blessing of the industrial revolution, the one that people still hasten to receive if they have the chance. This blessing is eagerly sought by people everywhere, and still motivates the industrialization of developing nations, thus making all of the other democratic freedoms possible and worthwhile.

Philosophical discussion of freedom has taken two primary forms. One seeks to find a way in which human will itself might be free of causal determinism, and the second seeks a definition of political freedom that prevents the freedom of one person from restricting the freedom of others. It seems pretty clear that the first investigation has revealed indeterministic freedom of the will to be a contradiction-ridden metaphysical fiction (Dennett 2003). In any case, it is not at issue here. Political freedom is closer in spirit to the sort of freedom that people are seeking, although, as the previous paragraphs imply, it too is not in itself quite what people are looking for. Political freedom legally protects us from being completely controlled by our fellow citizens. Crucial as this sort of freedom may be, philosophical discussions of it ignore the fact that it is valueless without the material prosperity required for *natural* freedom.* Because these discussions neglect the material basis of freedom, they tend to lose touch with reality and to venture into utopian visions of universal equality and the elimination of differences between people. By not first ensuring the material basis of freedom, they risk undermining the very thing they seek.

The most efficient definition of what we seek is relief from the mythical curse that was placed upon humankind for stealing fire from the gods (in its pagan form) or for disobeying God (in its biblical form). Whatever the narrative may be that explains and justifies the curse, the curse itself is the same: work, suffering, and death. In the most common form of the curse today, the biblical form, as God casts us out of the very embodiment of nature beatified, the Garden of Eden, he commands that we shall henceforth gain our bread only by the sweat of our brow, that our children will be born only through the labor pains of their mothers, that suffering and death will follow us all the days of our life, and so on. What we ultimately seek, in its

---

* John Rawls, whose work (1971) has defined the contemporary discussion of political freedom, separates the *worth* of our political freedoms from the *fact* that we have them—and then goes on to ignore the first in favor of the second. This fundamental error is part of the romantic industrial-counterrevolutionary attitude that typifies modern philosophy and political science.

full perfection, would be the lifting of this curse: freedom from work, suffering, and death. By providing us a degree of freedom from work, science, technology, and industrialization have given us the basis to pursue (with some success) freedom from suffering and death as well. Nature and human nature cooperated to subject us to hundreds of thousands of years of slavery, suffering, and death. They have also provided us the intelligence and motivation to escape this natural bondage. Because it is natural for us to seek and obtain this escape, we may call it *natural* freedom.

Science and technology have generally been embraced by humankind because they provide human beings a substantial increase in natural freedom, and where they are resisted they are resisted by those in power who gain from keeping their fellow human beings enslaved. Industrial civilization has reduced the amount of work that people have to do, has reduced their suffering from injury and disease, and has tripled their normal expected lifespan from about 25 years to 75 years. Reduced disease and increased lifespan are well-known aspects of the rise of modern life and are well documented, well studied, and highly prized. They have been gained by the gradual reduction of natural hazards to human well-being from the elements, predatory animals and microbes via improved housing and public sanitation on the one hand, and the gradual improvement of disease prevention and treatment via the advance of medical science on the other. They also depend on partial freedom from work. Work is required for us to gain the necessities of life, in particular food, clothing, and shelter. Modern industrialized life has permitted us to gain these necessities much more easily and safely. Just as freedom from excessive work is at the core of the standard political freedoms (or natural rights, as they are sometimes, somewhat misleadingly, called), it is also at the core of the other natural freedoms as well. Our increased freedom from suffering and death are in large measure possible only because of our increased freedom from work.

Until industrialization freed us from muscle power as our only source of energy, humankind generally could not afford the generosity of spirit that makes freedom and tolerance possible. Freedom from the struggle with other human beings for the means to survival was the first luxury of industrialization. We tend romantically to forget that evolutionary competition is equally intense within a species as between species. It is the members of our own species who fight with us for the very things we need in order to exist. Whereas we naturally sympathize with our next of kin in their struggles, it is not natural for our fellow-feeling to extend to people in general. Sympathy with our fellow human beings in general is a luxury, a political luxury, and one of the great boons of material prosperity.* This political luxury simply cannot be afforded until industrialization and the general technological advance of our struggle to survive are achieved.

Although the political rejection of slavery that swept the industrializing nations of the nineteenth century removed the legal mechanisms of slavery, it did not itself produce any natural freedom. As many recently freed slaves discovered, only a small

---

* Environmental sympathies are a further broadening of natural human sympathies and require further growth of freedom from work, suffering, and death in order that they may emerge.

increase in their natural freedom resulted when the legal institution of slavery, the ownership of one person by another, was rejected. Free, that is, unowned, people still required life's necessities—food, clothing, and shelter—and found that they were required to work to obtain them just as hard as before. They became wage slaves, or were reduced to effective labor serfdom in company farms, camps, or towns. Although the coal miner who spent his wages to rent a company house and to buy food from the company store was indeed free, politically speaking, to leave his job without any fear of punishment, he could do so only by losing his access to the food, clothing, and shelter that he and his family needed. This amounts to virtual slavery, even in the absence of legalized ownership of one human being by another.

Only a real reduction in the amount of human labor required to produce life's necessities could ever have permitted this virtual slavery to be overcome by humankind in general. Fortunately, this was precisely what industrialized life accomplished: real reduction in the amount of work required to make a living. It actually has brought us closer to the Biblical and romantic ideal of the birds of the field, which neither reap nor sow, and yet are fed, clothed, and sheltered. The Garden of Eden that never was may one day be achieved.

THESIS 19:   Natural freedom, the freedom from work, suffering, and death, is good and the basis of all other goods and freedoms.

Work, as understood by everyone who has ever had to work, consists of things that you have to do even though you are not inclined to do them. Children do schoolwork, adults work to make a living, and in both cases they rejoice when their work is done. Work is unique to our species, and would not exist on Earth were it not for *Homo sapiens*, although we also extend the curse of work to other species as well. To this day, countless animals are forced to do what they are not inclined to do in order to satisfy human needs: pulling wagons, dragging plows, herding sheep, and so on. Their workload has also decreased due to industrialization, and in industrialized countries horses are far more likely nowadays to be lazily ranging in pastures than working. However, were it not for humankind, animals would not know work at all. In their natural state, animals only do what they are inclined to do. They hunt for food when they are hungry, and when they find it, they eat it, as they are inclined to do. They run in flight or turn to fight as they are inclined. Usually, they feel no external compulsion to do what they do.

Human beings, by contrast, spend the bulk of their days doing things that they are quite consciously aware that they are not inclined to do, but *have* to do. From an early age children are forced to get up when the alarm rings, to wash, eat, and go to school, where they behave as instructed all day long—despite being inclined to do otherwise. We require this of children as part of their education, which in turn is required in order that they may get a good job—more work—when they grow up. The necessity for work is one of the chief forms of humankind's supposed alienation from nature that so fascinates the romantic mind. It is, however, a completely natural effect of our social nature coupled with our intelligence.

The natural human tendency toward organized work is worthy of attention. It places modern *Homo sapiens* in a rather unique category, since we spend so much of our time working, hence doing things that we would not do in the absence of social pressure. Work is just one aspect of the necessity to act against one's own inclinations that people generally face in civilized life. It seems plausible that the continual dislocation of our behavior away from what our instincts tell us to do does make it harder for us to be happy. We do not wake up when we have gotten enough sleep, but when the alarm clock goes off. We do not lie in bed musing about this and that, but get up and prepare ourselves to face the day. We do not eat because we are hungry, but because we know we need the nourishment to get through the day. We get our school-age kids up as well, and stifle their cries and complaints, telling them that they need to go to school or do whatever else that we've planned for them. Our deeper inclinations were fashioned long ago in our evolutionary past, and they are out of harmony with what we actually need to do now. Surely, the constant need to deny our inclinations is the source of a great deal of stress in modern life.

On the other hand, the romantic's proposed solution to the problem of inclination-denial, which they conceive as alienation from nature, is to head backward, toward the structures and practices of the past. This surely is not the solution. The way forward does not lie in the past. To begin with, people living in less modern cultures apparently suffer from inclination-denial at least as much as we do. Indeed, slavery was more, not less, common in pre-industrial societies, and only disappeared with the rise of industrialization. *Homo sapiens* is a paradigmatically social species. We are adapted to life in groups and to social control of behavior, including work. So the work that characterizes modern life is perfectly natural. Inclination-denial is part of human nature. We are born programmable, not programmed. Because we are big-brained organisms capable of sustaining rich cultural evolution in the arts and sciences, we have adopted technological innovations to gain a degree of freedom from work. This, too, is neither new nor unnatural. Work-saving technology is just another means whereby the tension between the inclinations of the individual and the inclinations of the group can be resolved more harmoniously. We do not want to reject technology, but to develop it in a way that continues to increase our natural freedom.

Natural freedom thus emerges as a core natural value for humankind. Natural freedom is freedom from excessive work, suffering, and death. No one would deny that escaping these things would be good in itself, although people could reasonably worry whether the consequences of escaping them might be bad. With few exceptions, nobody wants to work (assuming that work is doing what one is not inclined to do), to suffer, or to die. So freedom from these things is a natural good. But escaping or delaying death, for instance, might lead to intolerable population growth. But if we consider work, suffering, and death just in themselves and not in terms of their consequences, they are the sorts of thing that we would avoid. Sure, work may produce results that we desire. A person may willingly, even passionately, work to achieve a goal, putting aside mere pleasures in order to work harder, and so on. But if there were no possibility that our work could ever achieve our goals, we would not want to do it. We desire our goals, not the work itself. Since work is the sort of thing that we want to avoid, then freedom from work has a positive value for us. Its positive value, moreover, is completely *natural* (as opposed to supernatural). Work, suffering, and death are things that we observe among us and experience for ourselves. We can see for ourselves whether they are good or bad. In our view, not only are these things bad for us, they are bad for living things in general. Human freedom is a *natural good* in just the same way that work, suffering, and death are naturally bad: They are at opposite ends of the same spectrum, the spectrum of natural value.

10 Backward or Forward?

A man with a half volition goes backwards and forwards, and makes no way on the smoothest road; a man with a whole volition advances on the roughest, and will reach his purpose, if there be even a little worthiness in it.

—Thomas Carlyle (1899, p. 331)

The wind was whistling through the oaks when I took Sarah outside so she could urinate. There was just a glimmer of light in the southeast, so I knew it must already be around 6 A.M. As I attached the yard-lead to Sarah's collar, the freezing wind gusted, howling in the tree branches. Sarah grunted in disgust. She hates going out into the cold to pee first thing in the morning. But she is 15 years old now, old even for a cocker spaniel (which are long-lived dogs). She has an enlarged heart, liver disease, and kidney disease, which means she has to drink a lot and pee a lot. She cannot hold on until morning anymore, so she comes and wakes me up. She does not want to pee in the house, and she knows she can count on me to let her out. I've told her, by hand as it were, that she is welcome to come to me at anytime day or night and I will let her out. Whenever she wakes me up to be let out, I always pet her, no matter how sleepy and grumpy I may feel. I don't want her to pee in the house, which she surely will do, as she has done in the past, if she cannot get anyone up to let her out. I ducked back inside to watch her through the window, while the wind blew so hard outside that it ballooned her little pullover up over her shoulders. She got the pullover as a Christmas present from her nominal owner, my daughter, whose present she once was. The Christmas lights on some trees down the street shook and swayed as Sarah finished and came trotting up to the door, her tiny stump of a tail wagging with relief and joy to see that I was there to let her back in.

The scene was filled with tension between the natural and the artificial. On the one hand it was completely natural. To begin with the obvious, I am an animal, a largish ape, living on the urban steppes now that the African savannahs that bred my species are only a spot on a map for most of us today. It is natural for *Homo sapiens* to form symbiotic relationships with other organisms. Even as stone-age hunter-gatherers, *Homo sapiens* fed their scraps to their dogs and let them pee around the camp. We know that thousands of years ago, long before the dynamo and long before the wheel, our ancestors marked the winter solstice, perhaps with the giving of presents and a big, bright fire. On the other hand, their fires were not brightly colored electric lights, and if they gave a present to the dog, it most definitely was not a pullover made

Beyond Environmentalism: A Philosophy of Nature, By Jeffrey E. Foss
Copyright © 2009 John Wiley & Sons, Inc.

in Nepal. Is it possible that somehow, simply by following its instincts and using its own natural intelligence, *Homo sapiens* has become *unnatural*? Is our house not just a *really* fancy den? It is unlike any other animal's den: It is heated to a preset temperature, supplied with water and sewage, has doors and windows to look out of, and is decorated with Christmas lights. Could it be that our brains have become overdeveloped, hypertrophized by natural selection gone wild? Has *Homo sapiens* become an unnatural species: the postindustrial, alienated ape? Have we become *artificial*, products of our own cultural fascination with energy-burning technology, and in that sense not natural?

Certainly Sarah, my dog, is a completely artificial sort of canine: nothing like her exists in the wild, and for good reason. Her larger face, shortened muzzle, and frontward-facing eyes make her look more human than other dogs, and probably have a lot to do with the biological success of her specific genes, since they make her more attractive to some humans as a pet. These same features, however, would be handicaps in the wilderness—as would her cute, floppy ears. But Sarah's poor adaptation to the wilderness does not make her unnatural. Every plant or animal that ever existed was poorly adapted to some environment or other. Fish gotta swim, birds gotta fly, domestic dogs have to hang around humankind. True, Sarah's genetic makeup has been directly influenced by another species, *Homo sapiens*, but that does not make her unnatural either. Animals influence each other's genetic makeup all the time. The antelope is fast because the slowest antelopes are chased down and eaten by speedy predators. Thus, the coyote and the cougar are in the animal-breeding business, whether they know it or not, tirelessly working to improve the speed of antelope breeding stock. The antelope, for their part, return the favor by preferentially feeding the faster predators and thereby improving the speed of coyotes and cougars. So there is nothing unnatural in humans influencing the genotype of organisms in their environment. When other animals do it, we call it natural selection—how could it become *unnatural* selection just because we do it?

But was it not unnatural that this canine, Sarah, had been taken out of the environment it had been adapted to by *natural* selection? She was bred to be cute for humans, after all, bred, among other things, to fill the emotional niche left in the *Homo sapiens* behavioral repertoire when its offspring grow old enough to move out of the family shelter. She, a canine, was bred to fill a role in the social life of a different species. Is that not unnatural? Well, I for one will not contest the child-substitute theory of dog ownership, even though I doubt it is the full story since I have looked after dogs since I was a boy. Nevertheless, I confess that I baby talk to this little dog, just as my daughter did, perhaps the surest indication it plays the role of child substitute for me. I admit that I do feel a need to nurture someone or something, and that I have chosen to satisfy that need, in part, with Sarah. But I do not feel that this need is in any way unnatural. It is the most natural thing in the world.

My ancestors, noble savages that they were, began entering into symbiotic relationships with dogs many thousands of years ago. Dogs, like us, are social animals. They look you in the eye and listen when you speak. Their emotional tone is the same as ours, so there is intuitive understanding at the level of preverbal vocalization. Even small children can tell when a dog is happy and when it is angry, and puppies have the same innate understanding of human moods. The dog–human relationship goes beyond symbiosis on the purely biological level into the social realm. Household pets are domesticated to be companions. And, lest we forget, *Homo sapiens* is not the only animal that domesticates other species. I have watched the ants tending aphids in the garden, bringing them out of their nests in the morning on fair days to guardable spots on the flowers so they can suck back the sweet sap and make honeydew for the ants.

The only reasonable conclusion is that the artificial is *part of* the natural. It is natural for organisms of all sorts to engage in artifice, the artful making of things. The commonplace distinction between the artificial and the natural is an expression of a commonplace prejudice, like the signs on supermarket doors which say that no animals are allowed inside. If no animals were allowed inside, people would not be allowed inside either. The sign works because we are used to distinguishing ourselves from animals. Darwin's lesson has not been fully absorbed. If it were, we would see clearly that the artificial is not unnatural. There is nothing unnatural under the sun. To put this point loudly, the artificial is *perfectly* natural. And to bring the point home, we do not feel unnatural looking after domestic animals, and the fact that, unlike the ants, we do it not for food but rather for companionship only makes the relationship more genuine. We are connected to nature through our house pets and house plants. They are bits of nature that we nurture because we have affection for them. They are the media of our affection for the natural world. Like the Christmas lights, they mark our respect, our thanks, and our joy for the rhythm and harmonies of nature. Surely our caring for animals and plants is the furthest thing from our supposed alienation from nature.*

10.1   BACK TO NATURE: LEOPOLD AND CALLICOTT

Why, then, do environmental ethicists have such disdain for domestic plants and animals? To answer this question we can do no better than study the works of J. Baird Callicott. His is surely the purest, most philosophically complete and persuasive expression of the thesis that humankind must return to a closer relationship with nature. He has taken the writings of Aldo Leopold, the author of the famous *land ethic*, elaborated them into a complete ethical system, provided arguments in their favor, and defended them against criticism. Callicott's philosophy is classic environmentalism since it implicitly sets up the good of the environment as a transcendent objective (as defined in Chapter 3). Explicitly, he says that his is an "environmental ethic which takes as its *summum bonum* the integrity, stability, and beauty of the biotic community" (Callicott 1980, p. 62). In other words, the only thing that is good in itself, or intrinsically good (see the box "Nature and Intrinsic Value" in Section 9.1), is the health of the biological environment. All other things, insofar as they are good at all, are good only instrumentally, as a means to the unique intrinsic good: the integrity, stability, and beauty of the environment.

Now this in no way *entails*, for example, that such an environmental ethic will deny human rights or interests. It is logically possible, for instance, that human interests will converge smoothly with those of the biotic community. This is not a matter of

---

* Sarah died a few months after this passage was written.

logic alone, but of other facts and values as well. Whether Callicott's environmental ethics conflict with human interests depends on the relevant facts and values as he sees them. The stability and integrity of the biotic community, the first two-thirds of the *summum bonum*, are sufficiently well defined, perhaps, to count as mainly factual matters. The last third, by contrast, the beauty of the biotic community, is clearly a matter of values. So too, let us not forget, is his choice of the integrity and stability of the biotic community as the first two-thirds of the ultimate good. We have pretty good reason to believe, for example, that the evolutionary leaps that made the current state of nature possible required disruption of the integrity and stability of the natural order by such things as climate change or comets nearly destroying the Earth. Unless the old is destroyed, the new cannot emerge. This is ironic inasmuch as Callicott, Leopold, John Muir, Ruth Harrison, Dave Foreman, E. O. Wilson, Niles Eldridge, and uncounted other philosophers who lionize the wilderness are absolutely outraged by any stress placed on the current wild environments. They are arch-conservatives from an evolutionary point of view. Stability, not change, is their goal; the warm and familiar past, not the unknown and unsettling future, is their ideal.

As Callicott sees it, not only the industrial revolution, and before it the agricultural revolution, but civilization itself is contrary to the integrity and stability of the biotic community. According to his ethics, no one would ever let their pets live in the house, for neither the house nor the animal would be permitted to exist. It follows, as Callicott draws out in some detail, that human beings like those of us who read this book would cease to exist as well. Humankind must be transformed in order that the wilderness can be reborn, and the needed transformation will take us back, not forward.

Callicott's work is the most philosophically complete in part because he does not sugarcoat his vision of the environmental ideal. His honesty is bracing, and the clear, heroic outlines that he draws of the natural life are beguiling—indeed, they are romantic. He also takes time to address topics that other champions of the wilderness skirt. For example, we noted at the start of Chapter 5 that what is good for the prickly pear cactus is bad for the pronghorn antelope (the cactus prefers less rainfall, the antelope more), and that what is good for the lynx may be bad for the rabbit (the rabbit prefers not to be the meal desired by the lynx). To put it bluntly, the interests of animals within what Callicott calls the biotic "community" are hardly communal: Everything is either being eaten, or crowded aside, or both, all of the time. The paradigm of a community with the full panoply of biological, social, cultural, and artistic aspects is the human community, not the wilderness. In the wilderness survival *is* the struggle of living things with each other and with the elements. In this struggle there are no rules: Everything is fair. That is the definition of war—the polar opposite of community.

Environmentalists such as Callicott romanticize the intrinsic violence of nature, and praise the masculine virtues of living bravely within that violence without complaint or regret. They do their best to reconceive the constant hostilities as a form of community, although they cannot deny the fact that organisms are ever vigilant to take advantage of each other without the slightest care for each other's joy or suffering. Leopold himself was a hunter, and a sport hunter at that, since hunting (perhaps augmented by gathering) was not his sole source of food. That the animals he killed suffered is of no ethical interest to either him or Callicott. The suffering of individual animals is *irrelevant* from the point of view of the good of the environment as a whole.

The good of the community as a whole serves as a standard for assessment of the relative value and relative ordering of its constituent parts and therefore provides a means of adjudicating the often mutually contradictory demands of the parts considered for *equal* consideration (op. cit., p. 62).

The good of the biological "system" (for it is certainly not a community) is its integrity, stability, and beauty. Apparently, there is nothing beautiful about equality or equal consideration. If the suffering and death of rabbits and moles were in the balance against the unsuffering death of bees (they simply do not have the nervous systems required for pain), the bees should win. Callicott's argument for this is substantial and persuasive, as far as it goes: The bees play the important role of pollinating plants, the basic food supply of the entire biological system, while rabbits and moles do nothing as important. So much for integrity and stability, but can we not also recognize, at least, that the suffering and death of rabbits and moles is *ugly*? Apparently not. To the contrary, agonizing death is to be conceived as beautiful simply because it is natural. Nor is even human suffering to be seen as ugly. It, too, is natural and beautiful, and humankind has no claim to equal consideration either: "The land ethic, on the other hand, requires a shrinkage, if at all possible, of the domestic sphere; it rejoices in a recrudescence [sic]* of wilderness and a renaissance of tribal cultural experience" (op. cit., p. 66).

The domestic sphere is precisely the subsphere of Earth rearranged by human beings, including not just our houses, but all our buildings, highways, fields, plantations—and all of the domestic species of plants and animals. The sorts of plants that we grow in our gardens and buy in our markets are not found outside human cultivation. The sorts of animals that we farm are also not found in the wild. Sometimes the same species exists in the wild but not in the same form, as in the cases of the common carrot and the cocker spaniel. Sometimes the domestic species simply does not exist in the wild at all, as in the cases of wheat and sheep. Some species, like the common rat, are never found far from human habitation and have come to depend on human beings, although we see them as vermin, pests, weeds, or infectious agents. All of the apples, pears, roses, onions, pumpkins, pigs, chickens, and pickles that have been the pride of street markets and county fairs over the centuries are part of the domestic sphere. All of this "requires shrinkage." In fact, it would be best if it could disappear entirely, and we could return to our hunter-gatherer roots in a "renaissance of tribal cultural experience." Why? Because it would make more

---

* It is ironic in the extreme that Callicott chooses to use the word *recrudescence* at this point, a term which literally means the renewed breakout of a disease that had waned or been brought under control. Callicott presupposes that wilderness is hostile to human beings (perhaps because human beings have tamed, hence damaged, the wilderness), and that for wilderness to prosper, humankind must suffer. Apparently, he cannot fully and clearly conceive of a confluence of the health of the environment and human interests in which wilderness does not regain its ascendancy over humankind. To put it another way, it is not enough, given his ethics, that wilderness thrives—it is also necessary that humankind sink in submission to it. He views history as a battle between us and the wilderness which we have won through an unfair technological advantage. He wants us to surrender this advantage. He assumes that it is a matter of humanity versus the wilderness, and it is clear that his sympathies do not lie with us but with the wilderness.

room for wilderness, allowing it to reconnect as a single all-embracing entity, thus regaining its integrity, stability, and beauty. We have shattered the wilderness, and this wrong will be righted only when the wilderness regains its territory.

A host of questions arise. Why did these very same noble savages whom we are to emulate create a new stone age, agriculture, and civilization in the first place? If we become like them once again, why would we not do just what they did once again? If they are our ideal, why would we not accept their ideals, including a settled and civil form of life? What, aside from romantic notions, was so wonderful about "tribal" life? Preagricultural technology, culture, and lifestyle were, after all, technology, culture, and lifestyle in the *full* sense of the term. The suggestion seems to be that early humans were better than us because they were innocent (or nearly innocent) of the taint of technology. That is simply false. There were no pretechnological *Homo sapiens*. We had fire and stone tools right from the start. We had language, culture, religion, poetry, myth, explanation, and an intelligent interest in the world around us right from the get-go. So we cannot prefer the earliest ways of humankind to our own on the fanciful grounds that they were untainted by technology. Instead, we must show that their religion, poetry, myths, and explanations were better than ours. This is never done.

Instead, the physical closeness of earlier people to nature—which is nothing but their inability to defend themselves against natural predation, disease, and the elements—is taken as proof of their superiority. Their inability to overpower other contestants within the natural agonistic sphere, the sphere of the battle for existence, explains this "proof." It is the *inability* of our ancestors to have much influence on the wilderness that is their *virtue* in the eyes of the environmentalists. So human values are inverted within environmentalism: Weakness becomes our strength, and power becomes our weakness. This not only flies in the face of everything that one ought to teach one's children, but flies in the face of biology as well. Every organism, plant or animal, gets to play a role on the evolutionary stage only for as long as they remain strong enough not to be pushed aside. Little rabbits should be fast and quick to learn the wily ways of the lynx. Similarly, little human beings should be clever and quick to learn the state of the art in human technology. The question is not whether humankind will have an impact on nature, but what that impact will be. Although Callicott says that *less* impact is what we should aim for, really he wants us to make a *bigger* impact by withdrawing from the scene, which would result in a stunning resurgence of wilderness.

10.2   THE ENVIRONMENTALIST'S VILIFICATION OF DOMESTIC ANIMALS

It is a brute fact that all species have a significant influence on the form and genes (phenotype and genotype) of the species with which they come in contact. Why are we to believe that *every* change we have effected within nature, and in particular every change on the forms of life itself since we adopted farming and settled down in durable buildings, is wrong? This sounds more like a romantic curse than anything to do with biology. At bottom, the logic of pure environmentalists such as Callicott is that the natural is opposed to the artificial. Every single domestic strain and species is in some sense a witting or unwitting human artifact, and hence unnatural, hence an enemy of the environment. The lovely sweet-pea perfuming the evening air, the plump potato humbly hiding underground, and the pet canary are, one and all, Frankenstein creations. They are domesticated, hence tamed, hence unfree. Like us, they are alienated from nature and their own nature. We see this logic at work in Callicott's reference to the philosopher who devised the land ethic, the idea that human beings have recognized an expanding sphere of rights (e.g., my kin, freemen, slaves, women,) that would eventually expand to include *land itself*. The surface of the planet should be, and indeed would eventually be, recognized to have rights. Leopold mused about how this change might be realized or at least facilitated. Among the last philosophical remarks penned by Aldo Leopold before his untimely death in 1948 is the following: "Perhaps such a shift of values ... can be achieved by reappraising things unnatural, tame, and confined in terms of things natural, wild, and free" (Leopold, 1949, p. 64).

From this point of view, sheep and dogs are just as ugly an assault on nature as a fleet of all-terrain vehicles careening through the wilderness. Domestic animals are said to be only "half alive"! Callicott (1980, p. 64) praises John Muir for having devised this notion—although surely it makes no sense. Whatever problem the environmentalist has with the cow, the chicken, and the pet dog, it cannot be that they have only a half-measure of life—they are just as alive as the moose, the goose, and the wild dog. It is rather that their life is tainted by the influence of our species, an influence which by definition turns them into artifacts, hence making them unnatural. In Callicott's words, "There is thus something profoundly incoherent (and insensitive as well) in the complaint of some animal liberationists that the 'natural behavior' of chickens and baby calves is cruelly frustrated on factory farms. It would make almost as much sense to speak of the natural behavior of tables and chairs" (ibid.). Indeed, long before tables and chairs existed, humankind was already working its dark and diabolical magic of turning the natural into the unnatural: "Animals, beginning with the Neolithic Revolution, have been debased through selective breeding, but they have nevertheless remained animals. With the industrial revolution an even more profound and terrifying transformation has overwhelmed them. They have become, in Ruth Harrison's most apt description, 'animal machines'" (op. cit., p. 66).

Ruth Harrison and Mary Shelley obviously share an unnatural horror of the modern realization that life is indeed a matter of physics and chemistry. Like Byron, they are fascinated with the idea of the "undead," the quasi-life of such things as vampires and Frankenstein's monster. In their unscientific metaphysics the organic is diametrically opposed to the mechanical. When the undirected wanderings of natural chemistry—that is, evolution—animate the human form with life, it is genuine life, but when the same chemistry is influenced in any way by the intentional actions of human beings, we get undead monsters. There are only Franken-chickens and Franken-cows, there are no real chickens and cows. We have also "debased"—polluted—natural animals with our artificial, hence unnatural taint. There are no truly wild—hence truly good—animals anymore. The entire natural kingdom is under our curse, since our effects (like those of many other organisms) are not constrained within any particular boundaries. But at least the formerly wild animals are still animals as such. This (paradoxically) is not true of domestic animals, which, like Frankenstein, have only a semblance of life, the twilight-zone subsistence of the undead.

The metaphysical dichotomy between the natural and the unnatural cuts off domestic animals from the moral realm. Although it may seem that the horse laboring under the whip is our brother-or-sister-in-suffering, this is a dangerous illusion. Should we find ourselves sympathizing with their plight and making an effort to prevent cruelty to domestic animals, we should remedy this wanton spread of natural human affection by noting a *disanalogy* (or, rather, a metaphysical gap) between them and us: "Here a serious disanalogy . . . becomes clearly evident between the liberation of blacks from slavery . . . and the liberation of animals. . . . Black slaves remained, as it were, metaphysically autonomous. . . . But this is not true of cows, pigs, sheep, and chickens" (op. cit., p. 64). Those of us with any interest in autonomy or metaphysics are, upon reading this passage, anxious to hear all about metaphysical autonomy. There is a great divide running through nature separating those animals that have it and those that do not. On one side there is *Homo sapiens* and wild animals, all of whom are metaphysically autonomous, and on the other side are domesticated animals, all of whom are not. How disappointing to discover that this great divide consists merely of the fact that domestic animals "have been bred to docility, tractability, stupidity, and dependency" (ibid.). Who would have thought that metaphysical change was so easily achieved? Our ancestors wrought a metaphysical change simply by influencing the breeding of other animals. Wild animals are metaphysically autonomous, but when we domesticate any of them it becomes "literally meaningless to suggest they be liberated. It is, to speak in hyperbole, a logical impossibility" (ibid.). I suggest that once Callicott says it is literally meaningless that domestic animals be liberated, further hyperbole is clearly unnecessary—if it is even possible.

The romantic rejection of science rises to a crescendo in passages such as these. What has become clearly evident is Callicott's state of denial of a crucial discovery of biology: We are firmly, completely, irrevocably part of the animal order. The main reason that Darwin was so heatedly and loudly rejected by the moral authorities of the day is the *humble* place he envisioned for humankind: We do not transcend the natural order. We are part of it. We might expect that this rediscovery of pagan wisdom would be welcomed by environmentalists—but no. As far as freedom is concerned, the evidence shows that all complex life-forms are *chaotic*, and hence unpredictable. Whether they are wild or domesticated makes no difference. A domestic horse is no different from a wild horse in this respect: Both are unpredictable systems. A postindustrial, modern human is just as unpredictable as the noble savage of yore. To speak of wild animals and human beings as being "metaphysically autonomous" is simply to engage in metaphysical posturing. As far as the *evidence* is concerned, they are no different from us. Theoretical notions such as "metaphysical autonomy" may be advanced in order to draw a line between the good and the bad, between wild animals and humans on the one hand and between domestic animals and humans on the other, but advancing invisible marks of moral superiority is the visible mark of prejudice, no matter how it is dressed up.

ALDO LEOPOLD ON THE FREEDOM OF ANIMALS, WILD OR OTHERWISE

J. Baird Callicott is generally thought to be the foremost proponent of the environmental philosophy of the revered Aldo Leopold. Callicott's position on animal liberation, however, is plainly at odds with Leopold's. Callicott takes the view that while wild animals and human beings are autonomous, domesticated animals are not. Indeed, he avers that it is meaningless, even contradictory, to think that domesticated animals could ever be free, since they lack the autonomy that is the special preserve of humans and wild animals. Much as we may be inclined to think that a farm horse can be freed from its harness and set free to roam, Callicott says that the words "freed" and "free" are literally meaningless in this context: The farm horse lacks the sort of metaphysical autonomy required for freedom.

Leopold differs from his disciple Callicott, not because he thinks domestic animals *can* be liberated, but rather because he thinks humans and wild animals *cannot*. As Leopold sees it, *no* animals are capable of freedom—and that is a good thing. "I cannot imagine a worse jumble than to have the whole body politic suddenly 'adopt' all the foolish ideas that smolder in happy discontent beneath the conventional surface of society. There is no such danger" (Leopold 1949, pp. 174–175). Why not? "Nonconformity is the highest evolutionary attainment of the social animals.... Science is just beginning to discover what incredible regimentation prevails among the 'free' savages and freer mammals and birds" (ibid.). Nature requires regimented behavior, and what we call *freedom* is merely nonconformity, something that, if unchecked, leads to a jumble. Humans are clearly part of nature according to Leopold, not separate from it as environmentalists, including Callicott, aver. While nonconformity is functionally useful among human beings as the source of innovation and invention, it is strictly limited by the need for regimentation.

Freedom for Callicott consists in the metaphysical potential for autonomy, or self-rule, rather than in the ability to avoid suffering and death. It is a matter of fact, not metaphysics, that a domestic pig or domestic dog will run free if not restrained by a pen or a rope. Anyone who has sympathetically observed these supposedly artificial animals will see that they are no different from wild animals. Although they are more used to captivity than animals that do not experience it as their daily norm, they still do not like it. They do not voluntarily remain in captivity once the barbed wire and bonds are removed. It is a matter of fact that their suffering is practically the same as our own. Unless we are willing to discount human suffering itself as morally insignificant, consistent values would require that we do not discount their suffering either. Again to his credit, Callicott does not shy away from the unpleasant consequences of taking the good of the environment as the only intrinsic good. This does indeed entail dismissing the moral relevance of human suffering even as we dismiss the suffering of animals: "Pain and pleasure have nothing at all to do with good and evil if our appraisal is taken from the vantage point of ecological biology. Pain, in particular is primarily information..." (Callicott 1980, p. 65). While it is nice to see Callicott assume a more scientific point of view, he immediately jumps to an ethical conclusion, as though the logical distinction between fact and value did not even exist: "The doctrine that life is the happier the freer it is from pain and that the happiest life conceivable is one in which there is conscious pleasure uninterrupted by pain is biologically preposterous" (ibid.). Much as we may think that eating an apple or dancing just for the sheer pleasure of it is a good thing, this implies a biological absurdity. Preferring such pleasures to things like toothaches and tearing one's flesh on a thorn is preposterous according to "ecological biology." Pain is morally irrelevant, simply because it is information. The ethicists among us will be shocked and pained by this inference should it somehow turn out to be valid: Torture will be morally irrelevant simply because it is merely the information that we are being biologically damaged.

Of course, it is biologically preposterous that life have more pleasure than pain. We have already considered the reasons that evolution will design organisms in such a way that bodily pain (which does not include all pain, only pain that has a bodily location) outweighs bodily pleasure (Case Study 10, Section 9.1). Callicott, whose discussion of the pain experienced by animals presupposes that it is bodily pain he is talking about, agrees: Pain has a vital biological function, so it cannot be removed. And nobody, so far as I know, has suggested that happiness involves constant bodily pleasure. Happiness is certainly connected to bodily pain and pleasure, however, despite what Callicott avers, although the connection is complex. The way we use the words *pain* and *pleasure* (and cognate terms) indicates a continuum between bodily pain and emotional pain. News, for instance, can be very painful, although the pain is not felt in one's ears. It would be very *painful* for you to hear that your daughter was burned to death in an accident. Your pain is not identical with whatever unpleasant bodily sensations are caused by the emotional shock of learning this painful fact. You would eagerly trade ten times that pain if you could undo the horror that has occurred.

Being burned alive *is* a horror, and it is a horror precisely because of the enormous bodily pain it causes. Death is a mercy in such a case, although death in itself is bad enough. You *cannot* be happy with the fact that your child died in this way. Your child's painful death prevents your happiness. If she had been happy up to the point of death, certainly dying in such pain spoiled her happiness—and not just the happiness of her final moments, either. The happiness of her whole life is diminished by its intensely painful last minutes. It is now harder to say that she lived a happy life. A happy life is one that you would choose for yourself or your child, and who would choose a life that ends in screaming agony? The horror of your daughter's death is not just a matter of its painfulness, but also of the uselessness and injustice of that pain. Dying from burns suffered while rescuing a friend or fighting for a just cause may leave one's happiness intact. Suffering that is voluntarily accepted or risked may be tolerated as the cost of achieving one's goals and in that way is consistent with one's happiness. Dying painfully for a cause can complete one's life by bringing it to a meaningful conclusion. But your daughter's death was an accident. It did not complete her life's plans, but destroyed them. The meaninglessness of her death denies her life full meaning. Her potential was wasted.

So pain certainly does have something to do with happiness. Pleasure does, too, although the connection is more subtle, so we cannot look into that connection here. Pain also has something to do with good and evil. To see this, all we need do is change the previous example from the accidental death by burning of one's child to the intentional torture and murder of one's child. Such torture is evil, if the term *evil* is to have any use at all. Torture is the intentional infliction of pain. It is the infliction of *pain* that makes torture evil. Clearly, this evil would be inconceivable if there were no such thing as pain. So to say that pain and pleasure have nothing at all to do with good and evil is plainly false, since pain clearly has everything to do with the sort of evil we have just considered. Once again, pleasure, too, is connected with good and evil, although the connections are more subtle and cannot be looked into here. Of course, Callicott does not deny that pain and pleasure have anything to do with good and evil *at all*, but rather, that they have nothing to do with good and evil from the point of view of his environmental ethics. This is true, of course, and it implies that environmental ethics are incomplete, that they leave out entire dimensions of right and wrong, justice and injustice, good and evil. The only other option is to conclude that they are unacceptable, since taking the presumed good of the environment as a transcendent objective has horribly immoral consequences.

10.3   THE ENVIRONMENTALIST'S VILIFICATION OF HUMAN DOMESTICITY

Callicott himself does not admit that environmental ethics are incomplete. This would be inconsistent with his goal of increasing the domain of wilderness. He is arguing that we ought to retreat so that the wilderness may advance, and this is not easily done if he admits legitimacy to normal human ethics. Normally, we think that those who feed the poor, tend the sick, and bind the wounds of the suffering are doing a good thing. But once this sort of good and evil is taken seriously, it becomes obvious that increasing the extent of wilderness is but one good thing (assuming that it is a good thing) among others that call for our attention. It also becomes obvious that we must choose which goals are more important and that these goals may be in tension with each other. For example, enlarging the wilderness may increase hunger. Rather than engage in this very complex and difficult discussion, and rather than help us decide how to harmonize the tensions between different ethical dimensions, Callicott restricts himself to promoting the good of the environment, period. So he argues against human values and interests in general. People have attempted to exempt themselves from the life/death reciprocities of natural processes and from ecological limitations in the name of a prophylactic ethic of maximizing rewards (pleasure) and minimizing unwelcome information (pain) (Callicott, 1980, p. 65).

This insinuates that we civilized people are all shameful pleasure seekers without a care for what is really important in life. No doubt there are too many people wasting their lives on drugs and other sources of pleasure, but if Callicott means to imply that administering morphine to a burn victim is an ignoble attempt at "minimizing unwelcome information (pain)," this is nonsense. Identifying pain as "information" does not make it one bit less painful, so cannot transform it into something that has nothing to do with happiness or good and evil. From the scientific point of view, all states of consciousness whatever are information states of the brain. Assuming that this is true, it does not change anything. We are used to being told that we should be tougher and that we should not seek relief from pain, but surely enduring pain for no reason at all is irrational. An aging relative who is suffering agonies because his spine has collapsed due to osteoporosis deserves relief, not a lesson in bravery. Since the pain is largely intractable, and he is suffering it, he is the one to teach us lessons in bravery and toughness, not the other way around. If there is a pill that lessens his pain, he should have access to it. Denying him the medication is tantamount to torture.

Callicott disagrees. As he sees it, providing relief from pain is wrong because it alienates us from nature: "Civilization has insulated and alienated us from the rigors and challenges of the natural environment" (op. cit., p. 66). But what is this "alienation"? Yes, civilization has *insulated* us from wind, rain, cold, biting insects, famine, disease, suffering, and death—thank goodness! Certainly, this insulation appears to be a good thing, the same good thing discovered by the bees in their hives and the rabbits in their burrows. Is "alienation" something in addition to the insulation, and if so, what? In what mysterious way does the relief of our pain injure or insult the natural environment? To be frank, it is difficult to construct a story in which this mystery is revealed. It seems that the insulation *is* the alienation. To relieve someone's pain is to distance them from what would *naturally* occur, and this is the sin or crime of alienation. The relief of pain requires artifice, and the artificial is the opposite of the natural, hence is wrong.

This alienation is compounded when we extend our concern for the suffering beyond humans to include other animals. If, heaven help us, we should be so "soft" (a term Callicott uses to describe the civilized ethos) as to administer analgesics to a wild animal with osteoporosis or burns, we not only perversely persist in our own alienation, but we "impose" it on other innocent creatures, as though we were addicts spreading our addiction: "Rather than imposing our alienation from nature and natural processes and cycles of life on other animals, we human beings could reaffirm our participation in life as it is given without a sugar coating" (op. cit., p. 65).

How does this alienation work when we provide relief from suffering to the merely "half alive" domestic animals such as our farm animals and our pets? These animals themselves are embodiments of the unnatural and the alienated. Being so fully alienated from nature already, presumably they could not be further alienated by us when we give them analgesics. A pet dog undergoing a surgical operation is already so unnatural that not much further harm is done by administering an anesthetic during the operation.

We are lucky to have the works of Callicott, for they illuminate just what is implied in the classical environmentalist ethic, in which the good of the natural world (however that is conceived) is seen as intrinsically good in and of itself. It helps us to understand why environmentalists have pressed continually for the total elimination of DDT, even though this has resulted in a recrudescence of malaria and the deaths of millions of children. If DDT is bad for the "biotic community," it is bad, period. Thus, it must be eliminated. Millions of human deaths by malaria, on the other hand, are part of the "natural processes and cycles of life" and should be accepted "without a sugar coating." In any case, lower human population will advance the domain of wilderness, too. Within the ideal of the natural as good, there is the ideal of the natural human being. Natural human beings willingly accept the suffering and death meted out by the fickle finger of nature, as if they were, like other animals, unable to avoid them. This keeps their population small, and this is good for the environment. This ignores the fact that the ability to prevent suffering and death makes us responsible for them when we choose not to prevent them. It ignores this responsibility because human suffering and death are irrelevant to the *summum bonum* of the environmentalist. It is not even that such human costs are thought reasonable in order that the good of the environment may be achieved, but that they are not even *recognized* as costs at all, given that the only thing good in itself is the health of the environment.

DO ENVIRONMENTALISTS REMEMBER RIGHTLY?

In *Landscape and Memory*, Simon Schama argues patiently, and in great detail, that modern Western civilization has never been opposed to nature: ". . . to take the many and several ills of the environment seriously does not, I think, require that we trade in our cultural legacy or its posterity. It asks instead that we simply see it for what it has truly been: not the repudiation, but the veneration, of nature" (Schama 1995, p. 18). Environmentalists who, like Lynn White (1967), argue that the "environmental crisis" can only be solved by a radical reconstruction of Western culture (see Section 7.2), have a false consciousness of the past. Schama does sympathize with the environmentalist's readiness for such drastic action: ". . . today, the most zealous friends of the earth become understandably impatient with the shuffles and scuffles, the compromises and bargains of politics when the 'death of nature' is said to be immanent, and the alternatives presented as a bleak choice between redemption and extinction" (ibid.). However, Schama asks that we do not panic about the future, but instead remember the past as it really was, rather than as environmental zealotry would have us believe: "It is at this point, when environmental imperatives are invested with a sacred, mythical quality, which is said to demand a dedication purer and more uncompromising than the habits of humanity usually supply, that memory may help to redress the balance" (ibid.).

Since environmentalism considers the health of the environment to be an objective that outweighs all our other more transient concerns (that is, a transcendental objective; see Chapter 3), balance is precisely what needs to be restored. What Schama shows in *Landscape and Memory* "is that the cultural habits of humanity have always made room for the sacredness of nature. All our landscapes, from the city park to the mountain hike, are imprinted with our tenacious, inescapable obsessions" (ibid.). Which obsessions? Our obsessions with the beauty, awe, and sacredness of nature.

10.4   FORWARD: ACCEPTING OUR RESPONSIBILITY

In its boldest outlines, we face a choice between two alternatives: going backward to a state of submission to nature, or going forward to accept our responsibility for nature. I believe that we must go forward, not backward, both for our good and for the good of the planet. The most efficient way to define our responsibility is by way of an image: We are emerging as Earth's nervous system. Just as the nervous system of an animal has a specific role to play within and for the animal as a whole, we too have a specific role to play within and for the Earth as a whole. Indeed, one of the roles of the nervous system is to make a whole out of the parts of which the animal is constituted. The nervous system of an animal orchestrates the activities of the digestive system, circulatory system, musculature, sensory systems, and so on, in order that they act as a whole. The nervous system identifies the self of the animal whose existence it supports and partially comprises. At this moment in history, Earth is not a biological entity in its own right; Earth is neither a plant nor an animal. In a purely formal sense, the Earth contains or houses a biological system, but this system is nothing more or less than the sum total of all living things on the planet. But a total is not an entity. We have the unique potential to transform Earth into an entity with its own coherence and integrity—its own self. This is something we should do, both for our own sake and for the sake of the natural planet.

Making an organic entity out of the sum of living things on the planet is essentially a matter of resolving tensions between opposites. To that extent, it requires thinking through what to the casual glance may look like contradictions or paradoxes. For example, we are clearly only part of nature, and a tiny part at that, so it may seem impossible or paradoxical that we at the same time take charge of the whole. We can help ourselves see through this apparent impossibility by reminding ourselves that the nervous system of any animal is a mere fraction of the whole, and that the paradox is resolved in a harmonious resolution of tension. Biological entities are organic, organ*ized*: Their parts, all of which have interests of their own, come together to serve their various interests cooperatively. This cooperation presupposes a primary tension between the interests of the individual cells in the first place, as well as a way to resolve these tensions for the mutual benefit of all. Thus, the cells of the kidneys, for example, are both servants of the body as a whole, their job being to cleanse the blood, and are served by the body as a whole, which provides their food, water, and oxygen. Similarly, the cells of the nervous system are both served by the body and are servants of the body: both masters and slaves.

This is no magic involved in this; nothing supernatural is required. Nervous systems have provided animals from time immemorial with an advantage in the evolutionary race: the advantage of cooperation. Nevertheless, there is a tension. Seeing how a whole can be made of its parts requires looking past apparent clashes of interest to the underlying unities. Environmentalism thinks in terms of the clash between humankind and the rest of the natural order. Environmentalism has witnessed the manifold injuries inflicted on nature by humankind, and it recommends that we repent, assume a posture of humility, and try to make ourselves as small as possible (try to reduce our ecological footprint as much as possible), so that nature can reassert its power over us. In this vision one side or the other must triumph. Callicott does us the service of sketching what a victory of nature over humankind would look like. I suggest that as long as the conflict between human interests and nature remains at center stage, the conceivable solutions to our "environmental" problems (whether or not they amount to a crisis) will be restricted to victory of one side over the other. We must instead learn to see tensions, rather than clashes, between nature and humankind. We must look beyond victory for either side to a unity of the two sides. In this unity, there is a special role for us to play: we seek the harmonious resolution of tensions between humankind and nature. Just as a nervous system enables your billions of billions of individual cells to be the single entity that you understand yourself to be, so too we can become the unifying information-processing system of the planet as a whole.

CASE STUDY 11: SHOULD WE TRY TO TAKE CONTROL OF EARTH'S TEMPERATURE?

Years ago, while pondering climate change and wondering what, if anything, could be done about it, John Latham (1990) identified a mechanism whereby human beings could gain a large measure of control over climate: by controlling the amount of sunlight that Earth reflects back into space, that is, by adjusting its albedo. Global temperatures depend to a large extent on the fluffy cumulus clouds that float above us in the skies. They have a large cooling effect because they are very white and therefore increase Earth's albedo. This is readily seen in photographs of Earth from space: The big white patches are composed primarily of cumulus clouds. Since most of the global surface is ocean, and the ocean spawns clouds because of its moisture, most of the cumulus is over the oceans: marine cumulus clouds. Latham realized that it might be possible to influence both the amount and the albedo of Earth's marine cumulus, and that this would provide a way of getting a measure of control over global temperatures. In other words, Latham had identified a pressure point in the natural system that would enable us to subdue global warming.

The reflectivity of a cloud depends on the size of the cloud's water droplets: The smaller the droplets, the more sunlight it reflects (Twomey 1977). Droplet size depends on the number of cloud condensation nuclei, the microscopic bits of matter around which moisture condenses to form clouds. Stephen Salter, an environmental engineer who designed wind and tidal power generators (as well as a computer engineer who invented the touch screen computer), teamed up with Latham to devise a way to increase the numbers of cloud condensation nuclei available for marine cumulus cloud formation (Salter and Latham 2007). As it turns out, a main natural source of condensation nuclei is the evaporation of sea spray, which creates innumerable tiny particles of salt that are left suspended in the air and enable clouds to form.

So Salter went to work and designed a seagoing mechanism for creating huge numbers of sea salt condensation nuclei that would increase the amount and albedo of marine clouds. He also designed ships to carry these mechanisms that would be powered solely by the wind, and a global communications and control system to control the ships and direct them to just the right places to create superfine sea spray droplets and blow them upward above the layer of stagnant air near the sea surface. These microscopic droplets would then evaporate, creating condensation nuclei when and where they might be needed. The process this employs, moreover, is precisely the one that nature uses to make marine cumulus form in the first place. The Salter ships would not introduce a new and foreign process into the global climate, but merely enhance the natural process already occurring.

A change of 3% in the albedo of marine clouds would reverse the current level of warming estimated by global warming theory, and Salter calculates that "50 spray vessels costing a few million pounds each with a life of 20 years could cancel the effects of a one-year increase in world CO_2." This is an extremely cost-effective solution to the climate-change problem. By comparison, the direct costs of the Kyoto accords plan are generally reckoned to be on the order of $50 per ton, or some $450 billion per year. But that is just the tip of the Kyoto iceberg, so to speak. The Kyoto expense would not represent an investment in the economy, but rather a tax (an entire scheme of carbon taxes, to be more precise) placed on it. Rather than an economic engine the $450 billion that Kyoto costs would instead be a brake on the world economic engine, and a drag on world economic progress. Just how bad this braking effect might be is a matter of intense dispute, but decades of ongoing recession and depression are clearly possible.

Question: Should we investigate the Salter–Latham plan (SLP) with the intention of trying to realize their design goals and thereby gaining control of the climate?

No: The Environment Ministry of the United Kingdom has refused to provide funding to even test SLP, pointing out that it may cause more problems than it solves. For example, by increasing marine cumulus by some 3% or so, SLP probably will increase marine rainfall as well. It will also have effects on marine life by producing more shade over the oceans along with the increase in rain. We do not know just what these effects may be, but the chances are that they will be serious. Thus, SLP has the usual characteristics of the techno-fix, and should be rejected. The environmentally correct way to proceed here is for us to undo the harm we are already doing by our unrestrained use of fossil fuels. Trying to find some techno-fix for the problem that lets us continue dumping ever increasing amounts of CO_2 into the atmosphere means that we will eventually but inevitably have to face two sets of problems rather than just one: the as yet unforeseen environmental impacts of steadily increasing CO_2 and the unforeseen environmental impacts of SLP. That is why the UK Environment Ministry has correctly chosen to stick by its commitments to carbon trading (which are legally required by the Kyoto accords in any case) as a means of countering global warming.

Yes: If we grant for the sake of argument that global warming theory is right, then obviously we ought to be trying to make SLP a reality. The status quo response to the global warming problem has been to demand penitential action on the part of humankind, as though its use of fossil fuels has been a sin. We are being told that if the economy suffers, that is just what we deserve, since our greed and self-centeredness is what caused the problem in the first place. But all we have been trying to do is earn our food, clothing, and shelter without relying solely on muscle power. It is not greedy to want enough to eat, a place to live that is warm and dry, and a means of transport. Now that we have managed to achieve these entirely reasonable improvements in the human condition and at last have the chance of offering them to humankind as a whole, we are being told that the entire attempt has been immoral all along and ought to be abandoned. Granted, there is a down side to everything, and human progress is no exception. But let us not throw out the baby with the bathwater. The intelligent thing is to deal with the problems that arise and continue to improve the conditions not only for human beings but for all living things on the planet. SLP represents just that sort of intelligence.

No: This sort of thinking has gone on for too long. Every technological trick we have devised to evade the realities of biological life has created its own problems, problems which we then solve with another technological trick, and so on and so on, and our problems have just gotten bigger and bigger. Now that they have finally gotten so huge that they threaten global catastrophe, you suggest more of the same. This is nothing other than outright denial, the inability to face up to the nature of the problem, and an attempt to dodge responsibility once again. SLP just raises the stakes in a game of chance the human race has been playing and losing for generations. Rather than waiting for our luck to run out finally and catastrophically, we should quit playing this game altogether. Rather than gambling with the life of the planet as a whole, we should quit gambling altogether and go back to honest work and earning a modest, but honest, living.

Yes: This issue should not and cannot be decided at this level of grand and sweeping metaphor. Interesting as the environmental morality of the human species may be as a philosophical issue, the threat posed by global warming theory (GWT) is based on scientific facts, not global generalizations about human environmental ethics. Assuming that GWT poses a real problem, it does so because it has added up thousands of bits of data and has gotten its addition right. SLP is no different. Assuming that it presents us with a solution to the problem, it does so because it has added up its data and has gotten its addition right. The case we are studying is whether or not to go through the data more closely, with a fine-toothed comb as it were, to see whether it really has a chance of working.

No: And then do what? You are not proposing a program of mere scientific research, but to take control of global climate. You are proposing a massive and historically unprecedented intrusion into the planetary ecosystem, nothing less, and that goes beyond mere science to a style of thought and action that can only be characterized as arrogant.

Yes: Again, I urge us to stick to the facts rather than cast grand moral generalizations over the issue. And the fact is that GWT says that we are *already* controlling climate through our influence over CO_2 levels. The Kyoto accords, in fact, propose to control climate as well, by means of those same CO_2 levels. Assuming that GWT has identified one of the levers that controls temperature, the people behind the Kyoto accords are attempting to grab that lever themselves. So let us not pretend that the issue is the environmental morality of whether or not to control climate. The only question is how to control it.

No: In that case, the answer is perfectly clear: Put the CO_2 lever back where it was before we began to fiddle with it in the first place.

Yes: That is not at all clear. The fate of humankind depends on that lever—a fact that I am sure is not lost on the people behind the Kyoto protocols. If you can control the CO_2 lever, you can control humankind. We have every reason to worry about putting control of humankind in the hands of any particular political structure or regime. But even if we put those worries aside, putting that lever back where it was, given the present state of technology, will cause hardship for billions of our fellow human beings. What SLP does is identify another lever, one that can be adjusted without sad consequences for humanity—or for the other organisms that enjoy life on this planet.

No: Your fear seems like paranoia to me. In any case, even if we grant that one way or another controlling climate is the issue, the onus is upon SLP supporters to show that it will not have dire consequences for the environment. This simply has not been done.

Yes: Granted. That is why it is important to continue investigating SLP.

10.5 FORWARD: OUR UNIQUE ABILITIES

We are part of the totality of life on the planet, so total victory for humankind is impossible. Surely it is true that we have been in a contest with nature to survive and hopefully flourish. Like every other species on the planet, we have struggled to obtain the necessities of life and avoid the dangers that the world presents us. We cannot make the mistake of thinking that because we are engaged in this ongoing struggle, our goal is total victory or unconditional surrender of the planet to our desires. Instead, we must realize that there is a deeper unity between ourselves and the natural order, come to appreciate the tension between the unity and the struggle, and start seeking a harmonious resolution. Humankind comprises a system in an even stronger sense than the rest of the natural order. No man is an island, the poet tells us, and we can see the manifold ways in which that is true. We are social animals who have spun a global clan, driven by the instinct to communicate and cooperate in projects of mutual benefit. We are linked to each other in our cities, and our cities are connected to each other by flows of materials and information. We have very slowly begun to realize that we are a global village (McLuhan 1962, p. 32), that we have a global political identity. In fact, the current surge in global political activity centered around the threat of global warming is but one element of the emerging planetary polis, the Earth aware of itself as such.

For all of its beauty and awe, the order of nature considered in itself and separate from a relatively recent species, *Homo sapiens*, is an unconscious giant. We are by no means the only conscious organisms within nature. There is no doubt that lynxes and rabbits are conscious, as are thousands upon thousands of other species. Perhaps all organisms have some version of consciousness. Paul Churchland plausibly argues that life and consciousness differ only in degree, not in kind (1988, chap. 7). In this view, a conscious being is simply an organism equipped with sufficient information-processing and behavioral plasticity to be able to adapt to its environment within its own lifetime instead of having to wait for the geologically slow process of natural selection. The lynx is capable of learning to hunt different prey in different habitats offering different opportunities for shelter and reproduction, as is the rabbit.* But both the lynx and the rabbit are unconscious of the order of nature as such. They have no idea of, and no care whatever for, the health of the environment. They live and die in a world of much more pressing and immediate concerns of feeding, fleeing, fighting, and reproducing. Only human beings, alone among all of the species on Earth, has any comprehension of the natural order or any sense of its beauty and awe.

Rabbits are not in danger of extinction, but even if they were, the extinction of their species would not hurt any actual rabbit. Even if rabbits were disappearing from the face of the Earth, each individual rabbit faces life prospects that are perfectly normal for rabbits. If, for example, a plague were gradually killing all of the world's rabbits, each individual rabbit would then die of disease, not an uncommon end for a rabbit nor one that subjects a rabbit to uncommon hardship or pain. No rabbit would ever mourn the loss of its kind, or recollect the illustrious history of rabbit-kind, or feel its heart overflowing with grief that rabbits would nevermore burrow in the sweet soil of Earth. We human beings would do all of these things if we knew about the plight of the

---

* If we add to this Churchland's view that life and negentropic entities differ only in degree, not in kind, it seems to be implied that he has some sympathy with the Gaia hypothesis (Lovelock 1991, 1995). A negentropic entity is one that extracts order out of its environment, that is, one within which entropy decreases at the expense of increasing entropy in its environment. Physicists who study dynamical systems sometimes refer to this as the spontaneous creation of order within dissipative systems. In a dissipative system there is a continuous throughput of energy, resulting in the evolution of self-sustaining, even self-replicating processes. Common examples include such simple things as a candle flame and such complex things as a giraffe. The Earth itself is just such a negentropic system. Energy flows into it continuously from the Sun as white light and out of it into empty space as infrared light, and in the process creates the complex systems we know as life. Thus, Churchland envisions a continuum between continuously moving physical matter on one hand and consciously thinking intelligent organisms on the other. In between lie only faintly conscious living beings such as trees. Churchland is not alone in this respect, since philosophers of such different persuasions as Dennett (1991) and Chalmers (1996) recognize the fact and the import that there are no fissures running through the natural world to separate life from mere motion, or consciousness from mere life. Aristotle, I fancy, would be pleased.

rabbit species, although this would be entirely beyond the comprehension of rabbits. A species is an abstraction. A real rabbit has real ears and paws, but the species of rabbit has neither, and indeed is invisible. A species is not capable of digging or eating anything, and cannot possible suffer or enjoy anything. So neither the species of rabbit nor any individual rabbit cares one bit whether the species becomes extinct. We alone are capable of such subtleties. We can understand what a species is and the interlocking role of species in a system of life. We can grieve the extinction of a species—or take action to prevent it.

Our special abilities and potentials mean that there is a special role that we can play in the order of nature, a role that serves the entire order. Just exactly what that role may be is part of the question we face. Be assured—indeed, be warned— that we will indeed play a role of some sort or other. The real question is just what that role should be. I am presupposing that we must gain an understanding of ourselves and of nature in order to answer that question. I am proposing that one way to gain an appreciation of the best role for us to play is by taking a lesson from nature itself, in particular from the history of life on the planet. Given our nature, we can play the role of the information-processing system of the planet. Moreover, this would be a good thing to do. In particular, it would be what the enlightened environmentalist would want. What nature teaches us is that the role of the nervous system is threefold: to gather information, to understand it relative to the needs of the organism, and to coordinate the actions of the entire animal on the basis of that understanding.

Note well that the nervous system does not act on the basis of its own needs, but on the needs of the organism it enervates. Our role would be to serve as the eyes and ears of the planet, its power of understanding relative to its own needs, and the coordinator of its actions as an entity in its own right. Alone among the species of the world, we have the ability to play this much-needed role. As the romantic environmentalist insists, we can withdraw from this challenge, shunning the very thought as the pride that goes before a fall. Or we can humbly, cautiously, and with due circumspection realize that we must begin to serve not only the interests of *Homo sapiens*, but of life itself. Life itself arose before any of us was born, and we would like to see it continue after each of us has died. Human beings are capable of consciously serving a purpose that we know goes beyond the interest of any one of us, although it is indeed in the interest of all of us. This is not a matter of either–or: either choosing our own interests or choosing the interests of the totality of living things. It is, rather, the recognition of our common interests.

Although the natural order transcends us as individual beings, its good is not a transcendent objective, a goal that is to be achieved regardless of our good. Contrary to what environmentalists profess and presuppose, we are not to disappear or minimize our environmental impact or marginalize ourselves. Instead, we are to realize our goals as a component of the goals of the entire living system. We are to conceive of our own good as a strand in the good of the whole, and to conceive of the destiny of the planet in such a way that it fulfills our own destiny. We are to serve life, and to create an entity, something larger than ourselves, something that flows through us and into subsequent generations. This is both a blessing and a curse. It is a blessing to each of us because we each have life. It is a blessing to humankind because we collectively get to pursue our destiny as a species. It is a blessing because as a global clan we will witness and partake in the ongoing drama of life on Earth. It is a curse because our intelligence not only qualifies us for our special role, but forces us to know that each of us must die, that each of us has a responsibility simply by virtue of our unique nature as human beings, and that we never see more than a fraction of the whole.

THESIS 20:   We should not go backward to an earlier form of our relationship with nature. We should go forward to a relationship with nature that recognizes our unique abilities and embraces the unique responsibility they create.

CASE STUDY 12: SHOULD WE ACCEPT GLOBAL WARMING ON THE PRECAUTIONARY PRINCIPLE?

Yes: Whether or not a scientific hypothesis should be accepted is a function of what hangs on its acceptance. Suppose, for example, that you were driving your family to a movie and a funny noise made you suspect that a tire was going flat. How should you react to the hypothesis that you have a flat tire? Since it would be very dangerous for you to keep driving with a flat tire, and since the lives of your family members hangs on your decision to accept or reject the hypothesis that you have a flat tire, it would be completely rational for you to accept the hypothesis even if its probability was quite low. Even if the probability of the flat tire is only one in ten (0.10), you would be rational to accept it, pull over, and check the tire. This is a perfect example of the *precautionary principle*: The hypothesis of danger to someone or something that is precious (i.e., of very high value) can be accepted even when its probability is low, just to be on the safe side. An ounce of precaution is worth a pound of cure, as they say.

GWT is not simply a matter of pure theoretical science, but also a matter of global public policy concerning a deadly risk to something even more precious than any one person's family: the health of the environment. If GWT is true, the harm not only to human beings, but to all life on this planet, would be a disaster of such devastating proportions that it is perfectly rational to accept GWT even if its probability is only one in ten, or even one in a hundred. Human beings have a sorry history of environmental degradation caused by throwing caution to the winds in the greedy pursuit of economic growth. It is high time this historical trend is reversed, before we destroy the planet for our children and their children. It is high time to approach these issues more rationally, and rationality demands caution. So, for safety's sake—or, in other words, by virtue of the precautionary principle—we must accept GWT.

No: Your argument involves a dangerous error in logic. I agree that it is perfectly rational to pull over and check to see whether you have a flat tire, but this does not in any way require that you *accept* the hypothesis that you have a flat tire. You do not need to fool yourself that the tire really is flat in order to pull over. Even though you think that the chance of a flat tire is only one in ten, it is still rational to stop and check the tire, but this does not mean that you have suddenly come to believe that the tire is actually flat. You still believe that the probability the tire is flat is one in ten (0.10), but you still stop and check the tire, because the cost of not stopping should the tire actually turn out to be flat could be enormous: injury or death of your family. The logically correct way to think of the situation is that you should pull over because of the risk you face. The risk alone is enough to make you pull over, given its cost. At a first approximation, the cost of a risk is given by the product of its probability times the potential loss (i.e., multiplying its probability by its value, which in this case is negative). So even if the probability of a flat tire is one in a hundred (i.e., 0.01), the injury or death of your family would be such an enormous loss that even one one-hundredth of that loss (i.e., the product of that loss multiplied by 0.01) is a cost you quite rationally are not going to accept. So you should stop the car and check the tire, even though you do not believe the tire is actually flat, only that it might be.

Yes: I see what you are saying. I agree that you do not actually have to convince yourself that the tire is flat before it makes sense to pull over. But that does not make any difference in the case at hand. All that it means is that even if the probability of GWT is low, we should take precautionary action to protect ourselves and the planet against global warming. In fact, your logical clarification actually strengthens my case, for you have shown that we never have to accept GWT at all in order to take steps to stop global warming—and that this is perfectly rational.

No: I am gratified you see the logic of precautionary action more clearly. However, your application of this logic to GWT is hasty and incomplete. Suppose that you are not driving your family to see a movie, but are instead driving to the hospital because your child has a life-threatening injury and needs immediate medical attention. Now when you hear the funny noise that you fear is made by a flat tire, you do not immediately pull over to check, but instead keep driving to the hospital. You might slow down and grip the steering wheel more firmly, but you keep on going. This is also perfectly rational, and it shows that when you consider precautionary action you must take *all* of the significant costs and benefits into account. In this case you have to weigh the risk of the flat tire against the risk of your child not getting to the hospital in time. If, say, the probability of the flat tire is one in ten and the probability of your child dying because you get to the hospital too late is also one in ten, it is more rational to keep on driving to the hospital. A flat tire does not necessarily lead to an accident, after all, and an accident does not necessarily lead to a life-threatening injury. On the other side of the balance, your child already has a life-threatening injury and faces a high risk of death, so stopping to check the tire increases the risk to your child's life. So under these circumstances it is rational to keep on heading for the hospital despite the funny noise.

The situation with GWT is like the case where you are driving your child to the hospital: There are very serious risks on both sides. On the one hand, there is the risk of environmental and human harm associated with a temperature rise of 2 degrees over this century, while on the other hand there is the risk of harm to billions of the world's poorest people due to economic stagnation. The question comes down to the relative costs and the relative risks of these costs. As for the human costs, the IPCC itself does not predict widespread suffering or famine due to global warming (2007b). Instead, the results it foresees for us are mixed. Some agricultural lands will be lost in lower latitudes, but even more will be gained in higher latitudes; some areas will experience more drought, but overall there is an increase in rainfall that is good not only for human beings but also for the environment. So the effects for the environment as a whole are also mixed. The biggest problem is that some species face an increased risk of extinction, but it is also clear that some species face a decreased risk of extinction as well.

So we face two possibilities. On the one hand there is global warming, which brings both costs and benefits to us and the rest of nature gradually over this century. On the other hand there is the immediate reduction of fossil-fuel use and all of the human costs that will occur right away and which will have negative effects through the rest of the century, including negative effects on the environment as well. Poor and hungry people will not be much inclined to worry about the natural environment or work to improve it. Hunting wild plants and animals for food, or clearing forests to grow crops, will seem preferable to starvation for oneself or one's children. Under these circumstances, it seems there is more risk for the environment in abandoning fossil fuels than there is in global warming. We must conclude that the precautionary principle is actually in favor of maintaining our use of fossil fuels if that is what is required to maintain the health of the economy.

Yes: We have faced this choice before, and we do not want to make the same mistake again. We have chosen our own economic progress, our own comfort and luxury, over the health of the environment, and the result has been environmental devastation and, ultimately, damage to ourselves as well. Our bodies are laced with traces of every chemical and toxin known to chemistry—isn't that proof enough that the course you suggest—maintaining the *status quo*—is the wrong one?

No: I am suggesting no course. The question is whether we should accept GWT on the basis of the precautionary principle, and I am merely concluding that we should not. In any case, you have already admitted the most important point, which is that we can logically separate the question of the truth of GWT from the question of what we should do concerning that threat. Given that both global warming and stalling the economy in an effort to prevent it pose very serious threats, we can agree that the question of whether or not GWT is right is extremely important and must be taken very seriously. To do so, however, we must address the probability that the theory is right on its scientific merits alone. We cannot judge the risk GWT confronts us with until we have made our best estimate of its probability. This leaves us with no shortcuts. We cannot just accept the theory just to be on the safe side. We must instead face the scientific case for GWT head on, and make our best judgment of its credibility.

11 A Vision of the Future

If, as seems evident, the main business of the nervous system is to allow the organism to move so as to facilitate feeding, avoid predators, and in general survive long enough to reproduce, then an important job of cognition is to make predictions *that guide* decisions. *The better the predictive capacities, the better,* other things being equal, *the organism's chance for survival.*

—Patricia Churchland (2002, p. 40)

An early riser, I found our camp swathed in cloud when I got up. While my brother slept in the tent perched on the high mountain ridge, I started a fire in the tiny stone stove we had made and fed it chips of wood as the water in the pot above began to simmer. Cliffs fell away for thousands of feet on two sides of the camp, although I could see only a few feet through the fog in the dim dawn light. The stove, which was only slightly larger than the pot sitting on it, was built against a rock face that angled downward alongside a ridge broad enough to stand on. A number of hand-sized pockets had somehow been scooped out of the cliff face by ancient geological processes, and putting my face up to one of these, I discovered a microcosm inside. In a cup of soil there were dwarf grasses, miniature cinquefoil, tiny succulents, and mosses growing in their stone shelter. As the moist cloud collided with the cliff face, tiny beads of water condensed and ran down into the tiny world in the rock, supplying it with life-giving moisture. I fancied myself living in that diminutive world, looking out of the opening into the mysterious world outside.

Some ravens came up over the ridge riding the up-drafting breeze, croaking back and forth as they checked out the campsite, shaking me out of my reverie. I searched the mists to see them, but they floated on over the other side, never coming close enough to be seen. Those ravens, my brother, the mountain ridge, and I were also in our own microcosm, our own smaller world carved out of the larger universe around us. Our mountain ridge and all of the mountains and valleys for 50 miles in any direction were inside a large national park that had been protected from human incursion for over a century. Instead of a pocket in the rock, we were protected inside a pocket carved out of the economic landscape of humankind. For countless millennia our ancestors sheltered in caves against the wilderness. Now the wilderness found shelter against humankind in parks like this. Things evolve continuously, the tables

Beyond Environmentalism: A Philosophy of Nature, By Jeffrey E. Foss
Copyright © 2009 John Wiley & Sons, Inc.

turn continuously. Yesterday's competition becomes today's symbiosis; yesterday's invader becomes an essential part of today's body.*

From my unique human perspective the whole commanded admiration—and it got it. A golden glow suffused the fog as the Sun began to penetrate the mists. The awesome vertical slabs of stone above us gradually emerged against the painfully blue sky, and then the beautiful, green, tree-lined valleys were unveiled below.

11.1   OUR ONGOING ROLE REVERSAL WITH NATURE

The wilderness used to be the place where human beings were threatened with danger and death. It is now the place where we are the threat of danger and death. This represents a role reversal of magnificent proportions, a historical fact that we will be digesting for generations. In our urban and urbane romanticism of nature, we forget that nature is not our friend. It is totally indifferent to us. Nature is the arena in which we, along with other living things, have struggled to survive. Far from being a system, the natural world is a battle. Despite all of the poetic talk of ecosystems, or of *the* ecosystem, no system is made of such strife. Organisms, like all systems, are built of cooperation. And no organism can survive without carving out a space for itself and competing for access to the necessities of life with its own kind as well as with other kinds. True enough, in the end we all unwittingly play into the hands of the other competitors in the evolutionary contest, and to that extent there is a system. What the plants breathe out we animals breathe in, and vice versa, for example. But that does not stop us animals from eating plants. Indeed, ultimately all animal food comes from plants. Poor plants, they cannot even run away when lightning sets fire to the forest or the prairie. The body of a plant or an animal is a system precisely because it does not involve these sorts of internal battles. No part of your body eats another part, or infects it, or burns it up in a fire. To think of the ongoing drama within the cauldron of evolution as a system requires selective blindness, romanticization rather than real romance.

There is now a chance that nature may become a system, but only if we realize the role reversal that is taking place. It is not taking place for all of humankind at the same rate. Many people still do not have the upper hand over nature in the struggle for survival. From their point of view, nature does not have the romantic glow that it has developed for the city folk in the developed world who have gotten a bit of an upper hand over nature. The subsistence farmer who carves a cornfield out of the forest will not have the same sympathy for sparing the wilderness from human development as the city dweller will have. It is only because some of us have become secure in the struggle for existence that our sympathies can now extend to our erstwhile

---

* As we now know, thanks in large part to the work of Lynn Margulis, the mitochondria that are inside our bodily cells and are essential to our metabolism were once infectious agents that happened to choose a cooperative relationship with us rather than a competitive one, symbiosis rather than conflict. There is *evolutionary pressure*, to use the currently fashionable concept, for such symbiosis. We would do well to study and appreciate this pressure.

competitors. The role reversal of getting the upper hand over nature is necessary if nature is to be spared the depredations of human beings. In the end, human destiny cannot be divorced from the destiny of living things in general. But that can only be appreciated by those who are not facing death, suffering, or constant work just to survive. Nature will do best if we do best. National parks are promoted and protected by the prosperous.

11.2   WE ARE THE EMERGING NERVOUS SYSTEM OF THE PLANET

I am arguing that we should step forward to accept our responsibilities as nervous system of the planet instead of stepping back to let Earth continue colliding blindly with whatever fate happens to send its way. Whether or not you believe in a higher order, you cannot believe that we bear no responsibility for Earth's fate. My argument has two fundamental premises. The first is that we are *in fact* emerging as the sensory and motor system for the planet, whether we like it or not, and the second is that if we willingly engage in this process with our eyes open it will be better both for us and for the planet as a whole than it would be if we do not. By a nervous system I mean an information processing system with the three main functions of a nervous system. First it collects information about the state of the Earth (proprioception) and what the planet is likely to encounter in its trajectory through space. Second, it stores and processes information in order to learn how Earth's organisms and supporting systems work, the needs of those organisms and systems, their main threats, and how those needs may be met and those threats may be avoided. Third, it directs those behaviors it controls for the good of the whole. We can perform the same information processes for the Earth that the nervous system performs for an organism. In this sense the analogy between us and a nervous system is very close.

There are big disanalogies as well. For one thing, humans are not neurons, but entire animals. Each of us has our own mind, consciousness, intelligence, and ideas about what would be best for the environment. Neurons have no minds, no consciousness, no ideas about anything. Neurons are not smart cells, but electrically hyperactive cells. Neurons are just the sort of cells that might be used to construct an information processing system, as the process of natural selection has discovered, but they are just as unintelligent and unconscious as the cells of your liver or bones. Within the limits of their normal operating conditions, neurons behave more or less mechanically, firing when sufficiently stimulated by incoming signals, resting otherwise. The mindless consistency of the neuron is a good thing as far the nervous system is concerned, because information processing requires consistency. Human beings, by contrast, have minds of their own. They do not mindlessly fire when sufficiently stimulated. Instead, they receive information by seeing something or being told something. They may or may not pass on the information received, and if they do pass it on, they transform it in the process. If you see a dog you normally see an indefinite number of things in the process, such as the size and color of the dog, the location where you saw the dog, when, and what it was doing. You may or may not tell someone else about it, and if you do choose to pass on any information you must choose what to report of all the many things you perceived.

Nevertheless, despite this disanalogy, groups of human beings do collect, process, and use information. Indeed, we are adepts when it comes to information, unique among the species of the Earth. Human beings as a group are aware of everything from dinosaurs to the threat of global warming. As a group, humankind could perform the three planetary information functions outlined above: seeing, thinking, and acting. We could transform the planet by providing it with intelligence. And this brings us to another big disanalogy between us and nervous systems: we *could* provide the planet with senses and intelligence, but we have not done so yet. So far our most intelligent actions from the point of view of the planet as a whole have been to address the problems caused by our own waste products, our pollution, by reining in that pollution. These are humble beginnings, indeed, but they are beginnings nevertheless. We have shown that we are capable of seeing, thinking, and acting on behalf of all life on this planet. Indeed, we have shown this by acting as a simple and flawed nervous system. Now that the threat of global warming has been communicated to the global public, whether humankind chooses to act or not to act, it will as a matter of fact bear responsibility for the result. Thus we are *emerging* as the nervous system of the globe, although we are so far a feeble force for well-informed intelligence. If we can but see, collectively, what we are becoming, we can make the transition more efficiently and more advantageously for ourselves as well as for all the other living things on Earth.

So one argument is that we are emerging as the information and control system, or "nervous system," of the planet anyway, like it or not, so we would do best for all concerned if we squared up to this fact. Another argument is that environmentalists in general presuppose that we play the role of planetary information and control system, for they propose actions for the good of the planet on the basis of their analysis of the facts as they see them. Every proposal for action by every environmentalist group, including the Kyoto protocols, assumes that intelligent, well-informed action on behalf of the planet as a whole is possible. The main argument, however, is that it would be good both for us and for the rest of nature. In principle this is an argument that is accepted by environmentalists, but only in part since doing what is best for humankind is not their concern, and furthermore, they may disagree among themselves or with others about just what would be best for nature. Generally speaking, environmentalists think that the best thing we could do for the environment is to leave it alone. They picture us as having caused the "environmental crisis" in the first place by poking our fingers into natural mechanisms, and they recommend that we now take a strict *laissez-faire* approach. If only human beings had never existed, whatever state nature might be in would be not only good, but ideal. As environmentalists see it, we are the one and only environmental problem, and the solution to the problem is for us to remove ourselves from the scene and let nature once again follow its own course.

That is the way back, the return route to our primordial relationship with nature, whereas I am recommending a route forward in which we carefully, respectfully, take responsibility by influencing the course of events for the good of nature. People have differing ideas about what is good for nature, and widely differing ideas about what is good for humankind. Nevertheless, there is a fundamental, foundational level of good, which I have called *natural* good, that will enjoy much broader support. It is naturally good to satisfy the natural necessities (the need for nourishment, sunlight, water, shelter, etc.) of organisms. What makes nervous systems naturally good and favored by natural selection is their ability to provide the necessities of life for the organisms that have them. Similarly, as nervous system of the Earth, we would have the task of tending to the necessities of the totality of living things on Earth. No matter how much people might disagree about what is good for the environment, they can all agree that satisfying natural needs of the organisms living in the environment is a good thing. Indeed, unless these needs are met, we too will perish. Attending to these needs will be a good recognized by all environmentalists, with the exception of the needs of one species, our own. For environmentalists, *Homo sapiens* is a criminal species, the source of all environmental harm, and satisfying its needs is not necessarily a good thing. Indeed, satisfying the needs and desires of humankind is the original sin, the source of all environmental woes, so satisfying human needs is automatically assumed to be wrong, at least until it can be shown to be otherwise.

Fortunately, it can be shown to be otherwise. Moreover, one of the reasons that it is otherwise is already implicitly recognized by environmentalists: Human beings have a special nature, unique among the species of Earth, which enables them to appreciate the environmental state of health of the planet and to act on its behalf. Environmentalists implicitly recognize this in two ways: First, they themselves are human beings who appreciate the state of health of the environment and act on its behalf; and second, in seeking support for their cause among their fellow human beings they presuppose that other human beings have this potential as well. Environmentalists are apt to overlook the fact that *only* human beings have this potential. In their blind faith in the goodness of nature, they overlook the fact that other species are totally incapable of recognizing environmental facts and values, or acting on them. All other species are driven solely by instincts and desires that give their own species an advantage over other species in the continuous struggle for survival and reproduction. In this regard they are like human beings prior to their environmental awakening. There are no environmentalists among any other species other than our own. Every single environmentalist is a human being, and there is a crucial lesson in that fact. Whether or not it is part of some grand plan or merely an accident, human beings have a unique nature that enables them to be sensible of the value of nature. This sensibility gives them a special responsibility, a special role to play. Their needs and necessities therefore cannot be overlooked by environmentalists, despite environmentalists' tendency toward an antihuman bias.

The second reason is that the environmentalist's blind faith in the goodness of natural processes is misguided. Natural processes have repeatedly brought life on Earth to the brink of extinction, and they may do so again. Imagine that we knew a comet was going to collide with the Earth, causing terrific destruction, much like the destruction that was caused 65 million years ago when such a collision brought the age of the dinosaurs to an end. Suppose that we had the power to stop the collision. What would be the environmentally responsible thing to do in such a case? On the one hand, the environmentalist has faith in the goodness of nature and natural processes as long as they are not influenced by humankind. On the other hand, the massive extinction of species that would be caused by such a collision (at least 90% of all species would be exterminated) can only be seen as a bad thing, something that is to be avoided at all costs. The environmentalist cannot consistently condemn every species extinction caused by human beings while tolerating or praising those that would be caused by a comet smashing into Earth. Only sheer self-loathing at the species level could explain such an attitude, and nothing from an environmental point of view could justify it.

If we may assume that most environmentalists, if given the choice, would choose to avert this natural environmental disaster, we may conclude that their antihuman bias does not run as deep as their concern for the good of living things in general. When push really comes to shove, they are capable of appreciating the special role that human beings, and only human beings, can play in the protection of the environment. In principle, at least, environmentalists cannot help but admit that human interests are not implacably opposed to environmental interests. Regardless of whether environmentalists do recognize these things, the fourth argument for human beings serving as the nervous system of the planet is that doing so would recognize crucial facts: in particular, the fact that we are completely alone among all the species on Earth in our capacity to recognize environmental realities and values, and work on behalf of all living things.

THESIS 21: Humankind should accept its responsibilities for nature by providing information-processing functions for life on Earth of the sort that nervous systems perform for individual animals.

11.3   THE NATURAL VIRTUES AND VICES OF NERVOUS SYSTEMS

Nervous systems have arisen via the process of natural selection because they provide the animals that have them with advantages in the evolutionary struggle. These advantages may be placed under two headings: integration and intelligence. *Integration* includes all of the information functions required to make a single animal out of the millions or billions of differing cells from which it is composed. *Intelligence* includes all of the perception and thought involved in directing the actions of the animal so that they work for the benefit of the animal as a whole. Both of these functions require solving nontrivial problems of the general form of harmonizing drives or goals that are in tension. For example, it is essential that an animal not eat or attack its own parts, although sometimes the welfare of the whole will require that the welfare of a part be sacrificed. An integrated animal will recognize its own body as something it cannot eat no matter how hungry it is and no matter how nutritious its body might be. Although not eating or otherwise harming one's own body seems a perfectly obvious policy, it is not perfectly obvious just how a nervous system can ensure this form of integration. Fortunately, evolution has developed a number of workable solutions to this problem, although occasionally even human beings can get carried away chewing their nails, scratching themselves until they bleed, and so on. In tension with the need to protect the body, intelligence might require that one run through thorny bushes that damage one's body in order that a predator might be escaped, or that one burn one's hands opening a door to escape a fire. Again, nervous systems have evolved that have workable although not ideal solutions to these problems as well.

So the primary benefits we can provide the living things of the Earth are integration and intelligence. What this entails in the fullest sense would take us far beyond the scope of this book. In this full sense, only an ongoing investigation and discussion as events unfold can address the problem. However, if we restrict ourselves to natural goods, in particular survival and the necessities of survival, a number of crucially important basic goals of a planetary nervous system can be established. For example, the physical and chemical variables of the planet must be maintained within certain bounds. Extant species may be assumed to have the right to survive, unless they threaten the survival of other species or of the whole system. Just as an animal may rid itself of a harmful infection or parasite, so too the whole system of nature may rid itself of harmful infections or parasites. There is no assumption that everything within nature is sacrosanct. The good of the whole is primary, not the good of every single part without exception.

The human species is no exception to this rule. We too are not sacrosanct. Our right to exist depends on our being good for the system as a whole. Fortunately, nervous systems are generally so beneficial that they are accorded levels of support and protection within the animal that other components of the animal do not enjoy. When the animal is threatened by shortages of warmth, food, oxygen, and so on, supplies of these necessities are restricted for other parts of the body *before* they are restricted for the nervous system. If an animal is suffocating, the last part of the body to suffer loss of oxygen will be the central nervous system, the brain. Humankind stands to gain this special status within the Earth system *if* it accepts the function of information processing on behalf of the whole system. Our rights to special consideration within the system are only potential, not actual, at this point. We are *emerging* as the nervous system of Earth, but we have not achieved this status yet. This is yet another argument for taking on this task: It would justify our existence from the point of view of environmental values. In the next section we return to the special status we would have in an integrated system of nature.

One way to get a grip on what it would mean to perform the functions of integrating the natural systems of the planet and providing them intelligence is to see how nervous systems sometimes fail. We can get a better idea of the virtues of the nervous system by looking at its *vices*. These are too numerous to be detailed, including everything from imperfect perceptual, memory, and processing systems, through to tendencies of complex nervous systems to become unstable or "insane." But one very general form of nervous system failure is worth discussing if only briefly: addiction.*

---

* See Foss 1996a for more discussion of addiction.

Addiction always involves inflexibility of the nervous system, an inability to readjust for the good of the whole organism. A classic example is often described in evolution textbooks, although it is not recognized as a matter of addiction: the extinction of the Irish elk. Although it is impossible to tell at this time with certainty, it is usually thought that the Irish elk became extinct after the last ice age largely because its antlers became too large. Fossilized elk remains with antlers up to 4 m (13 feet) across have been found. Presumably these overlarge antlers made it more difficult for the elk to escape predators (including human predators)—an animal with such large antlers cannot easily hide in the forest, for instance. But why did the elk's antlers become so large? Why was there not selection pressure against such large antlers? We would expect that natural selection would favor animals with smaller antlers, and thus that the elk would have gradually developed smaller antlers. Why did natural selection fail in this case?

The answer lies in the inflexibility of the elk's nervous system, and hence its inability to learn. One essential role of the nervous system is to orchestrate reproductive activity. Females should find males sexually attractive, and males should return the favor. This requires a considerable amount of neural sophistication: the nervous system that must process incoming visual, olfactory, auditory, and tactile information in order that an animal can recognize sexually appropriate members of its species and initiate the proper behaviors. In addition, males and females should also be wired up neurally, so to speak, in such a way as to prefer healthier, stronger, faster mates. Among many species of deer, moose, and elk, antler size becomes the standard of sexual desirability. The general health of a male strongly correlates with the size of its antlers. In the case of the Irish elk, it is hypothesized that females preferred males with larger antlers, which led to runaway growth in antler size and thereby the eventual demise of the species. The problem was that the neural mechanisms of sexual preference were written in the genes of these elk, and changing these genes is a very slow process. Genetic change requires mutation and selection, and the first of these processes may be extremely slow. Neural structures underlying instincts, especially instincts essential to survival, tend to be robust and redundant: In short, they are inflexible. They cannot be adjusted very quickly. Because the elk were unable to *learn* new sexual behaviors within their lifetime, they had no choice but to follow the dictates of their instincts, even as these led to their extinction.

It is not mere metaphor to say that the Irish elk were addicted to antler size in their sexual behavior, since this behavior was in all essential respects just like other addictive behavior, such as drug abuse. Substances such as alcohol, nicotine, cocaine, and the opiates activate the natural reward systems of the body, strongly reinforcing maladaptive behavior. Because the neural systems of reward are deeply integrated into the human genome, it is very difficult to counter the use of these substances, even in the face of obvious damage to the body. Learning has only a superficial effect on the instinctive mechanisms of reward, and it is this inflexibility that leads to maladaptive behavior—just as in the case of the ill-starred Irish elk. Human beings currently suffer from a number of forms of addiction. Let us consider one of these: food addiction. Because periodic food shortages were the norm over most of human history, those of our ancestors who ate more food than was immediately needed built up stores of fat, and this enabled them to survive food shortages better than their less hungry, thinner friends. Thus, natural selection preferred big eaters over moderate eaters, with the result that most human beings will eat even when they are not hungry and the body has no need for nourishment. Since science and technology have largely eliminated food shortages, obesity has become a greater hazard to human beings than famine or undernourishment. Because the neural mechanisms controlling hunger are embedded in our genes, hunger is insensitive to learning. Even in the face of obvious bodily damage, the desire to eat may prevail.

This is relevant to the issues at hand in two ways. First, human beings consume more food than they need, so they engage in needless agricultural behavior. So when it comes to natural goods, it must be noted that food is not always a matter of need or necessity for human beings but is sometimes a matter of excess desire over and above what is necessary. Nor are our other desires and behavior immune to such addictive syndromes. Clothing, shelter, and mobility are also natural necessities that are subject to the development of addiction as well: Houses and clothing may be larger and more elaborate than needed. It is always a natural good to satisfy a desire, so it is possible that sometimes eating more than is necessary, wearing clothes that are nicer than is necessary, and so on, may be a good thing to do. For example, humans mark weddings and other with socially important events with ceremonial feasting and ceremonial dress. It is also possible that sometimes these things are bad things to do.

However, these are ethical issues that would take us beyond the scope of this book, and for that reason we must mainly restrict our attention to natural necessities: natural goods that are naturally desired and which are also required for survival or the avoidance of pain (or both). As we noted in Section 9.1, there is no sharp boundary between natural desires and natural necessities, but one gradually shades into the other. Thus, addictive behavior, one main vice of nervous systems, is a matter of degree. When biologists determine what the daily minimum food requirements are for a given species, they do not take the bare minimum required to permit survival but that level that allows the animal to grow to its full size, be strong, be active, be resilient to temporary shortages, and so on. Similarly, when we speak of natural necessities, we do not mean bare necessities for mere survival, but what is necessary for good health, activity, and development. A well-tuned nervous system inclines an animal to nearly-optimal levels of food intake, activity, rest, growth, and so on. A nervous system for the planet would do this for life on the planet in general, including human life as one important instance, particularly if humankind embodied that nervous system. One key to remaining well tuned is the ability to monitor sufficiency and excess as they crop up, and the ability to retune as needed.

Second, the widespread human addiction to overeating illustrates the need of nervous systems for flexibility and learning, especially when changes in behavior are required to handle changes in the world around us. For example, at this time there is a general tendency of environmentalists toward inflexible rejection of technology and technological solutions to the problems we face. Environmentalists are addicted to rejecting technology and blaming it for environmental problems. On the other side of the issue, lovers of technology, technophiles, tend to accept new technology inflexibly and to seek technological solutions to every problem. Rather than eating less, the technophile may expect scientists to develop low-calorie foods. As inflexible tendencies, both the environmentalist's rejection of technology and the technophiles demand for it are forms of addiction, and hence are species of vice. Of course, all nervous systems are imperfect and will make mistakes. The deer may not notice the wolves lying in wait at the edges of the waterhole, or the wolves may not realize that they are not well hidden and so can be spotted by the deer. To be susceptible to error is not a vice, or else all nervous systems would be riddled with vice. A vice is a particular form of error, an error to which a system is particularly prone.

To every vice there is a corresponding virtue. The opposite of the vice of addiction is the ability of nervous systems to learn, to be responsive to changes, to recognize failures and do something about them. Developing techniques to enable timely and efficient learning has been a main goal of philosophy right from the start. Energetic debate, or dialectic, has generally been the preferred method. In the dialectical method, a problem or question is identified, and someone states a thesis that solves the problem wholly or partially. Then someone vigorously tries to reveal the faults in the thesis. The thesis may be defended, revised, or abandoned, and the process then continues. Dialectic can be internalized in a form of meditation in which a single person internalizes dialectical debate by alternating between the two roles of advancing theses and criticizing them. Use of this method has solved many of our scientific, medical, ethical, legal, and political problems, and has led to specific forms of the dialectical method in each of these fields. Within the sciences, dialectical debate is carried on in universities, laboratories, conferences, and scholarly journals. A legal trial is a specific formalized version of the dialectical method developed to determine questions of guilt and innocence.

The philosophy of nature needs to develop its own set of techniques and discussions, with its own set of generally accepted results, its own crucial questions, its ongoing debates, its test cases, and so on. This book, including the theses it presents, is offered as a step in this direction. Among the theses it puts forward is the idea that humankind accept its environmental responsibilities by providing the Earth with the sorts of information-processing that nervous systems provide animals. If this proposed information-processing system is to succeed, it must avoid the trap of addiction, and so must be ready, willing, and able to learn at all times. I would further suggest, therefore, that the system employ the dialectical method.

11.4   WHEN IT'S HUMANKIND VERSUS NATURE

Now that we are going through this historical role reversal, we need to reconceive our struggle with nature. We would like to cooperate with nature, but we cannot cooperate with nature unless we have an idea where the limits of sacrifice on either side are to be drawn. For us to cooperate is for us to make an effort, to give of our own time and therefore our own lives, on behalf of nature. When push comes to shove, just how much should humankind sacrifice in order that nature as a whole may flourish? How much should nature as a whole sacrifice in order that we may flourish? Would it ever be right, for example, for another species to be sacrificed so that we can survive? Can every tension between our interests and those of other species be resolved harmoniously? Probably not. To say otherwise would be sheer romanticism or presumption. But if universal harmony between us and all other living systems cannot be assumed, how is the resulting conflict to be decided?

Let us grant that environmentalists are right about one thing, specifically that it is possible for human beings to harm nature. Does it follow that human interests can clash with the interests of nature as a whole? Sometimes people harm nature by following what they *think* are their interests. Human beings did pollute the air and the water in the past, and still continue to do so to some extent. But doing so was *not actually* in the interests of human beings as a group. Certainly, some humans may have benefited, but pollution hurt many humans as well. Pollution of nature by human beings is analogous to cigarette smoking in human beings. It is not good for the human being as a whole to smoke, although it is very satisfying to the nervous system. Partly because the lungs are poorly enervated and we do not feel pain in our lungs proportional to the damage that smoking causes, and partly because nicotine is experienced as a reward by the nervous system, human beings will smoke, even though they know it harms their body. Humans are similarly insensitive to damage to the body of nature and do not feel bodily pain when nature is harmed, so they polluted the water and air that they themselves drank and breathed. But when the smoker becomes ill, or when people begin dying in numbers, the mistake is realized: thinking that the interests of humankind and nature, or the brain and the body, could actually disagree in fact.

Humankind cannot exist without the body of nature, and a brain cannot exist without its body. The converse is possible, however. A body can exist without a complete brain, and nature can exist without human beings, as it indeed had done for billions of years before it gave birth to our species. There is something profoundly sad about a human body that exists without consciousness. The body of a person who is brain dead or in an irreversible coma is a waste, a tragic loss of what might have been. Is there nothing sad about the body of nature living on without any awareness of its own existence, no awareness of whether it is in good health or poor, no ability to sense danger or promise, no ability to take action on its own behalf? Human beings never have been Earth's nervous system. When human beings harm nature, they do not do it as Earth's nervous system involved in addictive behavior. They do it as one species among the many others that nature has created, a species that has to compete for survival within nature, contending not only with these other species (diseases, predators, competitors for food and shelter, thorns, poisonous insects, etc.) but the elements as well (wind, rain, cold, tidal waves, wildfires, etc.). We have taken our struggle within nature to extremes, with the result that we have clashed with nature itself. And we have come—or at least are coming—to the realization that we cannot hurt nature itself without hurting ourselves.

So it is a type truth (of the sort introduced in Section 7.5) that our interests cannot actually diverge from the interests of nature, and for the same reason that it is a type truth that the interests of the brain cannot actually conflict with the interests of the rest of the body: Neither can exist apart from the natural body to which they are matched. It is worth observing that a specific sort of brain goes with a specific sort of body. The brain of a fish cannot plug into our body, nor yours into it. Neither the metabolism nor the information processing will match. Indeed, your specific brain is matched to your specific body. Your brain knows how to walk in your body, not in another body. And what holds for walking also holds for talking and many other things as well. In just the same way, we are matched to the Earth, at least at the metabolic level. We are built to live in a specific mix of gases in a specific range of temperature and pressure and gravitational force, and to eat specific sorts of foods, and so on, all of which are present here on Earth. All of these elementary conditions and biological conditions of our existence are in turn connected to the physics and chemistry of this planet. So in the end we, too, are matched to Earth even in its physical and chemical nature, its distance from the Sun, its seasons, and so on. So we are deeply matched to Earth at the metabolic level. It is for this reason that we cannot survive without this planet.*

But unlike the match of brain and body, we are not matched to this planet in terms of information processing—although we could be. It is conceivable that we could become so valuable to nature that it would be sad for nature to go on without us, in the same way that it is sad for a body to go on living without its brain. We have so far spent our physical and intellectual energies struggling with other players on the natural stage to attain food, clothing, and shelter. We have struggled to gain a measure of natural freedom, freedom from excessive work, suffering, and death. With this freedom we now can contemplate a role that has been impossible up to the present moment: working for the good of nature, not as a sacrifice of our own interests and destiny, but as a fulfillment of them. Just as what is bad for nature as a whole is bad for us, what is good for nature as a whole is good for us—or would be if we were to become a new and useful system within nature as a whole: an information-processing system dedicated to the good of the whole. This is not a task to be lightly undertaken, for nervous systems can go wrong, as we know in our own case. Nervous systems can become motivated to do things that are not in the interests of the body and hence not in their own interest. The human nervous system is often inclined toward overeating, an understandable result of natural selection under conditions where famine was the main Malthusian check on our population. Understandable, yes, and arguably forgivable, but a flaw nevertheless in our very nature.

If we contemplate becoming Earth's information-processing system—and arguably this is a process that is as a matter of fact under way already—we must be on our guard against this sort of failure. We must learn from our mistakes. The current struggle of humankind with the threat of global warming presupposes both that we are becoming Earth's nervous system and that we are prone to mistakes which

---

* Of course, individual human beings can exist outside Earth, at least for a time, as astronauts currently do. But humankind—the entire species—cannot, at least for now. So it is a type-level truth that humans cannot survive without Earth because they are so deeply matched to it.

are similar (or functionally isomorphic) to addiction. Both those who warn that we must stop using fossil fuels and those who think the warning is alarmist base their stance upon information humankind has gathered about the climate. Both assume that we are capable of having significant effects upon nature for either good or ill. Both assume that we can use our intelligence and our information to act for the good if we choose to do so. Thus, everyone agrees that we can perform the three natural functions of a nervous system: Collect information relevant to Earth as a whole, process it in light of Earth's interests, and act on that basis. If that is correct, we must engage in an ongoing study of how such systems fail. We do not need a defeatist fascination with failure, however. We need to study how nervous systems fail in order to avoid these snares and pitfalls and thus succeed.

We can take a lesson from Immanuel Kant in our study of these snares and pitfalls. Kant conceived of human immorality as a form of irrationality. A person who steals something steals it because he or she wants to own it. That is obvious, yet it holds an implicit contradiction in values. On the one hand, the thief has no respect for personal property, as proven by the fact that the thief takes the property of others. On the other hand, however, the thief steals to enjoy the benefits of ownership. Bread is stolen so that it can be eaten, money is stolen so that it can be spent, and so on. Stealing works only because people generally respect property, so thieves can enjoy owning what they steal. Lying involves the same contradiction, for the lie is told in order that it will be believed. But lies are believed only because most of the time people tell the truth. If people lied all the time, lying itself would be impossible, for language would simply cease to have any meaning. If people stole things all the time, stealing would be impossible, for ownership would cease to have any meaning. So immorality is possible only when someone makes an exception to the rule in their own case, while expecting everyone else to follow the rule. Kant's advice to us is to ask whether the rule we follow in acting could be followed by everyone else, as though it were a law of nature. If it cannot, we know that the act is immoral.*

From the Kantian point of view, addiction is a normal case of immorality because it is irrational in the normal way. In addiction the nervous system persists in a behavior that clashes with the interests of the whole body, even while it depends of the rest of the body acting for the interests of the body as a whole. Among the systems of the body, it makes an exception to the rules in its own case, and the fruits of this exception can only be enjoyed because the other systems follow the rule. If all of the other systems clashed with the interests of the body as a whole, the body would die, and it would become impossible for the nervous system to gain its unfair advantage in the first place. At the level of type truths, the irrationality or immorality within a society is closely analogous (or functionally isomorphic) to the irrationality of addictive behavior. Plato explains this sort of behavior as a trick of perspective, of the close looking larger than what is farther away. Surely there is something in the idea that taking the longer, broader view is hard to do, although it is the right thing to do.

---

* I think that everyone will see the above as a relatively uncontentious reading of Kant, although there are details and problems in the interpretation of Kantian ethics that it does not address. What follows below, by contrast, are a series of implications that are not generally discussed in Kant scholarship.

Deferred rewards do not motivate us like immediate rewards, and this is not always irrational. "Seize the day!" sometimes saves the day. A bird in the hand is worth two in the bush. Hobbes thought that the stability of voluntary collectivities formed for mutual advantage was subject to systematic failures because cheating confers advantages. Surely there is some truth to this and to Hobbes's idea that cheating must be penalized.

What is not often recognized is that the laws against cheating only apply in a categorically well-defined world, where the boundaries between types are sharp. In fact, constant low levels of cheating are the norm in nature. Stealing and lying and addiction persist despite their irrationality. We always have thieves and liars among us. The cuckoos and the starlings continue to lay their eggs in the nests of songbirds, even though their behavior is inherently unstable and if too successful would eliminate the songbirds on which they rely.* And we do not want to realize that it would not always be a good thing if the laws were strictly enforced. Sometimes the poor must steal bread to survive so that they can farm the fields from which the bread comes. When your neighbor comes to you in a fit of anger, determined to murder his wife, it might be best for you to lie when he asks you whether you know where she is. We might do better to think of a tension that must be kept in harmony. Within nature, we see that the laws, although perfectly universal, do not produce simplistic effects. Although gravity is universal, it is not true that what goes up must come down. Gravity is but one force among many in nature's physical repertoire. Given sufficient acceleration, something can go into orbit and stay there just as long as you like. It can even achieve escape velocity and never return. The need for someone or some species to survive might sometimes overrule the general necessity for property or honesty.

If we think of the laws of natural collectivities, the sorts of laws that give rise to multicellular animals, for example, as type-level laws, we might do better. Certainly that is what observation shows us is nature's own way of doing things. Perhaps it is flawed. We have no guarantee that it is perfect. On the other hand, we would do well to at least observe to see whether it may not have something to teach us. What we see is dazzling complexity. The laws of nature may be expressed in a notationally brief way, but they give rise to the fractal complexity of the life we see around us. Life itself is the most complex phenomenon that we have witnessed in the universe. Virtually the entire observable universe is deadly boring. There are endless light years of empty space, occasionally punctuated by a star, which consists of mile after endless mile of hydrogen gas. Sometimes a star (of the second or third generation) will have a chemically complex planet circling it. Although the planet itself contains mile after mile of air or water or molten rock, a tiny, tiny fraction of the available volume is occupied by living cells, which are more complex than the world around them. These cells combine to form more complex things, multicellular animals, which in turn combine to form more complex things, such as kinship groups and countries, and so on. Although the laws are simple, their effect is anything but.

---

* It is interesting that one explanation of declining songbird numbers in eastern North America is the rise of cowbird numbers due to cattle farming. Like cuckoos and starlings, cowbirds trick other birds into tending their young.

Perhaps in this context, where simplicity breeds complexity, we should think of the simple laws of nature, whether physical or moral (the latter stemming from nature's rules for collectivities), as creating a harmonious tension. Nature's creativity seems to rely upon setting forces against each other in a natural dialectic. The dialectic itself is a process, an ongoing exchange marked by conception, birth, growth, maturity, and death. If Earth is to become an entity, a sort of multiorganismic animal on the model of multicellular organisms, it is now only at the stage of conception, or even pre-conception. Global entities of science, economics, and governance do exist, corresponding to the senses, metabolism, and motor control mechanisms of humankind. But global humanity is only a newborn entity, or a newly forming entity that is gradually assimilating its parts into something like a coherent whole. So we must not assume goals that are more appropriate to more mature stages of life—although we not only may, but must conceive them as we can.

CASE STUDY 13: WHAT IS THE IDEAL HUMAN POPULATION?

Whereas forecasts of human population growth used to reach 20 billion for the twenty-first century, it is now generally agreed that population will peak at around 9 billion in about 2075 (see Figure I.1 in the Introduction). It now appears that given material security and access to birth control, women on average tend to have fewer than two children each, which is well below the 2.2 children each required to sustain the population. Assuming that human population will be on the decline by 2100, how low should it go?

Zero. David Foreman, founding member of the environmentalist group Earth First! (a name that nicely expresses environmentalism's transcendent value), has been unusually candid in expressing just what humanity's environmental effect has been. Although environmentalists hate to admit it, and immediately get a harsh reaction when they do, the fact is that Foreman is right when he says "Humanity is the cancer of nature. . . . The optimum human population of Earth is zero" (Foreman 1998b). The idea that humanity is a cancer is not just a metaphor, and is often recognized by the best environmentalists. David Suzuki, a renowned environmentalist, says: "We are overrunning the planet like an out-of-control malignancy" (Suzuki 1994, pp. 18–19). Thomas Lovejoy, an environmentalist with impressive academic credentials, has observed:* "The planet is about to break out with fever, indeed it may already have, and we [human beings] are the disease. We should be at war with ourselves . . ." (Ray et al. 1990). Niles Eldredge, a renowned biologist and environmentalist, has

---

* "Dr. Thomas Lovejoy, a Yale University-trained biologist, is Science Advisor to Secretary of Interior Bruce Babbitt and project leader of the National Biological Survey, a comprehensive survey of the nation's biological resources being undertaken by the U.S. Department of the Interior. He has served as Assistant Secretary of External Affairs at the Smithsonian Institution and as Vice President for Science of the World Wildlife Fund" (the National Center for Public Policy Research, August 24, 1993; http://www.nationalcenter.org/dos7127.htm).

diagnosed this human disease that saps nature's strength and threatens its very existence: "Humans do not live with nature but outside it. *Homo sapiens* became the first species to stop living inside local ecosystems. . . . Humans do not live with nature but outside it. . . . Indeed, to develop agriculture is essentially to declare war on ecosystems" (Eldredge 2001).

We do not want to face this very unpleasant truth, but unpleasant truth is the truth nonetheless: Our existence is diametrically opposed to the health of the environment. We are the disease that afflicts the natural world. So if we truly care about the natural world and seek the health of the environment, we must cure this disease. What this entails, as unpleasant as it may sound, is that we must graciously exit from the scene. And it will not actually be painful to do so, after all. In any case, it is now clear that human beings do not want to go to all the trouble of having and raising children. As numerous population data show, given access to birth control, human fertility falls below the rate required for replacement. All we need to do is let this trend take its natural course. No one will be asked to give up their life. Everyone will live just as they are inclined to live. And we can all go to sleep at night knowing that we are doing not only the right thing, but the noble thing.

Not Zero. The often expressed idea that human beings are an out-of-control growth similar to a cancerous malignancy is a hasty conclusion adopted merely because it so nicely expresses environmentalists' antihuman bias. It is, however, based on two premises that are not usually made explicit, are never examined, and are false.

The first premise is that the human economy demands endless growth and this growth demands ever greater use of environmental resources. This idea is based on a fundamental mistake. Even if we assume that the human economy must grow in order to survive, economic growth refers to growth of economic *value*, not growth of the total amount of physical substance employed in the economy. For example, the value of your house may well go up (this is a quite common occurrence), and this would be economic growth, even though there was no physical growth in your house and no increase in the amount of environmental materials that compose it. As was discussed briefly toward the end of Case Study 8 in Chapter 7, much global economic growth involves such things as electronics, cell phones, personal computers, and music players, which have very high value but require very small amounts of physical materials.

The second premise of the idea that humanity is out of control is population growth itself. When Foreman, Suzuki, Lovejoy, and the rest put forth the human cancer hypothesis, it was very commonly believed that human population growth was out of control, a belief that they themselves did much to promote. Paul Ehrlich, another prominent environmentalist with impressive academic credentials, is the author of the famous and highly influential book *The Population Bomb* (Ehrlich 1968), which predicted apocalyptic consequences due to unrestrained population growth. This Malthusian prediction turned out to be false, as current population trends prove. The population bomb was a metaphor that failed to detonate. It is easy to be beguiled by the extrapolation of trends, and environmentalists are not immune to this weakness.

So there is no reason for us to commit collective suicide, no matter how easily this may be done.

Pre-agricultural Levels. Eldredge is still right about one thing: *Homo sapiens* is profoundly different from all other species that we see around us in the natural world—so different that we are, in effect, unnatural. It is because of our unnaturalness that we have created the ecological crisis that is going to destroy not only us but the rest of the life on this planet if we do not change our ways. He is also right that the crucial step that we took and which threw us off the natural path we should have followed was the development of agriculture. It was at that point that our destruction of the environment began, and it has continued ever since. The only conclusion we can draw is that we must reduce our environmental impact to the level it would have had if we had maintained our pre-agricultural form of life and our natural population.

Not Pre-agricultural Levels. That would require reduction of our numbers by something like a thousandfold (it is typically estimated that there were no more than 5 million people at the end of the last ice age, whereas there are currently some 6 billion of us). In addition, it would mean drastically reducing our use of environmental resources as well. All of the great cities of the world would have to disappear: Mexico City, New York, Rio de Janeiro, Paris, Peking, Tokyo, London, Toronto, Hong Kong, Singapore, Rome, Cairo. The rich culture these cities support—their art, their architecture, their dance, theater, festivals, sporting events, food—would also have to disappear or be reduced to some vestigial form. Science and technology would also have to disappear, since our exploration of the universe depends on large populations from which to draw and train expert personnel, and on the resources such populations provide. If we returned to pre-agricultural population levels, we would be reduced to curators of past riches, tending huge bodies of art, history, literature, music, and science, without the people and resources to comprehend it, let alone to keep it alive and developing.

Aside from the question of whether this is what we really want (and surely it is not), it does not even appear to be sustainable. In the long run our arts and our sciences would die out, and we would even forget our history as we once again became nothing more than a set of isolated groups that had little more in common with each other than mutual suspicion and hostility. The irony is that given human nature, our *natural* intelligence would once again assert itself, and we would simply begin the process that brought us to this point all over again. We would develop agriculture, technology, science, and finally would find ourselves in much the same situation as we are now.

1 or 2 Billion. We must aim for a population that is sustainable. According to most estimates, that is somewhere between 1 and 2 billion if we assume the level of resource consumption of the average U.S. citizen (Eldredge 1998, p. 184; Erickson 2000). This level would permit us to maintain some of our great cities, our culture, our science, and our technology without endangering the natural environment upon which it all depends.

Not 1 or 2 Billion. This level is the *maximum* carrying capacity of the Earth assuming that we maintain the resource consumption at current U.S. levels. But this scenario also continues to place the maximum possible stress on the natural environment indefinitely. This is not permissible from the point of view of environmental ethics. The natural environment must be permitted, indeed encouraged, to thrive, not just survive.

The Question Must Remain Open. Case studies are fine, and this discussion should be encouraged, but we must not assume that every question we can ask has a definitive answer once and for all. The hallmark of intelligence is *adaptability*. The intelligence of the robin hunting for breakfast for its brood does not consist in knowing in advance where the worms are. Instead, it employs a flexible strategy, or better yet, a flexible set of strategies, that will permit it to find the worms it needs under as yet undetermined circumstances. If there is a cat prowling about the spot where it normally gets its breakfast, it will be prepared to look somewhere else. If its own territory turns out to be unusable due to a flood, it must, and will, be ready to make a foray into the territory of the robin next door. If we are intelligent, we will not try to forecast every event before it happens. The robin does not need to know in advanced that a goshawk is hunting for it in order to duck the instant it catches a glimpse of it. We must be prepared to change our course in just this way.

Yes, we must achieve a population that is sustainable, but there is much debate about just what that means and just what the actual sustainable population is under any given definition. Joel E. Cohen, head of the Laboratory of Populations at The Rockefeller University, whose studies (Cohen 1995a, 1995b) of the Earth's carrying capacity are considered authoritative, comes to a very different conclusion from the 1 to 2 billion figure: "The fact is that no single number exists to answer 'how many people can the Earth support?' because human carrying capacity is dynamic and uncertain" (Cohen 2003). As we have seen earlier, sometimes science tells us that a particular quantity is unpredictable (see Section 6.6); the maximum human population is just such a quantity. Why? "The capacity depends on natural constraints and human choices, which are not captured by the ecological notions of carrying capacity we use for nonhumans" (ibid.). So the maximum human population depends on human choices. Which choices? "We must consider in our calculations the interactions of such constraints as food, water and livable land and choices about economies, environment, values and politics" (ibid.). Cohen is saying that sustainable population levels are not fixed in advance, but are largely chosen by us, which implies that they are a function of human values and so can best be determined by employing the sort of intelligence that is manifested in adaptability.

Yes, we must also sustain human culture, our science, technology, arts, architecture, dance, athletics—indeed, everything that makes us human. Contrary to what many environmentalists say, we are not unnatural and we have just as much right to exist in the ways we find natural as does any other species. Contrary to the gap between us and the natural world that is implicit in the distinction between humankind and the environment, and which is therefore implicit in environmentalism itself, *Homo sapiens* is, absolutely and without doubt, part of nature. We spring from the same source as every other species and continue to be sustained by it. Human nature is part of nature as a whole.

So we must support not merely the sustainability of humankind and the rest of living-kind, but our joint prosperity and flourishing. This means that we must keep our eyes and our minds open, and make the adjustments in-flight that are required, as they are required.

11.5   OUR RIGHT TO SURVIVE

For the newborn, survival is job one. We find ourselves, in Hobbes's charming words, in a state of nature. There is no harmony among the forces in contention within nature. There is no overall equilibrium, but a process of ongoing change. We have opened our eyes to discover that we are orphans in a jungle that is red in tooth and claw. Nature itself is indifferent to us, and places us in Earth's arena among millions of species that it has equally favored with existence. We have been blessed with quick wits and nimble hands, and this has given us a temporary advantage in fending off attacks and getting life's necessities. But every blessing is also a danger, and we have learned of good and evil by practical experience. We have begun to understand the advantages of transcending the warfare of each against all that characterizes the state of nature. We have begun to realize the kernel of truth in Hobbes' first law of nature, that even while we cannot surrender the advantages of warfare unilaterally, it is in our interest to join with our adversaries to create a mutually beneficial state of peace. In the process we must remember that at this point everything is fair, for there are no rules yet, and there is no peace. If we are wiped out or brought low by plague, famine, or meteorite impact, we cannot even lament that this is unfair, just so darned unfair. For us, survival is job one.

For this reason, we should first attend to our natural needs, the necessities of our existence. When we are more mature, we can move on to questions of virtue and higher goals. We have discovered that our necessities depend on the rest of nature and its necessities. So for now we must concern ourselves with simpler things. Those who propose that we concern ourselves with such things as the integrity, stability, and beauty of nature (Callicott's expression of the romantic ideal of pre-agricultural wilderness) have gotten way ahead of themselves. They have provided us with concepts that we are anxious to examine and must eventually turn this way and that so that we can understand them. But at this point it would be arrogant for us to change nature in such a way that it seems more beautiful to some of us. What is so special about a world in which there is nothing but wilderness?

If it is stability we seek, why not stabilize the current state of nature, complete with our farms, fields, cities, and space stations? These are no less natural than ant colonies, beaver ponds, or jungles, and are a product of the fully integrated processes of nature just as they are. Why should we think that the state of nature at the end of the last ice age, just before there were farms, fields, cities, and space stations, had more integrity than its current state? And if it did have such beauty, stability, and integrity, surely what flowed out of this state and this state alone must also have the same stability, beauty, and integrity. Or perhaps we should just drop our pretensions and admit that the only stable feature of nature is permanent evolution, that its beauty is in our eyes and is matched by its awe, and that its integrity is matched by its destructiveness. It destroys everything that it creates.

Our first concern, naturally, is that it not destroy us. This is not a matter of our rights, because nature does not give us rights—or anyone or anything else rights. It gives us existence and that is all, just as it does for everything else that exists. This existential equality is the full extent of our rights and the full measure of nature's concern for us. Our survival depends on the nature and viability of Earth's living systems. So if we are to survive, we must pay attention to this dependence. Our survival depends on seeing ourselves as part of the whole system of life on Earth. So our neonate need to survive pushes us toward a special role in the system: We can provide it with the same advantages that it has provided us. This is not a matter of treating everything in nature as sacrosanct. Survival requires destruction. Entropy can decrease locally only at the expense of at least as much increased entropy globally. Predators destroy their prey, forests advance by destroying the barren ground, and fields advance by annihilating the bare prairie. Even if we adopted the plants' strategy of rooting ourselves in one place, we would destroy the plants that would have otherwise lived on the spot that we appropriate for our own. Our business is survival, not the elimination of strife or suffering. One day, perhaps, in a distant future we might take on such lofty goals, assuming that they are not somehow illusory. In the meantime, we must play by nature's rules, whereby every act of creation is matched by an act of destruction. We must find our survival within this tension.

THE EMERGING GLOBAL CARBON REGIME

Like many other animals, human beings have increased their evolutionary fitness by working together to attain the necessities of survival and reproduction. Most animals—the bacteria, worms, and insects—go it alone from birth until death, making their way in the world as well as they might, cooperating only in the reproductive act itself, but most birds and mammals cooperate in many or all stages of life. There are exceptions to this general division on either side: Social insect species, the ants, termites, and bees, have reaped the benefits of social life, whereas male bears are aggressive even toward members of their own species and will even eat their own offspring. Human beings stand out from the rest of nature in many ways, and one of these is our high degree of sociality. Our history is one of ever-larger social groupings, from kinship groupings to tribes, from tribes to nations, from nations to international associations. The formal global structures that emerged in the twentieth century—the League of Nations, the United Nations, the General Agreement on Tarrifs and Trade—are part of a trend toward the formation of a global human social structure.

It is remarkable that one of the agents of human globalization in the twenty-first century is our general concern for "the environment," that is, for nature. The Kyoto accords have two main components. On the one hand there are mechanisms designed to reduce our CO_2 emissions, and on the other, mechanisms designed to reduce the disparities in material prosperity between different groups of human beings. Both of these are manifestations of our sympathy: Sympathy on the one hand with living things in general whether or not they are human, and sympathy on the other hand with other human beings whether or not they are from our national or ethnic group. The idea behind Kyoto is that a tax on CO_2 emissions will reduce and gradually eliminate them, while the proceeds can be used to aid underprivileged human beings. Laudable as this idea is in principle, there are a number of clear dangers in the plan in practice as presently conceived. The idea is to tax CO_2 emissions in prosperous, industrialized nations, and to transfer the income to industrializing nations, which will be allowed to emit CO_2 without any tax. The overall effect, then, will be to relocate, rather than reduce or eliminate, CO_2 emissions. Since industrializing countries tend to have poorer environmental standards, generally speaking, the effect may well be to increase pollution and environmental damage on a global level. So the Kyoto approach may well backfire as far as environmental harm is concerned.

It may also fail in terms of human prosperity and freedom. Increasing the central control of economies has generally depressed those economies so affected. The rise of democracy and human freedom has gone hand-in-hand with prosperity. Centrally controlled economies have tended to stagnate, whereas economies in which individuals are freer to buy, sell, and produce as they see fit have tended to grow. In political struggles, even in war whether hot or cold, countries with democratic freedoms have consistently triumphed over their foes. History teaches us that there is a symbiosis between political freedom and economic freedom, and that freedom in general comes with economic prosperity. Why is that? To put it in literary terms, those humans who have attained freedom have attained it by harnessing fire. In the days before the steam engine, the only source of power was muscle power. In those days, most humans and many animals as well had no choice but to live in slavery. By harnessing fire, people and animals could step out of their harnesses and achieve freedom.

The stated goal of Kyoto is to establish global control over fire by treating CO_2 as a pollutant. One problem of the process of harnessing fire, a process we know as industrialization, was pollution. Having achieved freedom from death, suffering, and oppressive work, those of us lucky enough to be in the industrialized world then successfully turned our attention to solving the problem of pollution. If implemented, Kyoto will establish a global carbon regime that will control every aspect of human behavior. Our food, our clothing, our housing, our leisure, our arts, our sciences—all are tied in various ways to our use of fire for heat, energy, and transportation. The growth of human freedom will be stopped or reversed. We are told that this is a necessary sacrifice, one that we must make for the sake of the environment and future generations of human beings. We are told that we must reduce our own strength, that we must diminish ourselves individually and collectively. This, we are told, is the way to solve the problem—for we are the problem. We are told that we must, insofar as it is possible, remove ourselves from the unfolding drama and once again let nature take its course, whatever that may be. We should take a final bow, and Earth should be left to the blind forces that have buffeted it over the course of billions of years.

Against this I offer a different vision. We should not choke off our own strength and intelligence, but let them grow. We should be calm in the face of the problems we face together with the natural world as a whole. We should recognize that even if we are causing a degree or two of global warming, this is a change of tiny proportions relative to the changes that the Earth has repeatedly undergone at the hands of nature over billions of years. We should recognize our own potential for good of a unique and transformative kind. For the first time in the history of the planet, it is possible to safeguard life itself. By employing the highest product of life, conscious intelligence, we can free Earth from the chain of cosmic accidents to which it has been subjected for so long. To do this, we must consolidate our power, not sap our strength. We must not lose our nerve through fear or intimidation. We and we alone are capable of understanding what dangers and challenges life on Earth faces; we and we alone are capable of facing up to them and overcoming them. We can do this only if we build on our strengths, not undermine them.

Playing on people's fears, chastising them for their pride and forcing them to assume a posture of submission were the only ways of achieving political organization back in the bad old days when muscles were the only source of power. Since the rise of science and technology, we have found a new way to unite even as we increase our freedom. We must not go back to the old days of fear and self-chastisement. Rather than self-doubt, we should claim the self-confidence justified by the actual record of history: We have dealt with the problems as they have arisen by moving forward, not back. We will do so again. Now is the moment to have faith in ourselves, a measured and sure-footed faith. Now is the moment to recognize our unique value and to continue building our strength, our unity, and our freedom. Now is the moment to increase the domain of human power, not to beat it into retreat.

11.6   OUR UNIQUE FEELINGS OF KINSHIP WITH OTHER SPECIES

The fact that individual *Homo sapiens* can feel natural sympathy for individuals from species other than their own, even for those which are troublesome to them, is a very rare ability. It makes it possible for us to perform a unique role for life on Earth. It also requires a bit of explanation.

The source of our sympathy is our social nature, the very thing that has led to global human structures (whether economical, political, or ecological) that are influencing the natural environment, and which have the potential for great harm or great good in the coming century. As discussed in Section 11.5, human beings cooperate in order to survive and reproduce, and the way this is achieved is by means of instinctive feelings. Because it gives us an advantage in the evolutionary process, a mother will fight to protect her young. We are often accused of being a fallen race, of having a natural inclination to evil. In particular, we are often said to be selfish. This is true, but it overlooks the fact that we are usually selfish not for ourselves alone but for those we love, in particular our children. Indeed, we are naturally prone to a variety of errors, but on the face of it unselfishness seems to trump selfishness. A father who wins a kingdom but loses his son and alienates his daughter will feel a huge gap in his happiness. A mother who refuses to donate money to charities will happily spend it on her daughter's wedding dress. A son will go to war to protect his family. Indeed, if it were not for the ready willingness of people to fight to the death for the ones they love, it would be difficult to raise armies ready and willing to wage war.

Our unselfishness, like our selfishness, is the product of an evolutionary struggle in which the need to survive and prosper sometimes required us to cooperate and sometimes to compete with our fellow human beings. Both of these needs were partially satisfied by giving us *feelings*, feelings of sympathy and feelings of selfishness. Generally speaking, our feelings of sympathy are strongest for those who are closest to us, because over the millennia those families that felt like sticking together have survived and prospered much more than those that did not. We are generally closest to our mother, father, brothers, and sisters. When they cry, we feel pain, when they laugh we feel like laughing even if we do not know what it is that is supposed to be funny. Our friends become like family to us, and we will take pains to save them pain. When times are good, our sympathies broaden, until they include our neighbors, our town, our country. In good times there is no need for violent competition with our neighbors, townsmen, countrymen, so we cooperate in a larger group and reserve our pooled violence for those who would attack us from the outside. Their pain feels like our pain, their joys are our joys. The twentieth century saw this pattern of ever-larger groups fighting ever-larger wars come to a nuclear stalemate.

We still take pride when our country wins gold medals in the Olympic games, but we would feel ashamed if we thought our country was sacrificing its sons and daughters in an unjust war. At the moment, the countries armed with nuclear weapons spend much, if not most, of their political and military effort on defusing the risk of war in nonnuclear conflicts of smaller scale. This may well be only a passing phase, and the risk remains of larger, more damaging wars than even those of the twentieth century. It should not be just a passing phase, and we should do whatever we can to secure peace among humankind. This is something that we owe not only to ourselves, but to Earth as a whole. This is enlightened selfishness, the selfishness of cooperation. It is entirely in our interests to keep the life of this planet healthy, vibrant, resilient, and safe. Safety requires intelligence, and intelligence requires data. We need to be the eyes and ears and brain of the planet. We need to develop our power to act on behalf of us all, not just ourselves but living things in general. We need to become the hands and feet of the Earth and become ready to walk or work as needed. We need to focus our competitive feelings on the largest and most worthy adversary we have ever faced: the uncaring physical universe.

The uncaring physical universe is nothing other than nature as a whole. On the one hand, the physical universe brought this living Earth into existence, and on the other, it has visited a series of catastrophes upon it as well. The relationship of the Earth to nature as a whole mirrors the relationship of any species to the entire system of life in which it finds itself. On the one hand, nature gives birth to us, and on the other, we must struggle with both the elements and with other living things in order to survive. Within the system of life on Earth, it is eat or be eaten, crowd out the others or be crowded out yourself. Each plant must either reach for the Sun or else be shaded by those that do. Competition is everywhere, even with your own kind. Your own kind, after all, wants the same things that you do. Their food could be your food if you can somehow eat it before they do. Unless, of course, there is plenty, in which case the spirit of cooperation prevails. Sympathy for the man or woman who lives far away in a country unlike your own comes easiest on a full stomach after a good night's sleep. We must secure our own necessities and the necessities of our family before we can worry about people whom we have never met or seen.

So humankind must secure its own necessities before it can care properly about nature as a whole. Our struggle with nature must be won before we can sympathize with our former competitors. We are on the verge of securing natural freedom for nearly all of humankind. Freedom from death, suffering, and excessive work is a necessity if our concern for nature is to be effective. Environmentalists have broad sympathies because they can afford broad sympathies. Poaching wildlife and secretly cutting wood in the forest will remain as long as people have no other way to get food, shelter, or fuel. Everyone needs fire, if only for cooking, warmth, or light, especially if they would like to share and develop sympathy with species other than their own. When human beings are free from the necessity to struggle just to stay alive, when they have the food, clothing, shelter, and freedom from pestilence which permits them to embrace life rather than merely running from death, their sympathies expand in an awe-inspiring way. Unlike dogs or cats or wolves or lions, humans can sympathize with their prey and find themselves unable to kill it. What is so wonderful, we may ask, in killing a deer? Generally, we do not allow hunters to kill does with fawns. Hunters are required to eat the deer they kill, and not just part of it either.

It is easy to say that writers and cartoonists have personified deer, and that is why we are able to sympathize with them. The question remains why it is that Bambi *can* capture our sympathies, and the answer lies in our ability to see ourselves in others, to identify with them, to *feel* the way that they do. Human beings have the intelligence and imagination not just to recognize, but to *feel*, the deep similarities between themselves and other species. Robert Burns, author of the "Ode to a Mouse," was not the only person who could visualize the joys and sorrows of animals other than the human animal. People can even feel sorry for their houseplants when they forget to water them. Aldo Leopold sympathized with the living land itself, and he was not alone in this, either. This is an *ability* of *Homo sapiens*. It is by no means a necessity, but it can be done. Indeed, it can even be overdone, although this too is not a necessity. Sympathy simply is feeling what others feel, which requires that they feel something in the first place. Given that other beings have the capacity for pain or joy, we are not only able to sympathize, but inclined to sympathize—assuming that we are not too emotionally exhausted from struggling to survive. Other species do not have this capacity, but we do. This is a unique ability, and arguably it gives us a unique responsibility to realize the good that this makes possible.

In any case, it provides us with far and away the best qualification of any species for the job of Earth's nervous system. As Descartes pointed out in his sixth meditation (1641), our minds are not lodged in our bodies like a captain in a ship, because the captain does not feel pain when part of the ship is damaged. We feel pain when our body is damaged, and we feel it where the damage occurs. We also have the ability of the social animal to feel the pain of others, as though they were extensions of our body, as though their hurt were our hurt. We can feel hurt not only in our own body, but in the body of someone we care about. We instinctively say "Ow!" when our children scrape their knees. This is not a matter of faith, but fact. The concept of humankind as Earth's nervous system, by contrast, is merely a possibility that can be expressed in terms of an analogy with natural processes of integration, in particular the banding together of cells to form multicellular animals capable of seeing, moving, and thinking. Perhaps this possibility is also analogous to becoming wise gardeners of the entire Earth, although we grow in the same garden we tend. Perhaps it is not so different from being wise stewards of the Earth, with the exception that we are self-employed and are among the things we tend. Perhaps it is like being housekeeper and maintenance crew on Spaceship Earth, except that we have no captain to give us orders, so must decide on our own what needs to be done. Even so, we are not the Earth's captain but merely one system within its body, the one system capable of seeing, thinking, acting, and *feeling* not just for ourselves but for the entire body.

11.7   A CODE

It is all well and good to recognize that we should accept our responsibility for nature, at least so far as we are able to affect it. It is even permissible within the field of values to conceive of possible futures—no, it is necessary, for that is the only way that any of us can make long-term plans, and we do need a long-term plan. But adding these two pieces together still leaves us basically with an argument to take charge of the planet for its own good. Why should we do this? What is in it for us? Are we to assume that whatever is good for the planet will necessarily be good for us? Maybe doing what is good for the planet will be bad for us. So are we being asked once again to sacrifice ourselves for the good of the environment? If not, what are we being asked to do?

We human beings should flourish, that is what we should do. We should continue to increase our natural freedom, our liberation from death, pain, and drudgery. We should become healthy, wise, powerful, peaceful, and good. We should advance our science and technology. We should seek understanding of the history of life on this planet and how nature works. We should continue to gather whatever news we deem worthy, whether trivial or important, whether from near or from afar, and keep ourselves informed about what is happening in the world we inhabit. We should tell stories, make movies, sing, dance, and play. In addition to all of these wholesome activities, whether scientific, artistic, or athletic, we should eat, drink, and be merry in the proper measure. To sum it all up, we should prosper in the fullest sense of the word and become *wealthy* in the fullest sense of the word. If we can do all of the things listed above, we will indeed be wealthy in this broad sense—which would be not only to our own good but to the good of life itself.

As we do these things, we must remember that we are here for nature, just as nature is here for us. To balance what we owe nature from what nature owes us, we need a code. I offer one for your consideration. It will consist of six theses which, fortunately, follow directly from those that went before.

THESIS 22:   Nature can be improved.

Thesis 16 states that nature is not good. This implies that nature can be improved, given that we have the power to act in ways that improve it. It could be, for instance, that we will learn that the elimination of a species of mosquito would actually improve the overall health of the environment by increasing the health and viability of other living things. Only a romantic belief in the ideal balance of nature (contrary to Thesis 12), or the natural perfection of all ecosystems (contrary to Thesis 13), would suggest that this could not be the case. Once we free ourselves from these romantic delusions, it is obvious that not every state of nature all through the history of the universe was ideal. To believe otherwise requires believing that every species that became extinct disappeared at precisely the right instant, and that every new species emerged at precisely the right instant, in a chain of cosmic coincidences that defies common sense. If life is a good thing, many events in Earth's history were bad since they were dangerous for life itself.

THESIS 23:   We have a right to do as we choose within the natural world.

This follows from Thesis 20, that we should embrace our responsibility for the state of nature. Given that we embrace responsibility, it follows that we must have a corresponding freedom of action. Thesis 23 states this freedom in a manner that evokes the contemporary resonance of the concept of rights, just to make its point very clear.* Our actions can only be guided by our own sense of values—and this entails

---

* It may be objected that using the terminology of rights in this context is incoherent, since rights have to be granted by actual political entities. On the other hand, however, it could be argued, along with Hobbes (1651), that in a state of nature, prior to any political contract whatever, everything is permitted and nothing obligated: Everyone has the right to do whatever they please. The role of political contracts can only be to restrict this native freedom. Humankind has no contract with the rest of nature, so is under no obligations to nature as a whole. So it does make sense, at least, to speak of our rights in this context. In any case, I am using the word *rights* in the popular sense, to indicate a moral freedom, the freedom to act, which itself is required in order that our actions are either right or wrong. Thesis 23 states that it is morally correct for us to do as we choose. Unless we do as we choose, our actions fall in the domain of the amoral, and are neither right or wrong. Since we want to take responsibility for our actions, we need to grant Thesis 23.

an enormous responsibility that we must take very seriously. The phrase "do as we choose" in Thesis 23 must not be interpreted as the suggestion that we act arbitrarily or capriciously. On the other hand, Thesis 23 does imply that our right to act is not in any way impeded by any obligations other than those we impose upon ourselves. Because we act as we choose, what we do falls in the domain of ethics and morality and must be seen as either right or wrong. Only by accepting Thesis 23 can we go on to judge the actions of humanity as either right or wrong when it comes to their effects on nature—and of course it is imperative that we do make these judgments. This thesis also says that there is no universal political or moral regime that has imposed any obligation on us with respect to the natural world. Any obligations that we may have to nature can only be self-imposed. As frightening as this fact may be, it is nevertheless true. We must face up to our awesome responsibility, and we can only do this if we realize its dimensions.

THESIS 24:   We should favor the living over the nonliving.

In Section 9.2 we discussed briefly the value of life itself and our solidarity with it. The value of life itself does not derive from the supposition that life is always good. Far from it. Life as we know it on Earth tends to generate more pain than pleasure, hence more bad than good, naturally speaking. On the other hand, life is the basis for anything of any value at all. The possibility of any good at all therefore depends on the existence of life. Our preference for life over the nonliving may be said to be biased in favor of ourselves since we are living beings, but so be it. This bias is perfectly reasonable given the necessity of life for the very possibility of goodness. If we are faced with the choice between either preserving certain geological features or else preserving certain life-forms, we should give preference to the life-forms, other things being equal.* If we are faced with the choice of destroying a stone or destroying a kitten, we should destroy the stone and save the kitten. Hopefully, this is obvious enough.

THESIS 25:   We should favor those that care over those that do not.

---

* If destruction of the geological feature hurt living things in some way, other things would not be equal, and we would be called upon to determine the overall best course to follow. It is quite conceivable that even the mere aesthetic satisfaction of human beings would justify the preservation of a geological feature (e.g., a range of mountains or a river) at the expense of the extinction of a species (e.g., a species of insect). We already consider it morally justifiable to destroy the bacterium that causes smallpox in order that human beings not suffer from the horrible disease of smallpox. This presupposes the principle that life-forms are not sacrosanct. Thus, it is at least an open question whether a life-form might be expendable in order to achieve specific human goods, one that cannot be answered by fiat in advance of the facts in any particular case.

Thesis 14 observes that pain and pleasure are the basis of natural value, and Thesis 15 notes that only beings capable of pain and pleasure may be harmed or helped. Thesis 25 draws out the central implication of these two prior theses for our values and ethics. To make it plain that more than just bodily pain and pleasure is at issue, Thesis 25 is stated in terms of caring. No event or state of affairs is of any value one way or the other unless someone (i.e., some conscious being) *cares* about it. The most direct and obvious caring of this form is bodily pain or pleasure, but as we have seen in Sections 9.2, pains and pleasures often extend far beyond their basic bodily forms. Because it is physically painful to be cut by a knife, being threatened with a knife will cause us pain as well, even though the sort of pain it causes is not felt in a part of the body. Since in the end all value will be embodied in pain or pleasure of one form or another, this thesis enjoins us to do the right thing by increasing the good over the bad by paying attention first and foremost to the welfare and interests of those organisms that are capable of pain and pleasure. If we are faced with a choice between preserving a tree, which cannot feel either pain or pleasure, with preserving an animal which can feel pain and pleasure, we must save the animal.

THESIS 26:   We should favor ourselves over other living things.

To put it crudely, if we were presented with a choice of preserving either our species or another species, it would be right, other things being equal, to preserve our own. This is not selfish or arrogant, but an obligation that we recognize given our special place in the natural order. Not only do we human beings have the widest range of pains and pleasures, we are also the only organisms capable of comprehending the beauty and awe of nature. Although there are many, many sentient animals capable of pain and pleasure, we are the only ones capable of understanding what a species is, or of recognizing its value. Only human beings can recognize the loss of a species, or take steps to protect it. We are the threads that sew the experiences of other animals together into an organic whole. If it were not for us, the beauty and awe of nature would be wasted possibilities. Just as colors cannot exist unless some being sees them, so too beauty and awe would not exist if we did not see them. We hear the melody that runs through the various and varied sounds of the natural world. We enable ranges of value to emerge that otherwise would not exist. We and we alone are capable of recognizing the significance of nature and of life, and give them meaning. We and we alone have risen above the struggle for existence and can champion nature as a whole and life itself. We and we alone can protect and nurture Earth and its precious cargo of life.

To put it picturesquely, after a long childhood struggle to survive in an indifferent universe, we now find ourselves alone, the only adults on Spaceship Earth. Although there are millions of other species of life on the ship, and whether we like it or not, we are the only species with any chance of figuring out how the ship works. We can do everyone on board a big favor by judiciously learning how to handle the controls and then setting the best course we can. Just what that course should be is a philosophical question that must be answered, and then asked again, and then reanswered and reasked, our understanding unfolding in parallel with the unfolding of the natural universe itself.

THESIS 27:   At this point in history, we are Earth's destiny.

References

Abrams, David. (1991) The mechanical and the organic. In *Scientists on Gaia*, Stephen Schneider and Penelope Boston, eds. Cambridge, MA: MIT Press.

———— (1994) Scattered notes on the relation between language and the land. In Burks (1994, pp. 119–130).

Adams, J. M., and H. Faure (1998) *Review and Atlas of Vegetation*. Oak Ridge, TN: Quaternary Environments Network, Environmental Sciences Division, Oak Ridge National Laboratory. http://www.esd.ornl.gov/projects/qen/adams1.html.

Alvarez, Walter (1997) *T. Rex and the Crater of Doom*. Princeton, NJ: Princeton University Press.

AR4, the IPCC's own abbreviation for IPCC (2007a), also known as Assessment Report 4. See IPCC (2007a).

Arthus-Bertrand, Yann, ed. (1999) *La Terre: Vue du Ciel*. Paris: Éditions de La Martinière.

Attaran, Amir, and Rajendra Maharaj (2000) DDT for malaria control should not be banned. *British Medical Journal*, vol. 321 (Dec. 2), pp. 1403–1404.

Avundo, Astrid, and Frédéric Marchand (1999) Auteurs des légendes. In Arthus-Bertrand (1999).

Bacon, Francis (1620) *The New Organon: Or True Directions Concerning the Interpretation of Nature*. See, e.g., http://www.constitution.org/bacon/nov_org.htm.

Baliunas, S. L., and W. H. Soon (1995) Are variations in the length of the activity cycle related to changes in brightness in solar type stars? *Astrophysical Journal*, vol. 450, pp. 896–901.

Beck, Ulrich (1992) *Risk Society: Towards a New Modernity*, Mark Ritter, transl. London: Sage Publications.

Bentham, Jeremy (1789) *Introduction to the Principles of Morals and Legislation*. London.

Bentley, Charles R. (1997) Rapid sea-level rise soon from West Antarctic ice sheet collapse? *Science*, vol. 275 (Feb. 21), pp. 1077–1078.

Berman, Morris (1981) *The Reenchantment of the World*. Ithaca, NY: Cornell University Press.

Berner, Robert A. (1993) Paleozoic atmospheric CO_2: importance of solar radiation and plant evolution. *Science*, vol. 261 (July 2), pp. 68–70.

———— (1997) The rise of plants and their effect on weathering and atmospheric CO_2. *Science*, vol. 276 (Apr. 25), pp. 544–546.

Bond, Michael, (2000) Dr. Truth. *New Scientist*, Dec. 25. http://www.greenspirit.com/logbook/the_log.cfm?booknum=4.

Bond, Michael et al. (2001) Persistent solar influence on North Atlantic climate during the Holocene. *Science*, vol. 294, no. 5549 (Dec. 7), pp. 2130–2136.

Beyond Environmentalism: A Philosophy of Nature, By Jeffrey E. Foss
Copyright © 2009 John Wiley & Sons, Inc.

Bookchin, Murray (1987) *Philosophy of Social Ecology*. New York: Black Rose Books.

Booth, A. L., and H. M. Jacobs (1990) Ties that bind: Native American beliefs as a foundation for environmental consciousness. *Environmental Ethics*, vol. 12, pp. 27–43. Reprinted in Armstrong and Botzler (1993, pp. 519–526).

Botkin, Daniel B., Edward A. Keller, and Isobel W. Heathcote (2005) *Environmental Science: Earth as a Living Planet*. Mississauga, Ontario, Canada: Wiley.

Brohan, P., et al. (2006) Uncertainty estimates in regional and global observed temperature changes: a new data set from 1850. *Journal of Geophysical Research*, vol. 111, p. D12106.

Brown, James Robert (1994) *Smoke and Mirrors: How Science Reflects Reality*. New York: Routledge.

Bugnion, V. (1999) *Changes in Sea-Level Associated with Modifications of the Mass Balance of the Greenland and Antarctic Ice Sheets over the 21st Century*. MIT Joint Program on Science and Policy Report 55. Cambridge, MA: MIT.

Burks, David Clarke, ed. (1994) *Place of the Wild*. Washington, DC: Island Press.

Caillon, N., et al. (2003) Timing of atmospheric CO_2 and Antarctic temperature changes across termination III. *Science*, vol. 299, pp. 1728–1731.

Callicott, J. Baird (1980) Animal liberation: a triangular affair. *Environmental Ethics*, vol. 2.4 (winter). As reprinted in Pojman (1997, pp. 57–68).

Carlyle, Thomas (1899) *Critical and Miscellaneous Essays*, vol. 1. New York: Charles Scribner's Sons.

Carson, Rachel (1962) *Silent Spring*. New York: Houghton Mifflin.

Carter, Bob (2008) The IPPC: on the run at last. *Canada Free Press*, Mar. 25. Based on data provided by John Christy and Roy Spencer, University of Alabama, Huntsville, AL. http://canadafreepress.com/index.php/article/2352.

Cartwright, Nancy (1999) *The Dappled World: A Study of the Boundaries of Science*. Cambridge, UK, Cambridge University Press.

Chalmers, David J. (1996) *The Conscious Mind: In Search of a Fundamental Theory*. New York: Oxford University Press.

Chapman, W. L., and J. E. Walsh (2007) A synthesis of Antarctic temperatures. *Journal of Climate*, vol. 20, pp. 4096–4117.

Charlson, Robert J., and Tom M. L. Wigley (1994) Sulfate aerosol and climatic change. *Scientific American*, vol. 270, no. 2 (Feb.), pp. 48–57.

Charlson, Robert J., et al. (1992) Climate forcing by anthropogenic aerosols. *Science*, vol. 255, p. 423.

Chen, Junye, Barbara E. Carlson, and Anthony D. Del Genio (2002) Evidence for strengthening of the tropical general circulation in the 1990s. *Science*, vol. 295, no. 5556 (Feb. 1), pp. 838–841.

Ching, Francis D. K., et al. (2007) *A Global History of Architecture*. Hoboken, NJ: Wiley.

Chiras, Daniel D. (1988) *Environmental Science: A Framework for Decision Making*. Menlo Park, CA: Benjamin-Cummings.

Christy, John R., and William B. Norris (2004) What may we conclude about global tropospheric temperature trends? *Geophysical Research Letters*, vol. 31, no. 6, p. L06211.

———— (1997) How accurate are satellite "thermometers"? *Nature*, vol. 389, p. 342.

Churchland, Patricia S. (2002) *Brain-Wise: Studies in Neurophilosophy*. Cambridge, MA: MIT Press.

Churchland, Paul M. (1988) *Matter and Consciousness*, 2nd ed. Cambridge, MA: MIT Press.

Clark, P. U., et al. (2002) The role of the thermohaline circulation in abrupt climate change. *Nature*, vol. 415, pp. 863–869.

Cohen, Joel E. (1995a) *How Many People Can the Earth Support?* New York: W.W. Norton.

——— (1995b) Population growth and Earth's human carrying capacity. *Science*, vol. 269, no. 5222 (July 22), pp. 341–346.

——— (2003) News release, The Rockefeller University, November 14, 2003; Internet: http://www.rockefeller.edu/pubinfo/jecAAAS.nr.html (January 15, 2004).

Comrie, Andrew C. (2000) Mapping a wind-modified urban heat island in Tuczon, Arizona. *Bulletin of the American Meteorological Society*, vol. 81, no. 10 (Oct.), pp. 2417–2431.

Dawkins, Richard (2006) *The God Delusion*. Boston: Houghton Mifflin.

Dennett, Daniel C. (1991) *Consciousness Explained*. Boston: Little, Brown.

——— (1996) *Kinds of Minds: Towards an Understanding of Consciousness*. New York: HarperCollins.

——— (2003) *Freedom Evolves*. New York: Viking.

Descartes, Rene (1641) *Meditations on First Philosophy*. Translated by John Veitch, 1901. Classical Library, 2001. http://www.classicallibrary.org/descartes/meditations/index.htm.

Drengson, Alan, and Yoichi Inoue, eds. (1995) *The Deep Ecology Movement: An Introductory Anthology*. Berkeley, CA: North Atlantic Books.

DSNY (2004) Recycling economics. In *Processing and Marketing Recyclables in New York City: Rethinking Economic, Historical, and Comparative Assumptions*. The City of New York Department of Sanitation, Bureau of Waste Prevention, Reuse and Recyling. http://www.nyc.gov/html/nycwasteless/downloads/pdf/pmrnyc03.ch1.pdf (accessed Apr. 17, 2008).

EC (European Commission) (2004) Reprinted in *The Times Concise Atlas of the World*, 10th ed. London: Times Books, 2006, p. 30.

Eddy, J. A., and H. Oeschger (1993) *Global Changes in the Perspective of the Past*. New York: Wiley.

Ehrlich, Paul (1968) *The Population Bomb*. New York: Ballantine Books.

——— (1988) The loss of diversity: causes and consequences. In Wilson (1988a).

——— (2005) *One with Nineveh: Politics, Consumption, and the Human Future*. Washington, DC: Island Press.

Ehrlich, Paul, and Anne Ehrlich (1998) *Betrayal of Science and Reason: How Anti-environmental Rhetoric Threatens Our Future*. Washington, DC: Island Press.

Einstein, Albert (1950) *The Meaning of Relativity*. Princeton, NJ: Princeton University Press.

——— (1954) *Ideas and Opinions*. New York: Bonanza Books.

Eldredge, Niles (1998) *Life in the Balance: Humanity and the Biodiversity Crisis*. Princeton, NJ: Princeton University Press.

——— (2001) The sixth extinction. http://www.actionbioscience.org/newfrontiers/eldredge2.html.

Enger, Eldon D., and Bradley F. Smith (1995). *Environmental Science: A Study of Interrelationships*. Dubuque, IA: Wm. C. Brown.

EPA (U.S. Environmental Protection Agency) (2006) Air emissions summary, 2005. http://www.epa.gov/airtrends/2006/emissions_summary_2005.html (accessed Aug. 21, 2007).

——— (2007a) Air quality and emissions: progress continues in 2006. http://www.epa.gov/airtrends/econ-emissions.html (accessed Aug. 21, 2007).

———(2007b) National trends in sulfur dioxide levels. http://www.epa.gov/airtrends/sulfur.html#sonat (accessed Aug. 21, 2007).

——— (2007c) National trends in nitrogen dioxide levels. http://www.epa.gov/airtrends/nitrogen.html#nonat (accessed Aug. 21, 2007).

——— (2007d) National trends in CO levels. http://www.epa.gov/airtrends/carbon.html#conat (accessed Aug. 21, 2007).

———(2007e) National trends in ozone levels. http://www.epa.gov/airtrends/ozone.html (accessed Aug. 21, 2007).

——— (2007f) National trends in lead levels. http://www.epa.gov/airtrends/lead.html#pbnat (accessed Aug. 21, 2007).

——— (2007g) National trends in particulate matter levels. http://www.epa.gov/airtrends/pm.html (accessed Aug. 21, 2007).

Erickson, Dell (2000) Sustainable populations using footprint data. http://www.mnforsustain.org/erickson_d_determining_sustainable_population_levels.htm (accessed May 1, 2008).

Esper, Jan, et al. (2005) Climate: past ranges and future changes. *Quaternary Science Reviews*, vol. 24, pp. 2164–2166.

European Science Foundation (2006) *Strategic Plan 2006–2010*. Strasbourg, Germany: IREG. http:/www.esf.org/fileadmin/be_user/publications/Plan20062010final.pdf.

Fagan, Brian (2000) *The Little Ice Age: How Climate Made History, 1300–1850*. New York: Basic Books.

Falkowski, Paul G. (2002) The ocean's invisible forest. *Scientific American*, vol. 287, no. 2, pp. 54–61.

Feyerabend, Paul (2001) *Conquest of Abundance: A Tale of Abstraction Versus the Richness of Being*. Chicago: University of Chicago Press.

Fischer H., et al. (1999) Ice core records of atmospheric CO_2 around the last three glacial terminations. *Science*, vol. 283, pp. 1712–1714.

Foreman, Dave (1985) *Ecodefense: A Field Guide to Monkeywrenching*. Tucson, AZ: Nedd Ludd.

——— (1987) Quotation from Bookchin (1987, p. 159).

——— (1991) *Confessions of an Eco-Warrior*. New York: Crown Publishers.

——— (1998a) Where man is a visitor. In Burks (1994, pp. 225–235).

——— (1998b) Inverview. *Sarasota Herald-Tribune*, Jan. 17.

Foss, Jeffrey (1992) Introduction to the epistemology of the brain: indeterminacy, microspecificity, chaos, and openness. *Topoi*, vol. 11, pp. 45–57.

——— (1996a) A scientific fix for the classical account of addiction. *Behavioral and Brain Sciences*, vol. 19, no. 4 (Dec., p. 579).

——— (1996b) Masters in our own house: a reply to brown. *Dialogue*, vol. 35, no. 1 (winter), pp. 165–175.

——— (2000) *Science and the Riddle of Consciousness: A Solution.* Dordrecht, The Netherlands: Kluwer.

——— (2006)The environmentalist faith. *In the Agora: The Public Face of Canadian Philosophy.* Toronto, Ontario, Canada: University of Toronto Press.

Friis-Christensen, E., and K. Lassen (1991) Length of the solar cycle: an indicator of solar activity closely associated with climate. *Science,* vol. 254, pp. 698–700.

Gaffen, Dian J., Benjamin D. Santer, James S. Boyle, John R. Christy, Nicholas E. Graham, and Rebecca J. Ross (2000) Multidecadal changes in the vertical temperature structure of the tropical troposphere. *Science,* vol. 287, no. 5456 (Feb. 18), pp. 1242–1245.

Galilei, Galileo (1615) Letter to the Grand Duchess Christina. In *Discoveries and Opinions of Galileo,* Stillman Drake, transl. New York: Doubleday, 1957.

Giere, Ronald N. (1999) *Science Without Laws: Science and Its Conceptual Foundations.* Chicago: University of Chicago Press.

Gleick, James (1987) *Chaos: Making a New Science.* New York: Viking Penguin.

Goodess, G. M., J. P. Palutikof, and T. D. Davies (1992) *The Nature and Causes of Climate Change: Assessing the Long-Term Future.* London: Belhaven Press; Boca Raton, FL: Lewis Publishers.

Gore, Al (2006) *An Inconvenient Truth: The Planetary Emergency of Global Warming and What We Can Do About It.* Emmaus, PA: Rodale Books.

GORP (2007) Fire: Friend or foe to Garry Oak ecosystems? In *GO Restore! Newsletter of the Garry Oak Restoration Project in Saanich.* Issue 7 (Fall 2006/Spring 2007), p. 1. Internet: http://www.gorpsaanich.com/pdf/fallwinter06.pdf (Sept. 13, 2008).

Goudie, Andrew (1993) *The Human Impact on the Natural Environment.* Oxford, UK: Blackwell.

Gribbin, John (1973) *Forecasts, Famines, and Freezes.* London: Wildwood House.

Guggenheim, Davis, director (2006) *An Inconvenient Truth.* Movie, produced by Lawrence Bender; Scott Z. Burns, Laurie David, and Lesley Chilcott, co-producers. Los Angeles: Paramount Classics.

Gunter, Lorne (2008) Perhaps the climate change models are wrong. *National Post,* Mar. 24, p. A8.

Hancock, D. W., and G. S. Hayne (1996) http://osb.3.wff.nasa.gov/topex/OscDrift.html.

Hansen, J. (2007) The real deal: Usufruct & the gorilla. http://www.columbia.edu/~jeh1/distro_realdeal.16aug20074.pdf (accessed Oct. 29, 2007).

Hansen, J., et al. (2001) A closer look at United States and global surface temperature change. *Journal of Geophysical Research,* vol. 106, pp. 23947–23963.

Harding, Sandra (1991) *Whose Science? Whose Knowledge? Thinking from Women's Lives.* Ithaca, NY: Cornell University Press.

Hartmann, Dennis L. (2002) Climate change: tropical surprises. *Science,* vol. 295, no. 5556 (Feb. 1), pp. 811–812.

Hartman, Dennis L., and Marc L. Michelsen (2002) No evidence for Iris. *Bulletin of the American Meteorological Society,* vol. 83, no. 2 (Feb.), pp. 249–254.

Hay, Simon I., Jonathan Cox, David J. Rogers, Sarah E. Randolph, David I. Stern, G. Dennis Shanks, Monica F. Meyers, and Robert W. Snow (2002) Climate change and the resurgence of malaria in the East African highlands. *Nature,* vol. 415 (Feb. 21), pp. 905–909.

Hendricks, J. R., R. R. Leben, G. H. Born, and C. J. Koblinsky (1996) Empirical orthogonal function analysis of global Topex/Poseidon altimeter data and implications for detection of global sea level rise. *Journal of Geophysical Research*, vol. 101, no. C6, pp. 14131–14146.

Hitchens, Christopher (2007) *God Is Not Great: How Religion Poisons Everything*. Lebanon, IN: Twelve Books, Hachette Book Group.

Hobbes, Thomas (1651) *The Leviathan*. London.

Huxley, Aldous (1932) *Brave New World*. London: Chatto & Windus.

Indermühle A., et al. (2000) Atmospheric CO_2 concentration from 60 to 20 kyr BP from the Taylor Dome ice core, Antarctica. *Geophysical Research Letters*, vol. 27, pp. 735–738.

Innanen, Kimmo (2006) Solving Laplace's lunar puzzle. *Science*, vol. 313, no. 5787 (Aug. 4), pp. 622–623.

IPCC (Intergovernmental Panel on Climate Change) (1990) *Climate Change: The IPCC Scientific Assessment*, J. T. Houghton et al., eds. Cambridge, UK: Cambridge University Press. http://www.ipcc.ch/pub/reports.htm.

——— (1992) *The Supplementary Report to the IPCC Scientific Assessment*, J. T. Houghton et al., eds. Cambridge, UK: Cambridge University Press. http://www.ipcc.ch/pub/reports.htm.

——— (1995) *IPCC Second Assessment Climate Change 1995*, J. J. Houghton, et al., eds. Cambridge, UK: Cambridge University Press. http://www.ipcc.ch/pub/sa(E).pdf.

——— (1996) *Technologies, Policies and Measures for Mitigating Climate Change*. IPCC Technical Paper I, R. T. Watson et al., eds. Geneva, Switzerland: IPCC. http://www.ipcc.ch/pub/IPCCTP.I(E).pdf.

——— (1997a) *Stabilization of Atmospheric Greenhouse Gases: Physical, Biological and Socio-economic Implications*. IPCC Technical Paper III, J. T. Houghton et al., eds. Geneva, Switzerland: IPCC. http://www.ipcc.ch/pub/IPCCTP.III(E).pdf.

———(1997b) *Implications of Proposed CO_2 Emissions Limitations*. IPCC Technical Paper IV, J. T. Houghton et al., eds. Geneva, Switzerland: IPCC. http://www.ipcc.ch/pub/IPCCTP.IV(E).pdf.

——— (2001a) *Climate Change 2001: The Scientific Basis*. Contribution of Working Group I to the IPCC Third Assessment Report. Internet: http://www.ipcc.ch/ipccreports/tar/wg1/index.htm; http://www.grida.no/climate/ipcc_tar/wg1/index.htm.

——— (2001b) *Climate Change 2001: Impacts, Adaptation, and Vulnerability*. Contribution of Working Group II to the Third Assessment Report of the IPCC, J. J. McCarthy et al., eds. Cambridge, UK: Cambridge University Press.

——— (2001c) *Climate Change 2001: Mitigation*. Contribution of Working Group III to the Third Assessment Report of the IPCC, B. Metz et al., eds. Cambridge, UK: Cambridge University Press.

——— (2001d) *Summary for Policymakers* of *Climate Change 2001: The Scientific Basis* (2001a). Bracknell, UK: IPCC/WMO/UNEP, UK Meteorological Office.

——— (2007a) *Climate Change 2007: The Physical Science Basis*. Contribution of Working Group I to the Fourth Assessment Report of the Intergovernmental Panel on Climate Change, S. Solomon, et al., eds. Also referred to as AR4. Cambridge, UK: Cambridge University Press. http://ipcc-wg1.ucar.edu/wg1/wg1-report.html (accessed Aug. 19, 2007).

———— (2007b) *Climate Change 2007: Impacts, Adaptation and Vulnerability*. Contribution of Working Group II to the Fourth Assessment Report of the Intergovernmental Panel on Climate Change, M. L. Parry et al., eds. Cambridge, UK: Cambridge University Press, pp. 7–22.

———— (2007c) Climate change and cities: the IPCC case for action. Jean-Pascal van Ypersele, Vice-Chair of IPCC Working Group II, ed. Cambridge, UK: Cambridge University Press.

James, William (1897) The will to believe. In *The Will to Believe and Other Essays in Popular Philosophy*. Cambridge, MA: Harvard University Press.

Jaworski, Zbigniew (2003) Solar cycles, not CO_2, determine climate. *21st Century Science and Technology*, winter 2003–2004, pp. 52–65. http://www.21stcenturysciencetech.com/Articles%202004/Winter2003-4/global_warming.pdf.

Jaworski, Z., T. V. Segalstad, and N. Ono (1992) Do glaciers tell a true atmospheric CO_2 story? *Science of the Total Environment*, vol. 114, pp. 227–284.

Johnston, Josée, Michael Gismondi, and James Goodman (2006) *Nature's Revenge: Reclaiming Sustainability in an Age of Corporate Globalization*. Peterborough, Ontario, Canada: Broadview Press.

Kay, David A., and Eugene B. Skolnikoff (1972) *World Eco-Crisis: International Organizations in Response*. Madison, WI: University of Wisconsin Press.

Keigwin, Lloyd D. (1996) The little ice-age and medieval warm period in the Sargasso Sea. *Science*, vol. 274, pp. 1504–1507.

Keigwin, L. D., W. B. Curry, S. J. Lehman, and S. Johnsen (1994) The role of the deep ocean in North Atlantic climate change between 70 and 130 kyr ago. *Nature*, vol. 371, pp. 323–326.

Kerr, Richard A. (2006) Climate change: a worrying trend of less ice, higher seas. *Science*, vol. 311, no. 5768 (Mar. 24), pp. 1698–1701.

Kirchner, James W. (1991) The Gaia hypotheses: Are they testable? Are they useful? In *Scientists on Gaia*, Stephen. Schneider and Penelope. Boston, eds. Cambridge, MA: MIT Press.

Krugman, Paul (2003) Good news on poverty. *International Herald Tribune*, Nov. 29–30, p. 8.

Kuhn, Thomas (1962) *The Structure of Scientific Revolutions*. Chicago: University of Chicago Press.

Kuo, C., C. R. Lindberg, and D. J. Thornson (1990) Coherence established between atomospheric carbon dioxide and global temperature. *Nature*, vol. 343, pp. 709–714.

Lamb, Hubert H. (1965) The early medieval warm epoch and its sequel. *Paleoclimatology, Paleoecology*, vol. 1, pp. 13–37.

———— (1982) *Climate, History, and the Modern World*. New York: Methuen.

———— (1988) *Weather, Climate and Human Affairs*. New York: Routledge.

Landscheit, T. (1983) Solar oscillations, sunspot cycles, and climatic change. In *Weather and Climate Responses to Solar Variations*. Boulder, CO: Associated University Press.

Landsea, Christopher (2005) Chris Landsea leaves IPCC. See "Open Letter to the Community from Chris Landsea." *Prometheus*, January 17, 2005. Internet: http://sciencepolicy.colorado.edu/prometheus/archives/science_policy_general/000318chris_landsea_leaves.html.

Landsea, Christopher W., Neville Nicholls, William M. Gray, and Lixion A. Avila (1996) Downward trends in the frequency of intense Atlantic hurricanes during the past five decades. *Geophysical Research Letters*, vol. 23, no. 13 (June 15), pp. 1697–1700.

Lanzante, J. R., S. A. Klein, and D. J. Seidel (2003) Temporal homogenization of monthly radiosonde temperature data. Part II: Trends, Sensitivities and MSU comparison. *Journal of Climate*, vol. 16, pp. 224–240.

Latham, John (1990) Control of global warming? *Nature*, vol. 347, pp. 339–340.

———— (2002) Amelioration of global warming by controlled enhancement of the albedo and longevity of low-level marine clouds. *Atomospheric Science Letters*, doi: 10.1006/Asle.2002.0048.

Lau, K.-M., and H. T. Wu (2003) Warm rain processes over tropical oceans and climate implications. *Geophysical Research Letters*, vol. 30, pp. 2290–2294.

Laudan, Larry (1981) A confutation of convergent realism. *Philosophy of Science*, vol. 48, no. 1 (Mar.): pp. 19–49.

Lee, M. I., et al. (2008) A moist benchmark calculation for the atmospheric general circulation models. *Journal of Climate*, forthcoming.

Leopold, Aldo (1949) *A Sand County Almanac*. New York: Oxford University Press. Page references from reprint by same publisher, 1966.

Lewis, Trevor J., ed. (1992) Climatic change inferred from underground temperatures. *Global and Planetary Change*, vol. 6, pp. 71–281.

Lewis, Trevor J., and Kelin Wang (1992) Influence of terrain on bedrock temperatures. *Global and Planetary Change*, vol. 6, pp. 87–100.

———— (1998) Geothermal evidence for deforestation induced warming: implications for the climatic impact of land development. *Geophysical Research Letters*, vol. 25, no. 4 (Feb. 15), pp. 535–538.

Lewis, Trevor J., H. Villinger and E. E. Davis (1993) Thermal conductivity measurment of rock fragments using a pulsed needle probe. *Canadian Journal of Earth Sciences*, vol. 30, pp. 480–485.

Lin, B. et al. (2002) The iris hypothesis: a negative or positive cloud feedback? *Journal of Climate*, vol. 15, pp. 3–7.

Lindzen, R. S. (1999) The greenhouse effect and its problems. In *Climate Policy After Kyoto*, Tor Ragnar Gerholm, ed. Brentwood, UK: Multi-Science Publishing, pp. 98–110.

———— (2007) Taking greenhouse warming seriously. *Energy and Environment*, vol. 18, no. 7/8, pp. 937–950.

Lindzen, R. S., and C. Giannitsis (2002) Reconciling observations of global temperature change. *Geophysical Research Letters*, vol. 29, no. 12, pp. 1583–1586.

Lindzen, R. S., et al. (2001) Does the Earth have an adaptive infrared iris? *Bulletin of the American Meteorological Society*, vol. 82, pp. 417–432.

Lomborg, Bjørn (2001) *The Skeptical Environmentalist: Measuring the True State of the World*. Cambridge, UK: Cambridge University Press.

Lorenz, Edward N. (1991) Chaos, spontaneous climatic variations, and detection of the greenhouse effect. In Schlesinger (1991, pp. 445–453).

Lovelock, James (1991) Mother Earth: myth or science. In *From Gaia to Selfish Genes: Selected Writings in the Earth Sciences*, Connie Barlow, ed. Cambridge, MA: MIT Press, pp. 3–19.

———— (1995) The Ages of Gaia. New York: W.W. Norton.

Lowry, William P. (1971) The Climate of cities. In *Man and the Ecosphere*. San Francisco: W.H. Freeman, pp. 180–188.

Lubick, Naomi (2002) Where biosphere meets geosphere. *Scientific American*, vol. 282, no. 1 (Jan. 28).

Luick, J. (2000) Seasonal and interannual sea levels in the western equatorial pacific from Topex/Poseidon. *Journal of Climate*, vol. 13, no. 3, pp. 672–676; 0894-8755.

Luick, J., and R. Henry (2000) Tides in the Tongan region of the Pacific Ocean. *Marine Geodesy: An International Journal of Ocean Surveys: Mapping and Sensing*, vol. 23, no. 1, pp. 17–30; 0149-0419/00.

Madigan, M. T., and B. L. Marrs (1997) Extremophiles. *Scientific American*, vol. 276, no. 4 (Apr.), pp. 82–87.

Malthus, Thomas (1798) An Essay on the Principle of Population, As It Affects the Future Improvement of Society, with remarks on the Speculations of Mr. Godwin, M. Condorcet, and other writers.

Manahan, Stanley E. (2007) *Environmental Science and Technology: A Sustainable Approach to Green Science and Technology*. Philadelphia, PA: Taylor & Francis.

Mann, Michael E., et al. (1998) Global-scale temperature patterns and climate forcing over the past six centuries, *Nature*, vol. 392, no. 6678, pp. 779–787.

——— (1999) Northern hemisphere temperatures during the past millennium: inferences, uncertainties, and limitations. *Geophysical Research Letters*, vol. 26, no. 6, pp. 759–762.

Margulis, Lynn (1998) *Symbiotic Planet: A New Look at Evolution*. London: Weidenfeld & Nicolson.

Marsh, Nigel D., and Henrik Svensmark (2000) Low cloud properties influenced by cosmic rays. *Physical Review Letters*, vol. 85, no. 23 (Dec. 4), pp. 5004–5007.

Mather, J. H., and T. P. Ackermann (1998) Analyses of measurements for the first year of operation of the Tropical Western Pacific Atmospheric Radiation and Cloud Station. *Proceedings of the Eighth Atmospheric Radiation Measurement* (ARM) *Science Team Meeting*, pp. 453–456. http://www.arm.gov/publications/proceedings/conf08/extended_abs/mather_jh.pdf.

McCluhan, Marshall (1962) *The Gutenburg Galaxy: The Making of Typographic Man*. Toronto, Ontario, Canada: University of Toronto Press.

McGee, Kenneth A., and Terrence M. Gerlach (1995) *Volcano Hazards Fact Sheet*. Open File Report 95-85. Washington, DC: U.S. Geological Survey, U.S. Department of the Interior. http://vulcan.wr.usgs.gov/Projects/Emissions/vgas_fsheet.pdf.

McIntyre, Stephen, and Ross McKitrick (2003) Corrections to Mann et al. (1998) proxy data base and northern hemispheric average temperature series. *Energy and Environment*, vol. 14, no. 6, pp. 751–771. http://www.multi-science.co.uk/mcintyre_02.pdf.

——— (2005a) The M&M critique of MBH98 Northern Hemisphere Climate Index: update and implications. *Energy and Environment*, vol. 16, no. 1, pp. 69–100.

——— (2005b) Hockey sticks, principal components, and spurious significance. *Geophysical Research Letters*, vol. 32, p. L03710.

McKitrick, Ross (2005) The Mann et al. northern hemisphere "hockey stick" climate index: a tale of due diligence. In Michaels (2005, pp. 20–49).

McKitrick, Ross, and Patrick J. Michaels (2007) Quantifying the influence of anthropogenic surface processes and inhomogeneities on gridded climate data. *Journal of Geophysical Research–Atmospheres*, vol. 112 (Dec. 14), p. D24S09.

McMichael, A. J., and M. Y. Beers (1994) Climate change and human population health: global and south australian perspectives. *Transactions of the Royal Society of Southern Australia*, vol. 118, pp. 91–98.

Meadows, Donella H. (1972) *The Limits to Growth*. Calcutta, India: Signet Publishing.

Meadows, Donella H., Dennis L. Meadows, and Jorgen Randers (1993) *Beyond the Limits: Confronting Global Collapse, Envisioning a Sustainable Future*. White River, VT: Chelsea Green Publishing Company.

——— (2004) *Limits to Growth: The 30 Year Update*. White River, VT: Chelsea Green Publishing Company.

Michaels, Patrick J. (2005) *Shattered Consensus: The True State of Global Warming*. Lantham, MD: Rowman & Littlefield.

Milankovitch, M. M. (1941) *Canon of Insolation and the Ice Age Problem*. Belgrade, Yugoslavia: Königlich Serbische Academie. English translation by the Israel Program for Scientific Translations, U.S. Department of Commerce and the National Science Foundation, Washington, DC.

Minster, J.-F., C. Brossier, and P. Rogel (1995) Variation of the mean sea level from Topex/Poseidon data. *Journal of Geophysical Research*, vol. 100, no. C12, pp. 25153–25162.

Moberg, A., et al. (2005) Highly variable northern hemisphere temperatures reconstructed from low and high resolution proxy data, *Nature*, vol. 433, no. 7026, pp. 613–617.

Monaghan, Andrew J., et al. (2006) Insignificant change in Antarctic snowfall since the International Geophysical Year. *Science*, vol. 313, no. 5788 (Aug. 11), pp. 827–831.

Monnin, E., et al. (2001) The Atmospheric CO_2 and temperature records of Dome Concordia, Antarctica. *Science*, vol. 291, pp. 112–114.

Montaigne, Michel de (1575) Of the institution and education of children; to the Lady Diana of Foix, Countesse of Gurson. In *Essays*, Chap. 25.

Morris, Henry M., and Gary E. Parker (1982) *What Is Creation Science?* El Cajon, CA: Master Books.

Mudelsee, M. (2001) Variations in atmospheric CO_2, temperature and global ice volume derived from the Vostok ice core. *Quaternary Science Reviews*, vol. 20, pp. 583–589.

Myers, Norman (1979) Sinking the Ark. New York: Pergamon Press.

Myers, Norman, and Andrew H. Knoll (2001) The biotic crisis and the future of evolution. *Proceedings of the National Academy of Science*, vol. 98, no. 10 (May 8), pp. 5389–5392, For excerpts, see http://www.actionbioscience.org/newfrontiers/myers_knoll.html#Primer.

Naess, Arne. (1995) The systematization of the logically ultimate norms and hypotheses of Ecosophy T. In Drengson and Inoue (1995, pp. 31–48).

NAS (National Academy of Science) (2001) *Climate Change Science: An Analysis of Key Questions*. Washington, DC: National Academies Press.

NASA (2007a) Earth's energy budget (graphic). http://eosweb.larc.nasa.gov/EDDOCS/images/Erb/components2.gif (accessed Nov. 14, 2007).

——— (2007b) A delicate balance. http://earthobservatory.nasa.gov/Study/DelicateBalance/balance2.html (accessed Sept. 10 2007).

National Forestry Database Program (2004) Canadian Council of Forest Ministers. http://nfdp.ccfm.org.

Neihardt, John G. (1975) *Black Elk Speaks: Being the Life Story of a Holy Man of the Oglala Sioux*. New York: Pocket Books.

Nemani, Ramakrishna, et al. (2003) Climate-driven increases in global terrestrial net primary production from 1982 to 1999. *Science*, June 6, pp. 1560–1563.

Nerem, R. S. (1995a) Global mean sea level variations from Topex/Poseidon altimeter data. *Science*, vol. 268, no. 5211 (May 5), pp. 708–710.

Nerem, R. S. (1995b) Measuring global mean sea level variations using Topex/Poseidon altimeter data. *Journal of Geophysical Research*, vol. 100, no. C12, pp. 25135–25152.

Nerem, R. S. (1997) Global mean sea level change: correction. *Science*, vol. 275 (Feb. 21), p. 105.

Newell, R. E., et al. (1974) *The General Circulation of the Tropical Atmosphere and Interactions with Extratropical Latitudes*, vol 2. Cambridge, MA: MIT Press.

Newhall, Chris, James W. Hendley II, and Peter H. Stauffer (1997) *The Cataclysmic 1991 Eruption of Mt. Pinatubo, Philippines*. Fact Sheet 113–97. Washington DC: U.S. Geological Survey. http://wrgis.wr.usgs.gov/fact-sheet/fs113-97/.

Newton, Isaac (1687/1947) *Philosophiae Naturalis Principia Mathematica*. London, 1687. Translated by Florian Cajori as *Sir Isaac Newton's Mathematical Principles of Natural Philosophy and His System of the World*. Berkeley CA: University of California Press, 1947.

NIAID-NIH (National Institute of Allergy and Infectious Diseases–National Institutes of Health) (2004) The history of malaria. http://www.niaid.nih.gov/publications/malaria/history.htm.

NRC (National Research Council) (2000) *Reconciling Observations of Global Temperature Change*. Washington, DC: National Academies Press.

——— (2003) *Understanding Climate Change Feedbacks*. Washington, DC: National Academies Press.

——— (2005) *Improving the Scientific Foundation for Atmosphere–Land–Ocean Simulations: Report on a Workshop*. Washington, DC: National Academies Press.

NSF (National Science Foundation) (2007) U.S. R&D increased 6.0% in 2006 according to NSF projections. http://www.nsf.gov/statistics/infgrief/nsf07317/ (accessed June 22, 2007).

NSSTC (National Space Science and Technology Center) (2007) Publicly accessible satellite data. http://vortex.nsstc.uah.edu/data/msu/t2lt/tltglhmam_5.2 (accessed Nov. 11, 2007).

Offman, Craig (2008) Jail politicians who ignore climate science: Suzuki. *National Post*, vol. 10, no. 87 (Feb. 7), pp. A1–A8.

Oke, T. R. (1973) City size and the urban heat island. *Atmospheric Environment*, vol. 7, pp. 769–779.

O'Neill, Terry (2008) By any means necessary. *National Post*, vol. 10, no. 87 (Feb. 7), p. A15.

Parkinson, Claire L. (2002) Trends in the length of the Southern Ocean sea-ice season, 1979–99. *Annals of Glaciology*, vol. 34, pp. 435–440.

Petit, J. R., et al. (1999) Climate and atmospheric history of the past 420,000 years from the Vostok ice core, Antarctica. *Nature*, vol. 399, pp. 429–436.

Pojman, Louis P. (1998) *Environmental Ethics: Readings in Theory and Application*. Belmont, CA: Wadsworth.

Ponte, Lowell (1976) *The Cooling*. Englewood Cliffs, NJ: Prentice-Hall.

Ramanathan, V., and W. Collins (1991) Thermodynamic regulation of ocean warming by cirrus clouds deduced from observations of the 1987 El Nino. *Nature*, vol. 351, pp. 27–32.

Rapp, Anita D., et al. (2005) An evaluation of the proposed mechanism of the adaptive infrared iris using TRMM VIRS and PR Measurements. *Journal of Climate*, vol. 18 (Oct.), pp. 4185–4194.

Rawls, John (1971) *A Theory of Justice*. Cambridge, MA: Belknap Press of Harvard University Press.

Ray, Dixy Lee, and Louis R. Guzzo (1990) *Trashing the Planet*. Washington DC: Regnery Publishing.

——— (1993) *Environmental Overkill: Whatever Happened to Common Sense*. Washington, DC: Regnery Publishing.

Rees, William E., and Mathis Wackernagel (1996) *Our Ecological Footprint: Reducing Human Impact on the Earth*. Gabriola Island, Canada: New Society Publishers.

Revkin, Andrew C. (2007) Arctic melt unnerves the experts. *New York Times*, Oct. 22. http://query.nytimes.com/gst/fullpage.html?res=9402E3DD1231F931A35753C1A9619C8B63.

Richards, John (1990) Land transformation. In Turner et al. (1990, pp. 163–180).

Roberts, Dave (1998) Eukaryotes in extreme environments. Natural History Museum, Deparment of Zoology. http://www.spaceref.com/redirect.html?id=0&url=www.nhm.ac.uk/zoology/extreme.html.

Roberts, D. R., L. L. Laughlin, P. Hsheih, and L. J. Legters (1997) DDT, global strategies and a malaria control crisis in South America. *Emerging Infectious Diseases*, vol. 3, pp. 295–302

Roberts, D. R., S. Manguin, and J. Mouchet (2000) DDT house spraying and re-emerging malaria. *The Lancet*, vol. 356, no. 9226 (July 22), pp. 330–332.

Roberts, Neil (1998) *The Holocene: An Environmental History*. Oxford, UK: Blackwell.

Robinson, Arthur B., Sallie L. Baliunas, Willie Soon, and Zachary W. Robinson (2002) Environmental effects of increased atmospheric carbon dioxide. http://www.oism.org/pproject/s33p36.htm.

Ruddiman, William (2003) The anthropogenic greenhouse era began thousands of years ago. *Climatic Change*, vol. 61, no. 3 (Dec.), pp 261–293.

Russell, Bertrand (1912) *The Problems of Philosophy*. London: Oxford University Press.

Russell, Peter (1983) *The Global Brain*. New York: J.P. Tarcher.

——— (1995) *The Global Brain Awakens: Our Next Evolutionary Leap*. Palo Alto, CA: Global Brain.

Ryckman, D. P., D. V. Chip Weseloh, and C. A. Bishop (2005) Contaminants in herring gull eggs from the Great Lakes: 25 years of monitoring levels and effects. http://www.on.ec.gc.ca/wildlife/factsheets/fs_herring_gulls-e.html (Aug. 21, 2007), see Fig. 3; http://www.on.ec.gc.ca/wildlife/factsheets/images/glfs_herring_fig3.jpg (Aug. 21, 2007).

Salter, Stephen H., and John Latham (2007) The reversal of global warming by the increase of the albedo of marine stratocumulus cloud. http://www.theengineer.co.uk/assets/getAsset.aspx?liAssetID=25650 (accessed Jan. 30, 2008).

Salter, Stephen H., and Graham Sortino (2007) Sea-going hardware for the implementation of the cloud albedo control method for the reduction of global warming. http://www.theengineer.co.uk/assets/getAsset.aspx?liAssetID=25649 (accessed Jan. 30, 2008).

Santer, B. D., T. M. L. Wigley, G. A. Meehl, M. F. Wehner, C. Mears, M. Schabel, F. J. Wentz, C. Ammann, J. Arblaster, T. Bettge, W. M. Washington, K. E. Taylor, J. S. Boyle, W. Brüggemann, and C. Doutriaux (2003) Influence of satellite data uncertainties on the detection of externally forced climate change. *Science*, vol. 300, no. 5623 (May 23), pp. 1280–1284.

Schama, Simon (1995) *Landscape and Memory*. Toronto, Ontario, Canada: Random House of Canada.

Schlesinger, M. E., ed. (1991) *Greenhouse-Gas-Induced Climatic Change: A Critical Appraisal of Simulations and Observations*. New York: Elsevier Science.

Seidel, D. J., et al. (2004) Uncertainty in signals of large-scale climate variations in radiosonde and satellite upper-air temperature datasets. *Journal of Climate*, vol. 17, pp. 2225–2240.

Shaviv, Nir J., and Jan Veizer (2003) Celestial driver of phanerozoic climate? *GSA Today*, vol. 13, no. 7, pp. 4–10.

Shelley, Percy Bysshe (1820) *Prometheus Unbound*.

———— (1821) *A Defense of Poetry*. First published in 1841.

Shelton, James Gary (1998) *Bear Attacks: The Deadly Truth*. Holt, MI: Partners Publishing Group.

Sinha, Ashok (1995) Relative influence of lapse rate and water vapor on the greenhouse effect. *Journal of Geophysical Research*, vol. 100, no. 3, pp. 5095–5103.

Smith, A. G. (2000) How toxic is DDT? *The Lancet*, vol. 356, no. 9226 (July 22), pp. 267–268.

Smith, Carlton (2003) Grooming an elf: how Tre Arrow turned Jake Sherman into an eco-terrorist. *Willamette Week Online*, Nov. 26. http://www.wweek.com/flatfiles/News4575.lasso.

Smith, Eddie (1999) Atlantic and East Coast hurricanes 1900–98: a frequency and intensity study for the twenty-first century. *American Meteorological Society Bulletin*, vol. 80B (July–Dec.), pp. 2717–2720.

Snyder, Gary (1995) Ecology, place, and the awakening of compassion. In Drengson and Inoue (1995, pp. 237–241).

Soon, W. (2005) Variable solar irradiance as a plausible agent for multidecadal variations in the Arctic-wide surface air temperature record of the past 130 years. *Geophysical Research Letters*, vol. 32, p. L16712.

Soon, W., and S. Baliunas (2003) Proxy climatic and environmental changes of the past 1000 years. *Climate Research*, vol. 23, pp. 89–110.

Soon, W., et al. (2003) Reconstructing climatic and environmental changes of the past 1000 years: a reappraisal. *Energy and Environment*, vol. 14, pp. 233–296.

Soulé, Michael E. (1991) Conservation: tactics for a constant crisis. *Science*, vol. 253 (Aug. 16), pp. 744–750.

Spencer, Roy W., and John R. Christy (1992) Precision and radiosonde validation of satellite gridpoint temperature anomalies. *Journal of Climate*, vol. 5, pp. 847–866.

Spencer, Roy W., et al. (2007) Cloud and radiation budget changes associated with tropical intraseasonal oscillations. *Geophysical Research Letters*, vol. 34, p. L15707.

Standing Bear, Luther (1933). *Land of the Spotted Eagle*. Lincoln, NB: University of Nebraska Press.

Staniford, Stuart (2006) Living in the Emian. *The Oil Drum*, Feb. 20. Based on NOAA data. http://www.theoildrum.com/story/2006/2/3/0394/97545&h=476&w=692&sz=203&hl= en&start=2&um=1&tbnid=SLv3ZIpHyM38VM:&tbnh=96&tbnw=139&prev=/images% 3Fq%3Dholocene%2Btemperature%26um%3D1%26hl%3Den%26client%3Dfirefox-a% 26rls%3Dorg.mozilla:en-US:official%26sa%3DN (accessed Apr. 25, 2008).

Stott, Lowell, et al. (2007) Southern hemisphere and deep-sea warming led deglacial atmospheric CO_2 rise and tropical warming. *Science*, vol. 318 (Oct. 19), pp. 435–438.

Stove, David (2003) In *On Enlightenment*, Andrew Irvine, ed. New Brunswick, NJ: Transaction Publishers.

——— (1990) What is wrong with our thoughts. From Stove (2003, p. 56).

Stroeve, J., et al. (2007) Arctic sea ice decline: faster than forecast. *Geophysical Research Letters*, vol. 34 (May 1), p. L09501.

Sun, D.-Z., and R. S. Lindzen (1993) Distribution of tropical tropospheric water vapor. *Journal of Atmospheric Science*, vol. 50, pp. 1643–1660.

Suzuki, David (1990) Balance the books on global warming. *The London Free Press*, London, Ontario, Canada, May 12.

——— (1994) *Time to Change*. Toronto Ontario, Canada: Stoddart Publishers.

——— (1998) *Earth Time: Essays*. Toronto, Ontario, Canada: Stoddart Publishers.

Suzuki, David, and Holly Dressel (2004) *From Naked Ape to Superspecies: Humanity and the Global Eco-Crisis*. Berkeley, CA: Greystone Books.

Svensmark, Henrik, and Nigel Calder (2007) *The Chilling Stars: A New Theory of Climate Change*. Cambridge, UK: Icon Books.

Svensmark, Henrik, et al. (2007) Experimental evidence for the role of ions in particle nucleation under atmospheric conditions. *Proceedings of the Royal Society A*, vol. 463, pp. 385–396. http://www.journals.royalsoc.ac.uk/content/3163g817166673g7/fulltext.pdf.

Taubes, Gary (1997) Apocalypse not. *Science*, vol. 278, Nov. 7, pp. 1004–1006.

Taylor, George (2002) State climatologists skeptical of administrations global warming. http://www.sitewave.net/news/s49p628.htm.

Turner, R. Kerry, David Pearce, and Ian Bateman (1990) *Environmental Economics: An Elementary Introduction*. New York: Harvester/Wheatsheaf.

Twain, Mark (1897) *Following the Equator.*

Twomey, Sean (1977) The influence of pollution on the shortwave albedo clouds. *Journal of Atmospheric Sciences*, vol. 34 (July), pp. 1149–1152.

UNDESA (United Nations Department of Economic and Social Affairs) (2004) *World Population to 2300*. New York: United Nations. http://www.un.org/esa/population/publications/longrange2/WorldPop2300final.pdf (accessed Apr. 19, 2008).

USCCSP (U.S. Climate Change Science Program) (2006) *Temperature Trends in the Lower Atmosphere: Steps for Understanding and Reconciling Differences*. Thomas R. Karl et al., eds.

USDA (United States Department of Agriculture) (2000) *RPA Assessment of Forest and Range Lands*. Washington, DC: USDA Forestry Service. http://www.fs.fed.us/pl/rpa/rpaasses.pdf.

USDHHS-NIH (U.S. Department of Health and Human Services–National Institutes of Health) (2002) *Malaria*. NIH Publication. 02-7139, Sept. http://www.niaid.nih.gov/publications/malaria/pdf/malaria.pdf.

Veizer, Ján, et al. (2000) Evidence for decoupling of atmospheric CO_2 and global climate during the Phanerozoic eon. *Nature*, vol. 408 (Dec. 7), pp. 698–701.

Watts, Simon, and Lyndsay Halliwell (1995) *Essential Environmental Science: Methods and Techniques*. New York: Routledge.

Wegman, Edward J., et al. (2006) Ad hoc committee report on the "hockey stick" global climate reconstruction, executive summary. http://republicans.energycommerce.house.gov/108/home/07142006_Wegman_Report.pdf.

White, Lynn (1967) The historical roots of our ecological crisis. *Science*, vol. 155, pp. 1203–1207.

White, Margaret R., ed. (1985) *Characterization of Information Requirements for Studies of CO₂ Effects: Water Resources, Agriculture, Fisheries, Forests and Human Health*. Washington, DC: U.S. Department of Energy, Office of Energy Research.

White, M. R., and I. Hertz-Picciotto (1985) Human health: analysis of climate related to health. In White (1985, pp. 171–206).

White, W. B., and C.-K. Tai (1995) Inferring interannual changes in global upper ocean heat storage from Topex altimetry. *Journal of Geophysical Research*, vol. 100, no.C12, pp. 24943–24954.

WHO (World Health Organization) (2003) . New book demonstrates how climate change impacts on health. http://www.who.int/mediacentre/releases/2003/pr91/en/print.html.

Wielicki, Bruce A., et al. (2002) Evidence for large decadal variability in the tropical mean radiative energy budget. *Science*, vol. 295,. no. 5556 (Feb. 1), pp. 841–844.

Wigley, Tom (1998) The Kyoto protocol: CO_2, CH_4 and climate implications. *Geophysical Research Letters*, vol. 25, no. 13, p. 2285.

———— (2005) The climate change commitment. *Science*, vol. 307, pp. 1766–1769.

Wilson, Edward O. (1975) *Sociobiology: The New Synthesis*. Cambridge, MA: Belknap Press of Harvard University Press.

———— (1988b) The current state of biodiversity. In Wilson (1988a).

———— (1992) *The Diversity of Life*. Cambridge, MA: Harvard University Press.

———— (2001) *The Diversity of Life*. New York: Penguin Books.

———— (2002) *The Future of Life*. New York: Alfred A. Knopf.

———— (2007) *The Creation: An Appeal to Save Life on Earth*. New York: W.W. Norton.

Wilson, Edward O., and Dan L. Perlman (2000) *Conserving Earth's Biodiversity: With E. O. Wilson*. Washington, DC: Island Press.

Wordsworth, William (1798b) Words written in early spring.

———— (1800) The oak and the broom. In *The Complete Poetical Works of William Wordsworth*. London: Macmillan, 1888.

Wright, H. E., Jr., J. E. Kutzbach, T. Webb III, W. F. Ruddiman, F. A. Street-Perrott, and P. J. Bartlein, eds. (1993) *Global Climates Since the Last Glacial Maximum*. Minneapolis, MN: University of Minnesota Press.

WWF (World Wildlife Fund) (1999) *Persistent Organic Pollutants: Hand-Me-Down Poisons That Threaten Wildlife and People*. Issue Brief. Washington, DC: WWF, pp. 1–20.

Yokoyama, Y., et al. (2000) " Timing of the last glacial maximum from observed sea-level minima", *Nature*, vol. 406, pp. 713–716.

INDEX

Beyond Environmentalism: A Philosophy of Nature, By Jeffrey E. Foss
Copyright © 2009 John Wiley & Sons, Inc.